Electromagnetic Response of Material Media

Electromagnetic Response of Material Media

Yu. A. Il'inskii and L. V. Keldysh

Lebedev Physics Institute
Moscow University
Moscow, Russia

Translated by
Vitaly Kisin

Springer Science+Business Media, LLC

Library of Congress Cataloging-in-Publication Data

Il'inskiĭ, I͡U. A. (I͡Uriĭ Anatol'evich)
[Vzaimodeĭstvie ėlektromagnitnogo izluchenii͡a s veshchestvom. English]
Electromagnetic response of material media / Yu. A. Il'inskii and L.V. Keldysh ; translated by Vitaly Kisin.
p. cm.
Includes bibliographical references and index.

1. Electromagnetic waves. 2. Electromagnetic fields. 3. Matter--Effect of radiation on. I. Keldysh, L. V., 1931- . II. Title.
QC665.E4I413 1994
530.1'41--dc20 94-29103
CIP

This book was typeset using $\mathcal{A}\mathcal{M}\mathcal{S}$-TEX

DOI 10.1007/978-1-4899-1570-2

Originally published by Plenum Press, New York in 1994
MyCopy version of the original edition 1994

Preface

The textbook we offer to the reader is based on a two-term course of lectures, "Electromagnetic Response of Material Media," that the authors gave for a number of years to the final-year students of the Physics Department of Moscow University.

This course built on courses in quantum electronics, nonlinear optics and theoretical fundamentals of quantum radiophysics; students are assumed to have mastered the fundamentals of quantum mechanics, laser physics and nonlinear optics.

The essential core of the course, and hence of the book, is the current general theory of electromagnetic response of a nonrelativistic medium. The main aspects are presented in Chapters 1 and 2. The second part is devoted to more traditional topics which students learn in this course of lectures and also in the course "Condensed Matter Physics" for students who choose to major in radiophysics and laser physics; this course is also taught by the authors at the Physics Department.

This volume was intended as a text for students and, as such, does not cite original publications. We decided to provide a list of additional recommended literature, mostly of well known, easily accessible textbooks. Specific papers are cited only in a few places, when something is mentioned in passing in the text or if detailed calculations are omitted. In fact, as far as the first part of the book is concerned, we refer the reader to first chapter of Reference 1, where a historical survey of the evolution of the main concepts is presented, the concepts themselves are outlined in a compact form, and the necessary references to original publications are given.

Yu. Il'inskii
L. Keldysh

Moscow

Introduction

This course of lectures is devoted to a theoretical analysis of the interaction of the electromagnetic field (mostly radiation) with matter.

The term "matter" will stand for the ensemble of nonrelativistic particles, that is, electrons and nuclei that compose the plasma, molecules of the gas or a solid. Electrons and nuclei are assumed to be pointlike, which limits the applicability of the theory to distances greater than the characteristic nuclear scale, that is, to distance much greater than 10^{-13} cm.

Since the wavelength of the electromagnetic radiation is large in comparison with nuclear size even in the x-ray range ($\sim 10^{-8}$ cm), nuclear diameters can indeed be regarded as negligible.

The book deals with the microscopic mechanisms underlying the response of the material media to arbitrary electromagnetic fields. A general description of these phenomena is presented in terms of response functions, including as special cases linear and nonlinear susceptibilities, conductivity, screening of the Coulomb interaction, and so on. These mechanisms account in a natural way for the microstructure of the field in a medium at length scales comparable to or less than interatomic distances, and so appear to be adequate for describing phenomena in the x-ray range. The well-known problem of distinguishing between the mean and local (acting on the particles) fields, crucial for any microscopic theory of response, also finds a natural solution.

Representation of the linear and nonlinear response functions, in terms of the correlation functions of current density fluctuations, by Kubo-type formulas shows explicitly the connection between the response by the microscopic structure of the medium and its excitation spectrum; it thus appears to be the theoretical foundation of all kinds of spectroscopy. The general structure and properties of response functions are presented as well as a few of the most important and characteristic phenomena, such as two-photon absorption and scattering, dynamic Stark and Kerr effects, considered in more detail.

In the second part of the book some typical processes contributing to electromagnetic response are illustrated in the framework of simple model systems, such as free charges (plasma), atoms and molecules, lattice vibrations and band electrons in solids, etc.

The interaction of hard gamma radiation with matter is not considered in this course and matter is treated as nonrelativistic, so that we could ignore the processes of creation and transformation of particles, as well as the familiar problems of relativistic quantum theory.

In most problems of quantum radiophysics, one typically uses the so-called semi-classical method, in which matter is described in quantum terms and nonrelativistic quantum mechanics is used to the full, while the electromagnetic field is treated classically, via Maxwell's equations. This approach is widely used in the present book; it is quite adequate and sufficient, except in some problems where quantization of electromagnetic field is essential.

The problem of field–matter interaction must be treated in a self-consistent manner, that is, one must consider both the effect of the field on matter and also the effect of matter on the field. The latter effects are taken into account in electrodynamics by introducing material constants, such as the refractive index n, dielectric permittivity ε, conductivity σ, linear susceptibility $\chi^{(1)}$ (in linear theory) and nonlinear susceptibilities $\chi^{(2)}$, $\chi^{(3)}$ etc.(in nonlinear theory).

These constants and tensors, which are typically introduced in a phenomenological fashion, make it possible to solve problems involving the propagation of electromagnetic field (light). These problems are the subject of macroscopic electrodynamics, nonlinear optics, and wave theory. This book does not cover them. Our objective here is to study elementary (microscopic) processes of the interaction between electromagnetic field and matter and to analyze—using this stage as a basis—the physical nature and general properties of material constants.

The goal of this book is thus a description of microscopic processes that determine the response of matter to electromagnetic field. The description can be separated into two parts:

(a) a general theory which assumes that the spectrum and wave functions of matter are known and then solves the problem of calculating the response to the field (finding the linear and nonlinear susceptibilities);

(b) a description of specific media (plasma, molecules, solids) and of their interaction with the field (using simple models).

In other words, the book presents the microscopic theory of the effect of electromagnetic field on matter, and vice versa. If this problem is to be solved in a self-consistent manner, the material constants must be found in terms of only the fundamental constants: the charge of the electron e, the velocity of light c, Planck's constant $\hbar$, the electron mass m and the masses of nuclei M, and the atomic numbers N.

Their numerical values are:

$e = 4.80 \times 10^{-10}$ CGSE

$c = 3.00 \times 10^{10}$ cm/s

$\hbar = 1.05 \times 10^{-27}$ erg $\cdot$ s

$m = 9.11 \times 10^{-28}$ g

$m/M \sim 10^{-5}$–10^{-3}.

In this introduction and hereafter we are using the CGS system of units, since the SI system of units looks very incongruous for theoretical physics. This happens, first of all, because the dimensions of the electric and magnetic field strengths are different in SI, even though both enter the electromagnetic field tensor in a quite equivalent manner.

As a rule, the fundamental constants enter various expressions as certain combinations that determine the principal scales. These combinations determine the units of length, time and mass that are convenient in atomic physics (Hartree's system).

A convenient unit of length in atomic physics is the Bohr radius

$$a_0 = \hbar^2/me^2 \simeq 0.5 \times 10^{-8} \text{ cm}$$

This natural and unique combination of the constants e, m and $\hbar$, which are matter-related and enter the nonrelativistic equation of the motion of electrons in an atom, has the dimension of length. The Bohr radius a_0 is of the order of magnitude of atomic size and of interatomic distances in condensed media.

The natural unit of mass in atomic physics is the electron mass m; it is convenient to express the unit of time in terms of the atomic unit of energy

$$2I_0 = e^4m/\hbar^2 = 4.33 \times 10^{-11} \text{ erg} \simeq 27.2 \text{ eV}$$

($\hbar/I_0 \sim 10^{-16}$ s is the appropriate order of magnitude for atomic processes).

One half of the atomic energy unit, $I_0 = e^4m/2\hbar^2$, is called the Rydberg; it equals the energy of ionization of the hydrogen atom. The Rydberg is the natural unit and a scale of atomic energies.

The energy of a quantum of electromagnetic field is well known:

$$\hbar\omega = 2\pi\hbar c/\lambda$$

where ω is the frequency and λ is the wavelength. For estimates, it is convenient to use the formula

$$\hbar\omega \simeq \frac{1.24}{\lambda} \text{ (eV)}$$

where λ is assumed to be in micrometers.

Therefore, the wavelength of 1 μm (the near-infrared band) corresponds to the quantum energy of about 1 eV, the midpoint of the visible band ($\lambda \sim 0.5$ μm) corresponds to 2 eV, and the wavelength of 0.1 μm to the energy of about one Rydberg.

In a number of problems, the essential parameter is the temperature T, which can be measured in energy units. The correspondence with temperature expressed in more familiar Kelvins is given by the Boltzmann constant:

$$\mathrm{k} = 1.38 \times 10^{-16} \ \mathrm{erg/K}$$

Temperature on the order of room temperature, $T \sim 100$ K, corresponds to the energy of $\sim 10^{-14}$ erg $\sim 10^{-2}$ eV, so that kT can always be regarded as a small quantity on the atomic energy scale. As a rule, kT is much smaller than typical spacings between electron energy levels of a system, so that ordinary atomic systems at room temperatures are at the ground state (the lowest possible energy level). This is not true, however, for systems with closely spaced levels (vibrational and rotational levels of complex molecules) and for systems with continuous spectra (e.g. the plasma). Moreover, a plasma may even have k$T \gtrsim I_0$.

Contents

Chapter 1. General Theory of Interaction of Electromagnetic Fields with Matter 1
1.1. Classical Description of Electromagnetic Fields 1
1.2. Hamiltonian Formalism and Transition to Quantum Mechanics 7
1.3. Averaging of Maxwell's Equations 12
1.4. Expansion of Current in Powers of Field 17
1.5. Linear Susceptibilities and Their Properties 21
1.6. Relation of Dissipation of Energy in a Medium to the Anti-Hermitian Component of Linear Susceptibility 27
1.7. Nonlinear Susceptibilities 33
1.8. Susceptibilities of a Crystal 36
1.9. Mean Susceptibility of Crystals 43
1.10. Field Acting in the Crystal 48
1.11. Nonlinear Susceptibilities of Crystals 53

Chapter 2. Quantum-Mechanical Theory of Linear and Nonlinear Susceptibility 57
2.1. Density Matrix of a System 57
2.2. Representation of Interaction 65
2.3. Perturbation Theory for Density Matrix 68
2.4. Perturbation Operator for Problems of Field–Matter Interaction 71
2.5. Linear Response of a System to External Field 77
2.6. Explicit Form of Linear Response of a System to External Field 82
2.7. Fluctuation–Dissipation Theorem 87
2.8. Linear Susceptibility in Dipole Approximation 92
2.9. Quadratic Susceptibility. Symmetry Properties 102
2.10. Cubic Susceptibility. Degeneration Cases. Dynamic Kerr Effect 108
2.11. Dynamic Stark Effect and Balance Equations 112
2.12. Two-Level System Approximation 117

Chapter 3. Plasma in Electromagnetic Fields 125
3.1. Interaction between Free Charges and Electromagnetic Field 125
3.2. Elementary Theory of Dielectric Permittivity of Plasma 131
3.3. Kinetic Equation for Plasmas 137
3.4. Dielectric Permittivity of Plasmas, with Spatial Dispersion Taken into Account 142
3.5. Nonlinear Susceptibilities of the Plasma 147
3.6. Quantum Description of the Effect of Time-Dependent External Field on Collision Processes 153
3.7. Ionization in Intense Electromagnetic Field 158

Chapter 4. Atoms and Molecules in Electromagnetic Fields 163
4.1. Symmetry of Atoms and Molecules. Fundamentals of the Theory of Symmetry 163
4.2. Classification of Electron States of a Diatomic Molecule. Selection Rules 169
4.3. Atomic Spectra. Selection Rules for Atoms 174
4.4. Adiabatic Approximation 177
4.5. Types of Bonding in Molecules and Solids 183
4.6. Oscillational and Rotational Energy of Molecules 194
4.7. Optical Spectra of Molecules 202
4.8. The Placzek Approximation 210
4.9. Kerr Effect 216

Chapter 5. Interaction of Electromagnetic Radiation with Crystals 219
5.1. Model of a Crystal 219
5.2. Lattice Vibrations 222
5.3. Oscillations of Three-Dimensional Lattices. Quantization of Vibrations 232
5.4. Anharmonism. Phonon Lifetime 242
5.5. Interaction of Lattice Vibrations with Electromagnetic Field 250
5.6. Polaritons 258
5.7. Surface Polaritons 264
5.8. Model of a Lattice with Pointlike Ions 268
5.9. Electrons in Crystals. Bloch Theorem 277
5.10. Two Simplest Models for Calculation of Bloch Functions 281
5.11. Motion of Electrons in Crystals Placed in External Fields 287
5.12. Electric Conduction in Solids. Basics of Band Theory 289
5.13. Linear Susceptibility in Band Theory 295

5.14. Electron–Phonon Interaction 299
5.15. Mechanisms of Light Absorption in Solids 304

References 313

Index 315

Electromagnetic Response of Material Media

1

General Theory of Interaction of Electromagnetic Fields with Matter

1.1. Classical Description of Electromagnetic Fields

The electromagnetic field is described by two vectors $\mathbf{e}$ and $\mathbf{h}$, that is, by the macroscopic strengths of the electric and magnetic fields. These are the exact instantaneous values of the fields at a given point at a specific moment of time. We denote the averaged fields in the medium by $\mathcal{E}$ and $\mathcal{H}$.

The fields $\mathbf{e}$ and $\mathbf{h}$ satisfy the Maxwell–Lorentz equations

$$\begin{aligned} &\operatorname{curl}\mathbf{h} = \frac{1}{c}\frac{\partial \mathbf{e}}{\partial t} + \frac{4\pi}{c}\mathbf{j} \qquad \operatorname{curl}\mathbf{e} = -\frac{1}{c}\frac{\partial \mathbf{e}}{\partial t} \\ &\nabla\mathbf{e} = 4\pi\rho \qquad \nabla\mathbf{h} = 0 \end{aligned} \tag{1.1.1}$$

If the magnetic monopole exists, the righthand side of the second equation in (1.1.1) should be the density of magnetic charges, not zero. Modern physics accepts the possibility that the magnetic monopole exists; it may be found some day. At the present moment, however, this is a purely hypothetical possibility; taking it into account in equations (1.1.1) is unnecessary for the description of any phenomena physics has had to treat so far.

The righthand sides of equations (1.1.1) include $\mathbf{j}$ and ρ, that is, the macroscopic current density and charge density, which for pointlike electrons and nuclei take the form

$$\begin{aligned} \rho(\mathbf{r},t) &= \sum_i e_i\delta(\mathbf{r}-\mathbf{r}_i(t)) \\ \mathbf{j}(\mathbf{r},t) &= \sum_i e_i\dot{\mathbf{r}}_i\delta(\mathbf{r}-\mathbf{r}_i(t)) \end{aligned} \tag{1.1.2}$$

where $\mathbf{r}_i$ is the radius vector of the ith particle, e_i is its charge, and the summation is carried over all particles of the system.

At present, physics has no indication that the electron is not pointlike; however, the characteristic size of nuclei is of the order of 10^{-13} cm. Hence, as we have already discussed in the introduction, expressions (1.1.2) can be used only to describe phenomena at distances greater than 10^{-13} cm. Expressions (1.1.2) fail only for very hard γ quanta, of wavelengths of the order of, or less than, 10^{-13} cm. For nonrelativistic problems, the wavelength is much greater than the Compton wavelength:

$$\hbar\omega \ll mc^2 \qquad \lambda \gg 10^{-9}\ \text{cm}$$

Hence, the condition $\lambda \gg 10^{-13}$ cm is definitely satisfied.

Expressions (1.1.2) automatically satisfy the continuity equation

$$\frac{\partial \rho}{\partial t} + \nabla \mathbf{j} = 0 \tag{1.1.3}$$

which constitutes the law of conservation of the electric charge. In a finite region, this equation is equivalent to stating that the charge in this region changes only if a certain amount of charge is gained or lost across the boundary of the region.

Continuity equation (1.1.3) is not independent of Maxwell's equations (1.1.1). If we take the divergence of the first equation of (1.1.1) and take into account the third equation of (1.1.1), we arrive at the continuity equation; hence, the continuity equation is formally a corollary of Maxwell's equations. Actually, the charge conservation is a more fundamental notion and we should rather say that Maxwell's equations are constructed so as to automatically satisfy this law.

In order to close the set of equations (1.1.1) and (1.1.2), we need to complement them with the equations of motion of the particles. In the classical case, the motion in the electromagnetic field is described by Newton's equation with the Lorentz force in the righthand side:

$$m_i \dot{\mathbf{r}}_i = e_i \left(\mathbf{e}(\mathbf{r}_i, t) + \frac{1}{c} \left[\dot{\mathbf{r}}_i, \mathbf{h}(\mathbf{r}_i, t)\right] \right) \tag{1.1.4}$$

The set of Maxwell's equations (1.1.1) is conveniently rewritten by introducing the potentials in such a way that some equations in (1.1.1) are automatically satisfied. The last equation in (1.1.1) and the familiar theorem of calculus imply that there exists a function $\mathbf{A}(\mathbf{r}, t)$, known as the vector potential, such that

$$\mathbf{h} = \operatorname{curl} \mathbf{A} \tag{1.1.5}$$

The second equation of (1.1.1) then implies that

$$\operatorname{curl} \left(\mathbf{e} + \frac{1}{c} \frac{\partial \mathbf{A}}{\partial t} \right) = 0$$

whence we find, via another theorem of calculus, that there exists a function $\varphi(\mathbf{r}, t)$ such that

$$\mathbf{e} = -\operatorname{grad}\varphi - \frac{1}{c}\frac{\partial \mathbf{A}}{\partial t} \tag{1.1.6}$$

Therefore, expressions (1.1.5) and (1.1.6) are the electric and magnetic fields given in terms of the vector and scalar potentials. The second and fourth Maxwell's equations (1.1.1) are now automatically satisfied, and the first and third ones yield equations for the potentials $\mathbf{A}$ and φ:

$$\begin{aligned} \Box\mathbf{A} - \operatorname{grad}\left(\nabla\mathbf{A} + \frac{1}{c}\frac{\partial\varphi}{\partial t}\right) &= -\frac{4\pi}{c}\mathbf{j} \\ \Delta\varphi + \frac{1}{c}\frac{\partial}{\partial t}\nabla\mathbf{A} &= -4\pi\rho \end{aligned} \tag{1.1.7}$$

where $\Box = \Delta - \frac{1}{c^2}\frac{\partial^2}{\partial t^2}$ is the d'Alembertian operator.

The set of equations (1.1.7) is simpler than the original set (1.1.1), since it defines one vector and one scalar function (four components altogether) instead of two vector functions (six components altogether).

An important feature of expressions (1.1.5) and (1.1.6) is the invariance with respect to the gradient (gauge) transformations carried out by an arbitrary scalar function $\chi(\mathbf{r}, t)$:

$$\begin{aligned} \mathbf{A} \to \mathbf{A}' &= \mathbf{A} + \nabla\chi \\ \varphi \to \varphi' &= \varphi - \frac{1}{c}\frac{\partial\chi}{\partial t} \end{aligned} \tag{1.1.8}$$

A direct substitution demonstrates that

$$\mathbf{h}' = \operatorname{curl}\mathbf{A}' = \operatorname{curl}\mathbf{A} = \mathbf{h} \qquad \mathbf{e}' = -\operatorname{grad}\varphi' - \frac{1}{c}\frac{\partial\mathbf{A}'}{\partial t} = \mathbf{e}$$

The fact that the fields are invariant under this transformation of the potentials is known as the gauge invariance. Since only the fields $\mathbf{e}$ and $\mathbf{h}$ are observable (measured) quantities, not the potentials, the potentials $(\mathbf{A}, \varphi)$ and $(\mathbf{A}', \varphi')$ describe the same physical situation; in each specific problem, arguments of convenience determine which of the pairs of these potentials is to be used. The gauge invariance is a fundamental property of the theory of the electromagnetic field and, as became clear in recent years, of all other fundamental fields in nature.

A specific choice of potential and, hence, of the function $\chi(\mathbf{r}, t)$ is known as the gauge. A frequently used gauge is the Lorentz gauge, when an additional condition

$$\nabla\mathbf{A} + \frac{1}{c}\frac{\partial\varphi}{\partial t} = 0 \tag{1.1.9}$$

is imposed on the potentials.

By virtue of the relativity theory, the Lorentz gauge is invariant under a transition to a different inertial frame of reference. In view of this gauge, equation (1.1.7) for potentials acquire a simple form of wave equations:

$$\Box \mathbf{A} = -\frac{4\pi}{c}\mathbf{j} \qquad \Box \varphi = -4\pi\rho$$

In fact, the relativistic invariance is irrelevant for many problems in this course of lectures, since we use a specified reference frame — a frame at rest with respect to the system of particles, that is, to the medium. There is often a special convenience in the so-called Coulomb gauge, when the vector potential obeys the condition

$$\nabla A = 0 \tag{1.1.10}$$

It is not difficult to show that condition (1.1.10) can always be satisfied by using transformation (1.1.8). Indeed, let the potential $\mathbf{A}(\mathbf{r}, t)$ violate the Coulomb gauge condition:

$$\nabla \mathbf{A}(\mathbf{r}, t) = f(\mathbf{r}, t) \neq 0$$

By virtue of (1.1.8), the gauge invariance now permits the introduction of new potentials $\mathbf{A}'$ and φ'. If we demand that the new potential satisfy the condition $\nabla A' = 0$, we arrive at the following equation for the gauge function $\chi(\mathbf{r}, t)$:

$$\Delta\chi = -f(\mathbf{r}, t)$$

Its solution is

$$\chi(\mathbf{r}, t) = \frac{1}{4\pi}\int \frac{f(\mathbf{r}', t)}{|\mathbf{r} - \mathbf{r}'|}\, \mathrm{d}^3 r'$$

regardless of the form of the function $f(\mathbf{r}, t)$. Hence, we can always transform the equations from arbitrary initial potentials to potentials which satisfy the Coulomb gauge condition (1.1.10).

In the Coulomb gauge, expression (1.1.6) for the electric field automatically separates the field into potential and vortex components (the latter satisfying the condition $\nabla \mathbf{e} = 0$).

The equations for potentials (1.1.7) in this gauge are

$$\begin{aligned} &\Box \mathbf{A} - \frac{1}{c}\frac{\partial}{\partial t}\operatorname{grad}\varphi = -\frac{4\pi}{c}\mathbf{j} \\ &\Delta\varphi = -4\pi\rho \end{aligned} \tag{1.1.11}$$

The second equation has the same form as the electrostatic equation, and its solution can be written in the general form: as the sum of Coulomb's potentials (this is what gave the name to the gauge):

$$\varphi(\mathbf{r}, t) = \int \frac{\rho(\mathbf{r}', t)}{|\mathbf{r} - \mathbf{r}'|}\, \mathrm{d}^3 r' = \sum_i \frac{e_i}{|\mathbf{r} - \mathbf{r}'|} \tag{1.1.12}$$

In order to clarify the physical meaning of the Coulomb gauge, we expand the potential in plane modes:

$$\mathbf{A}(\mathbf{r},t)=\sum_{\nu}\exp(\mathbf{i}\mathbf{k}_{\nu}\mathbf{r})\mathbf{A}_{\nu}(t) \tag{1.1.13}$$

where the subscript ν enumerates the wave vectors and polarizations. Substituting this expression into (1.1.10), we find

$$\nabla\mathbf{A}=\sum_{\nu}\exp(\mathbf{i}\mathbf{k}_{\nu}\mathbf{r})i(\mathbf{k}_{\nu}\mathbf{A}_{\nu})$$

which is identically satisfied only if

$$(\mathbf{k}_{\nu}\mathbf{A}_{\nu})=0$$

This condition means that all modes of the vector potential are transverse.

If we similarly rewrite the scalar potential in the form

$$\varphi(\mathbf{r},t)=\sum_{\nu}\exp(\mathbf{i}\mathbf{k}_{\nu}\mathbf{r})\varphi_{\nu}(t) \tag{1.1.14}$$

then the electric field can be partitioned into

$$\mathbf{e}=\mathbf{e}_{\parallel}+\mathbf{e}_{\perp}$$

where

$$\mathbf{e}_{\parallel}=-\operatorname{grad}\varphi=-\mathrm{i}\sum_{\nu}\exp(\mathbf{i}\mathbf{k}_{\nu}\mathbf{r})\varphi_{\nu}(t)\mathbf{k}_{\nu}$$

is the longitudinal component of the field and $\mathbf{e}_{\perp}=-\frac{1}{c}\frac{\partial\mathbf{A}}{\partial t}$ is the transverse component of the electric field. The magnetic field is always a vortex field, that is, always transverse, and expressible in terms of the vector potential. We conclude, therefore, that the Coulomb gauge splits the field into a transverse component described by the potential $\mathbf{A}$ and a longitudinal one, described by the potential φ.

The use of the Coulomb gauge in nonrelativistic problems is convenient since the magnetic interaction between particles in a medium is much smaller than the electric interaction. As a result, the interaction between particles in a medium can be described by the potential φ. The field of radiation entering the medium is purely transverse and is described by the vector potential $\mathbf{A}$. In the medium, these fields are generally mixed; however, if the external field is small in comparison with the interatomic fields, the interaction between particles, which is the condition for the existence of the medium, remains purely Coulombic in the zero approximation. The Coulomb gauge is convenient in specific calculations since this interaction can be rewritten in explicit form (1.1.12), which immediately singles out the zero approximation of the problem — a system of particles in the absence of external fields.

Equations for potentials (1.1.11) look more complicated in the Coulomb gauge than in the Lorentz component; actually, this impression is misleading. We have already mentioned that the solution of the equation for the scalar potential can be written in a straightforward manner. The first equation of (1.1.1) is unwieldy because it includes both $\mathbf{A}$ and φ. In fact, these are two equations for the longitudinal and the transverse components:

$$\begin{aligned} \Box\mathbf{A} &= -\frac{4\pi}{c}\mathbf{j}_{\perp} \\ \frac{1}{c}\frac{\partial}{\partial t}\,\mathrm{grad}\,\varphi &= \frac{4\pi}{c}\mathbf{j}_{\parallel} \end{aligned} \tag{1.1.15}$$

where $\mathbf{j}_{\parallel}$ and $\mathbf{j}_{\perp}$ are the longitudinal and the perpendicular components of the current density. By definition,

$$\mathbf{j}_{\parallel} = \sum_{\nu} \exp(\mathrm{i}\mathbf{k}_{\nu}\mathbf{r})\,\frac{\mathbf{k}_{\nu}(\mathbf{k}_{\nu}\mathbf{j}_{\nu})}{k_{\nu}^2} \qquad \mathbf{j}_{\perp} = \mathbf{j} - \mathbf{j}_{\parallel}$$

where $\mathbf{j}_{\nu}$ are the Fourier components of the current density,

$$\mathbf{j} = \sum_{\nu} \exp(\mathrm{i}\mathbf{k}_{\nu}\mathbf{r})\mathbf{j}_{\nu}$$

The second equation of (1.1.15) can be ignored since it is implied by the continuity equation and Poisson's equation for the scalar potential φ (the second equation in (1.1.11)). Therefore, the equation for the vector potential has a simple form of the wave equation, as in the Lorentz gauge, with the current density in the righthand side replaced by its transverse component.

In what follows, we will frequently consider problems in which an atomic system interacts with an optical field or electromagnetic field of longer than optical wavelength. The dipole approximation is often helpful in this case, since the size of the system (of the order of a_0 for atoms and molecules) is much smaller than the wavelength λ. In the zero approximation in the parameter $a_0/\lambda \ll 1$, the electric field can be assumed as uniform within the system: $\mathbf{e} = \mathbf{e}(t)$. This field can be described in terms of only the vector potential,

$$\mathbf{A} = -c\int^{t} \mathbf{e}(t')\,\mathrm{d}t'$$

or only the scalar potential,

$$\varphi = -\mathbf{e}(t)\mathbf{r}$$

These two representations are related by the gauge transformation to the function

$$\chi = c\mathbf{r} \int^t \mathbf{e}(t')\, dt' = -\mathbf{A}(t)\mathbf{r}$$

The fact that the electric field is a function only of time signifies that the magnetic field plays a minor role, and is thus rightly neglected from the very beginning.

1.2. Hamiltonian Formalism and Transition to Quantum Mechanics

In nonrelativistic theory, the transition from the classical to the quantum description of a system of particles is performed in the framework of the so-called Hamiltonian formalism.

The basic equation of quantum mechanics, the Schrödinger equation

$$i\hbar \frac{\partial \Psi}{\partial t} = \mathscr{H}\Psi \tag{1.2.1}$$

contains the Hamiltonian $\mathscr{H}$ of the system. By virtue of the correspondence principle, the Hamiltonian of a particular system is obtained from the classical Hamiltonian function — the function of generalized coordinates and canonical momenta conjugate to them — by replacing the canonical momenta with the appropriate operators satisfying specific commutation relations.

The first problem of the quantum theory of motion of particles in the electromagnetic field is, therefore, to construct the Hamiltonian function. Let us consider the Hamiltonian (canonical) formalism for the equations of motion of charges and the field.

The Hamiltonian formalism assumes that a system is described by pairs of canonically conjugate variables p and q. Then there exists a Hamiltonian function $\mathscr{H}(p, q)$, such that the equations of motion can be recast to the form

$$\dot{p} = -\frac{\partial \mathscr{H}}{\partial q} \qquad \dot{q} = \frac{\partial \mathscr{H}}{\partial p}$$

In conservative systems, $\mathscr{H}$ is the energy expressed in terms of canonical variables. For the ensemble of field and particles, the energy (which is an integral of motion) is a sum of the kinetic energy of the particles and the energy of the field:

$$E = \sum_i \frac{m_i v_i^2}{2} + \frac{1}{8\pi} \int (e^2 + h^2)\, d^3r \tag{1.2.2}$$

We need not write here the energy of interaction since it is included in the energy of the field.

In fact, (1.2.2) is not yet the Hamiltonian function since the energy is given in terms of the velocities v_i, not in terms of the momenta.

The generalized momenta $\mathbf{p}_i$ are introduced by the relation

$$\mathbf{p}_i = m_i \dot{\mathbf{r}}_i + \frac{e_i}{c}\mathbf{A}(\mathbf{r}_i, t) \tag{1.2.3}$$

The quantity $\mathbf{p}_i$ is not gauge invariant; it is defined with the same degree of arbitrariness as the vector potential $\mathbf{A}$. Owing to this ambiguity, the generalized momentum $\mathbf{p}$ is physically not completely real. The observables are the coordinates $\mathbf{r}$ and the velocities $\dot{\mathbf{r}}$ of particles, not their generalized momentum. However, it is necessary to use generalized momenta in order to get the equations of motion (1.1.4) in the "canonical" form and in order to subsequently go over to the quantum description.

Having resorted to (1.2.3) in order to express the velocities in (1.2.2) in terms of the generalized momenta and the field $\mathbf{e}$ and $\mathbf{h}$ in terms of potentials, we obtain the Hamiltonian function

$$\begin{aligned}\mathscr{H} = &\sum_i \frac{1}{2m_i}\left(\mathbf{p}_i - \frac{e_i}{c}\mathbf{A}(\mathbf{r}_i, t)\right)^2 \\ &+ \frac{1}{8\pi}\int\left((\operatorname{curl}\mathbf{A})^2 + \frac{1}{c^2}\left(\frac{\partial \mathbf{A}}{\partial t}\right)^2\right) \mathrm{d}^3 r \\ &+ \frac{1}{4\pi c}\int \frac{\partial \mathbf{A}}{\partial t}\operatorname{grad}\varphi \, \mathrm{d}^3 r + \frac{1}{8\pi}\int (\operatorname{grad}\varphi)^2 \, \mathrm{d}^3 r\end{aligned} \tag{1.2.4}$$

The second term of (1.2.4) is the energy of the transverse field in the Coulomb gauge. This is essentially the energy of photons under field quantization. We need to additionally express $\partial\mathbf{A}/\partial t$ in this term via generalized momenta corresponding to the variables of the field. This term of the Hamiltonian function enables us to find the equations for the field (Maxwell's equations) but is unimportant for finding the equations of motion of particles because it does not carry their coordinates and momenta.

The third term in the Coulomb gauge ($\nabla\mathbf{A} = 0$) vanishes since it can be transformed as shown here:

$$\frac{\partial \mathbf{A}}{\partial t}\operatorname{grad}\varphi = \frac{\partial \mathbf{A}}{\partial t}\operatorname{grad}\varphi + \varphi\frac{\partial}{\partial t}\nabla\mathbf{A} = \nabla\left(\varphi\frac{\partial \mathbf{A}}{\partial t}\right)$$

and since the integral over the space can be reduced to the surface integral over an infinitely remote surface on which the fields vanish.

In order to transform the last term in (1.2.4), we make use of the identity

$$(\operatorname{grad}\varphi)^2 = \nabla(\varphi\operatorname{grad}\varphi) - \varphi\Delta\varphi$$

In the integration over space, the term including the divergence vanishes; if we now take into account Poisson's equation,

$$\Delta\varphi = -4\pi\rho$$

we can rewrite the last term in Hamiltonian function (1.2.4) as

$$\frac{1}{2}\int \rho\varphi \, \mathrm{d}^3 r = \frac{1}{2}\sum_i e_i \varphi(\mathbf{r}_i, t) = \frac{1}{2}\sum_{i \neq i'} \frac{e_i e_{i'}}{|\mathbf{r}_i - \mathbf{r}_{i'}|}$$

Ultimately, the Hamiltonian function of a system of particles in electromagnetic field is written in the form

$$\mathscr{H} = \sum_i \frac{1}{2m_i}\left(\mathbf{p}_i - \frac{e_i}{c}\mathbf{A}(\mathbf{r}_i, t)\right)^2 + \frac{1}{2}\sum_{i \neq i'} \frac{e_i e_{i'}}{|\mathbf{r}_i - \mathbf{r}_{i'}|} \tag{1.2.5}$$

The validity of expression (1.2.5) is supported by the fact that Hamilton's equations with Hamiltonian function (1.2.5) are correct equations of motion of particles.

The first set of Hamilton's equations yields

$$\dot{r}_{i\alpha} = \frac{\partial \mathscr{H}}{\partial p_{i\alpha}} = \frac{1}{m_i}\left(p_{i\alpha} - \frac{e_i}{c}A_\alpha(\mathbf{r}_i, t)\right)$$

which coincides with the definition (1.2.3) of canonical momentum. The subscript α =1, 2, 3 above and hereafter enumerates the Cartesian coordinates of vectors.

The second set of Hamilton's equations gives the rate of change of the canonical momentum:

$$\dot{p}_{i\alpha} = -\frac{\partial \mathscr{H}}{\partial r_{i\alpha}} = \frac{e_i}{m_i c}\left(p_{i\beta} - \frac{e_i}{c}A_\beta(\mathbf{r}_i, t)\right)\frac{\partial A_\beta(\mathbf{r}_i, t)}{\partial r_{i\alpha}} - e_i \frac{\partial \varphi(\mathbf{r}_i, t)}{\partial r_{i\alpha}}$$

The repeated subscript β in this and the subsequent equations implies summation.

In the resulting equation, we need to eliminate the momenta $p_{i\alpha}$ using the first set of Hamilton's equations or, which is the same, using definition (1.2.3). Differentiating (1.2.3) with respect to time and taking into account that $\mathbf{A}(\mathbf{r}_i, t)$ depends on t both explicitly and implicitly, via the function $\mathbf{r}_i(t)$, we obtain

$$\begin{aligned} m_i \ddot{r}_i + \frac{e_i}{c}\frac{\mathrm{d}A_\alpha(\mathbf{r}_i, t)}{\mathrm{d}t} &\equiv m_i \ddot{r}_{i\alpha} + \frac{e_i}{c}\frac{\partial A_\alpha(\mathbf{r}_i, t)}{\partial t} \\ + \frac{e_i}{c}(\dot{\mathbf{r}}_i \nabla)A_\alpha(\mathbf{r}_i, t) &= \frac{e_i}{c}\left(\dot{\mathbf{r}}_i \frac{\partial \mathbf{A}}{\partial r_{i\alpha}}\right) - e_i \frac{\partial \varphi(\mathbf{r}_i, t)}{\partial r_{i\alpha}} \end{aligned} \tag{1.2.6}$$

Let us make use of the vector formula

$$[\dot{\mathbf{r}}_i \operatorname{curl} \mathbf{A}]_\alpha = \left(\mathbf{r}_i \frac{\partial \mathbf{A}}{\partial r_{i\alpha}}\right) - (\dot{\mathbf{r}}_i \nabla) A_\alpha$$

Now (1.2.6) gives

$$m_i \ddot{r}_{i\alpha} = -e_i \frac{\partial \varphi(\mathbf{r}_i, t)}{\partial r_{i\alpha}} - \frac{e_i}{c} \frac{\partial A_\alpha(\mathbf{r}_i, t)}{\partial t} + \frac{e_i}{c} [\dot{\mathbf{r}}_i \operatorname{curl} \mathbf{A}]_\alpha$$

Taking into account the expressions for the fields $\mathbf{e}$ and $\mathbf{h}$ in terms of potentials, we find that in the vector form, the last equation is the correct classical equation of motion coinciding with (1.1.4). This is a proof that (1.2.5) is a correct expression for the Hamiltonian function. In fact, the canonical momenta were introduced by formula (1.2.3) precisely in order to arrive at this result.

The transition to a quantum description of a system is formally carried out by replacing the classical quantities in the Hamiltonian function (1.2.6) by the appropriate operators $\hat{p}$ and $\hat{r}$ which satisfy the commutation relations

$$[\hat{p}_{i\alpha}, \hat{r}_{i'\beta}] = -i\hbar \delta_{ii'} \delta_{\alpha\beta} \qquad [\hat{p}_{i\alpha}, \hat{p}_{i'\beta}] = [\hat{r}_{i\alpha}, \hat{r}_{i'\beta}] = 0$$

In the so-called coordinate representation, the coordinates continue to be ordinary numbers $\hat{r}_{i\alpha} = r_{i\alpha}$, and the operators of momenta are

$$\hat{p}_{i\alpha} = -i\hbar \frac{\partial}{\partial r_{i\alpha}}$$

This transition produces from (1.2.5) the Hamiltonian operator, or the Hamiltonian of the system:

$$\mathscr{H} = \sum_i \frac{1}{2m_i}\left(- i\hbar\nabla_i - \frac{e_i}{c}\mathbf{A}(\mathbf{r}_i, t) - \frac{e_i}{c}\mathbf{A}^{(e)}(\mathbf{r}_i, t)\right)^2 + \frac{1}{2}\sum_{i \neq i'} \frac{e_i e_{i'}}{|\mathbf{r}_i - \mathbf{r}_{i'}|} + \sum_i e_i \varphi^{(e)}(\mathbf{r}_i, t) \tag{1.2.7}$$

where we have singled out the field $\mathbf{A}$ connected with the particles and the external field $\mathbf{A}^{(e)}$, $\varphi^{(e)}$ created by sources that are external with respect to the particle system we consider. This is the standard form of the nonrelativistic Hamiltonian of the system of particles interacting with the electromagnetic field. In what follows, we regularly turn to Hamiltonian (1.2.7).

The external field $\mathbf{A}^{(e)}$, $\varphi^{(e)}$ in Hamiltonian (1.2.7) can be used in an arbitrary gauge, even if the field $\mathbf{A}$ connected with the particles of the medium is used in the Coulomb gauge.

The Hamiltonian formalism is very convenient for the transition to quantum mechanics. The "penalty" is the apparent lack of gauge invariance of the Hamiltonian itself and of the results derived from it. The gauge transformation of the potentials $\mathbf{A}^{(e)}$ and $\varphi^{(e)}$ changes the explicit form (1.2.7) of the Hamiltonian operator $\mathscr{H}$. However, this does not affect any physical observables.

Indeed, all quantum mechanical mean values are defined using the wave functions that satisfy the Schrödinger equation (1.2.1). Gauge transformation (1.1.8) which is performed by a function $\chi(\mathbf{r},t)$ modifies the Hamiltonian since

$$\Big(-\mathrm{i}\hbar\nabla_i - \frac{e_i}{c}\mathbf{A} - \frac{e_i}{c}\mathbf{A}^{(e)}\Big) \to \Big(-\mathrm{i}\hbar\nabla_i - \frac{e_i}{c}\mathbf{A} - \frac{e_i}{c}\mathbf{A}^{(e)} - \frac{e_i}{c}\nabla_i\chi\Big)$$

$$\varphi^{(e)} \to \varphi^{(e)} - \frac{1}{c}\frac{\partial\chi}{\partial t}$$

We easily see that a change in the Hamiltonian,

$$\mathscr{H} \to \widehat{\mathscr{H}'}$$

can be completely compensated for by changing the phase of the wave function. If ψ satisfies the Schrödinger equation (1.2.1), then the wave function

$$\psi' = \psi \exp\left\{\mathrm{i}\sum_i \frac{e_i}{\hbar c}\chi(\mathbf{r}_i,t)\right\}$$

satisfies an identical equation with the Hamiltonian $\widehat{\mathscr{H}'}$.

In quantum mechanics, the wave function is not gauge-invariant but neither is it an observable quantity. The observables are the mean values of physical quantities,

$$\overline{L}(t) = \int \psi^* \widehat{L}\psi \, \mathrm{d}^3 r$$

where $\hat{L}$ is an operator corresponding to a physical quantity L. For example, the probability distribution of particles in the configurational space $|\psi|^2$ is invariant since the phase factor vanishes. As a result, the mean value of any function of coordinates is gauge-invariant.

At the same time, the average value of the canonical momentum

$$\widehat{p_{i\alpha}} = -\mathrm{i}\hbar\frac{\partial}{\partial r_{i\alpha}}$$

is a function of gauge since the wave function is transformed. There is nothing surprising in it because by virtue of (1.2.3), the canonical momentum in the classic theory is equally not a gauge-invariant quantity. However, the mean value of the velocity of a particle,

$$\hat{\dot{\mathbf{r}}}_i = \hat{\mathbf{v}}_i = \frac{1}{m_i}\left(-i\hbar\nabla_i - \frac{e_i}{c}\mathbf{A} - \frac{e_i}{c}\mathbf{A}^{(e)}\right)$$

is gauge-independent; this can be confirmed by straightforward calculation. The same is true for any power of velocity and, hence, for any operator which is expandable in a series in powers of $\mathbf{r}$ and $\dot{\mathbf{r}}$.

1.3. Averaging of Maxwell's Equations

The true values of the fields $\mathbf{e}$ and $\mathbf{h}$ in a medium vary in a very complicated manner in space and time owing to the uninterrupted motion of all particles; this motion generates rapidly fluctuating local densities of charges and currents. Moreover, if the description is at a microscopic level, where quantum effects are essential, it is not possible at all to assign to each particle definite values of coordinates and velocities; one can only operate with the probability for particles to have specific coordinates and velocities. Strictly speaking, it is simply meaningless to try to determine the instantaneous local values of the fields $\mathbf{e}$ and $\mathbf{h}$. Any sensible theory can operate only with the values, averaged over fluctuations, of fields and with their correlations at different spacetime points, or with the probability distribution for various field configurations.

The Maxwell–Lorentz equations (1.1.1) are linear and thus allow direct averaging, which reduces to simple replacement of $\mathbf{e}$ and $\mathbf{h}$ by their mean values $\mathcal{E}$ and $\mathcal{H}$. The charge and current densities are then also replaced with the mean values. Obviously, the relations between currents and fields (the so-called material equations) can be nonlinear; however, this is a different problem which has no direct bearing on the problem of averaging of the field equations in the form (1.1.1).

The averaged Maxwell's equations are written in the form

$$\begin{aligned}
\operatorname{curl}\mathcal{H} &= \frac{1}{c}\frac{\partial\mathcal{E}}{\partial t} + \frac{4\pi}{c}j \\
\operatorname{curl}\mathcal{E} &= -\frac{1}{c}\frac{\partial\mathcal{H}}{\partial t} \\
\nabla\mathcal{E} &= 4\pi\rho \\
\nabla\mathcal{H} &= 0
\end{aligned} \tag{1.3.1}$$

The same notation as in nonaveraged equations (1.1.1) is used above for the averaged values of the current density $\mathbf{j}$ and charge ρ.

Owing to the linearity of equations (1.1.1), they can be averaged by physically very different methods. Thus the procedure of averaging the

Maxwell–Lorentz equations over the so-called physically infinitely small volume has been known for about a century now, since Lorentz's time. It is assumed in this procedure that the concept of field at each point must be discarded, and that the long-wavelength field (with $\lambda \gg a_0$) can be regarded as uniform within volumes whose linear size is much less than λ. If the chosen, physically infinitely small volume contains a large number of particles, the volume can be treated as macroscopic. The statistical fluctuations of the field averaged over a volume are small and one can regard the field averaged over a physically infinitely small volume as the statistically mean field. In this way, one can construct the macroscopic electrodynamics and optics.

This approach is not satisfactory for our purposes, because

1. The wavelength in the medium may be considerably reduced (owing to a high refractive index); moreover, there exist qualitatively new phenomena, such as gyrotropy, which are defined only by the drop of the field over molecular-scale distances,
2. The condition $\lambda \gg a_0$ is not satisfied at all in the x-ray range, but the most important reason is that
3. To find the motion of particles in a medium and their response to fields, one needs to know not the average field but the true field at points where particles are.

The averaging over a physically infinitely small volume immediately excludes from analysis the field acting on a particle and forces one to forego the response of the medium to the field; after this, one has to be satisfied with taking only a phenomenological account of this response.

Therefore, averaging over a physically infinitely small volume has to be dropped from a microscopic theory of matter's response to electromagnetic field and one has to resort to the method, standard for the statistical physics, of averaging over the ensemble of the possible states of the medium, for example, over the Gibbs distribution. In the quantum case, this approach implies also the averaging over wave functions of particles. In fact, charge and current densities are thereby averaged, and the corresponding mean fields are automatically found from equations (1.3.1). This averaging is carried out over fluctuations, not over volume. By virtue of the ergodic hypothesis, this also eliminates temporal fluctuations since the statistical averaging is equivalent to averaging over time.

This does not mean that the information on fluctuations is lost completely. We can still write equations for correlation functions of the type $\overline{e_\alpha(\mathbf{r},t)e_\beta(\mathbf{r}',t')}$ when the mean fields are zero, as in the case of thermal radiation. Again, some averaged characteristics of the fluctuation ensemble are taken into account.

It is important that there is no spatial averaging and the fields are referred to a point (to within a length of the order of 10^{-13}, as we have mentioned above).

It is thus sufficient, owing to the linearity of Maxwell's equations, to average their righthand sides that contain $\mathbf{j}$ and ρ; note that the charge and current densities are averaged by averaging over the Gibbs ensemble. If external fields are also present and $\mathbf{j}^{(e)}$ and $\rho^{(e)}$ are their sources, then $\mathbf{j}(\mathbf{r},t)$ and $\rho(\mathbf{r},t)$ represent the medium's response to these external fields. In their turn, these also induce fields in the medium, so that the next problem is to establish the relation of the induced currents $\mathbf{j}$ and charges ρ to the external fields. As the next step, averaged equations

$$\begin{aligned}
\operatorname{curl}\mathcal{H} &= \frac{1}{c}\frac{\partial\mathcal{E}}{\partial t} + \frac{4\pi}{c}\mathbf{j} + \frac{4\pi}{c}\mathbf{j}^{(e)} \\
\operatorname{curl}\mathcal{E} &= -\frac{1}{c}\frac{\partial\mathcal{H}}{\partial t} \\
\nabla\mathcal{E} &= 4\pi\rho + 4\pi\rho^{(e)} \\
\nabla\mathcal{H} &= 0
\end{aligned} \tag{1.3.2}$$

must be used to find the relation between the currents $\mathbf{j}$ and the mean fields in the medium,

$$\mathbf{j} = \mathbf{j}(\{\mathcal{E}\})$$

The relation between the currents $\mathbf{j}$ and the field $\mathcal{E}$ in the medium dictates such characteristics of the medium as its linear and nonlinear susceptibilities.

The refusal to average fields over physically infinitely small volumes results in a significant restructuring of the entire approach to describing the electromagnetic field in matter. The standard procedure of macroscopic electrodynamics is to single out the physically different components in the current $\mathbf{j}$,

$$\mathbf{j}(\mathbf{r},t) = \mathbf{j}_{\mathrm{f}} + \mathbf{j}_{\mathrm{b}} + \mathbf{j}_{\mathrm{m}}$$

where $\mathbf{j}_{\mathrm{f}}$ is the current of free charges (conduction electrons, etc.), $\mathbf{j}_{\mathrm{b}} = \partial\mathcal{P}/\partial t$ is the polarization current, or the displacement current (by definition, $\mathcal{P}$ is the medium's polarization vector) due to the motion of bound charges, and $\mathbf{j}_{\mathrm{m}} = c\operatorname{curl}\mathcal{M}$ is the magnetization vortex current (by definition, $\mathcal{M}$ is the magnetization vector). The mean value of the vector $\mathbf{h}$, denoted above as $\mathcal{H}$, is usually referred to as the magnetic induction $\mathcal{B}$; also introduced are the electric induction vector

$$\mathcal{D} = \mathcal{E} + 4\pi\mathcal{P}$$

and the magnetic field strength vector

$$\mathcal{H} = \mathcal{B} - 4\pi\mathcal{M}$$

As a result, we obtain ordinary equations of macroscopic electrodynamics for the vectors $\mathcal{E}$, $\mathcal{H}$, $\mathcal{D}$, and $\mathcal{B}$, which must be appended with two phenomenological relations between these vectors.

This procedure is justified in the range of quasistationary currents but is not valid for the optical and still shorter wavelength range, where unambiguous decomposition of current into $\mathbf{j}_\mathrm{f}$, $\mathbf{j}_\mathrm{b}$, and $\mathbf{j}_\mathrm{m}$ is not possible and thus becomes meaningless.

Indeed, we can only distinguish between free and bound charges, or distinguish between closed (vortex) and open currents, if they are considered in a volume that is greater than the region of localization of the bound charges or of closed current loops. At each individual point, however, these contributions to current are essentially identical to one another: they are all contained in the common expression (1.1.2).

We shall illustrate this general statement with several simple examples. At low frequencies, the separation of charges into free and bound types is obvious: free charges are displaced by external fields to macroscopic distances, while bound charges stay within their respective atoms or molecules, and their small displacements within these bounds produce only a local polarization. However, this difference vanishes at high frequencies. Indeed, the classical equation of motion of free charges in a field $\mathbf{E}$ at a frequency ω,

$$m\ddot{\mathbf{r}} = e\mathbf{E}\cos\omega t$$

gives the oscillation amplitude of the order of $eE/m\omega^2$. Even in the strong field of a ruby laser ($\omega \simeq 3\times 10^{15}\,\mathrm{s}^{-1}$), with the electric field strength of the order of $E \sim 10^5$ CGS $= 3\times 10^7\,\mathrm{V\,cm}^{-1}$, the oscillation amplitude is of the order of 10^{-8} cm, and is even lower in the weak field of ordinary light. A bound electron is characterized by a very similar amplitude of motion (of the order of one Bohr radius). Consequently, the currents of free and bound electrons cannot be separated.

Another example is the photoelectric effect, in which bound electrons are transferred to the free state as a result of absorption of a quantum of light. At high frequencies, light quanta knock out bound electrons from atoms and thus convert them from bound into free particles. The question is this: to which current should we assign the contribution of these transition processes, that of the free or that of the bound charges? This example again demonstrates the arbitrariness of separating the current into distinct parts.

We find the same situation with the magnetization current, which typically involves the motion of charges along closed trajectories. At low frequencies, this motion is driven by the magnetic field but if electrons are subjected to circularly polarized light, they move along circular trajectories of radius $eE/m\omega^2$, which, as shown above, can be of the order of a_0. Free charges in electric field then contribute to the magnetization current just as bound electrons do. It is thus clear that it is impossible and physically meaningless to strictly separate the magnetization current and the current of free charges (even though one can always identify the total vortex component of the current; it is not determined by magnetic field alone).

The only way left to us in the case of high-frequency fields is to operate with the total current $\mathbf{j}$ which describes all processes. The equations for the mean fields $\mathcal{E}$ and $\mathcal{H}$ then take the form (1.3.1) or (1.3.2). These equations are written identically to those used in the electrodynamics of metals but $\mathbf{j}$ and ρ are now not the current and charge densities of free particles but the total current and the corresponding charges.

Sometimes the total polarization $\mathcal{P}(\mathbf{r}, t)$ is used instead of the total current $\mathbf{j}(\mathbf{r}, t)$; formally, it is defined as

$$\mathbf{j} = \frac{\partial \mathcal{P}}{\partial t} \qquad \mathcal{P}(\mathbf{r}, t) = \int_{-\infty}^{t} \mathbf{j}(\mathbf{r}, t')\mathrm{d}t'$$

that is, it contains the contributions of both free and bound charges and of magnetization currents. The lower bound of integration must be taken as $t = -\infty$ (we assume that neither fields nor polarization existed in the distant past).

Now we can also introduce the total induction

$$\mathcal{D}(\mathbf{r}, t) = \mathcal{E}(\mathbf{r}, t) + 4\pi\mathcal{P}(\mathbf{r}, t)$$

just as this is done for bound charges.

The equations can now be rewritten as

$$\begin{aligned} \operatorname{curl}\mathcal{H} &= \frac{1}{c}\frac{\partial \mathcal{D}}{\partial t} + \frac{4\pi}{c}\mathbf{j}^{(\mathrm{e})} \\ \operatorname{curl}\mathcal{E} &= -\frac{1}{c}\frac{\partial \mathcal{H}}{\partial t} \\ \nabla\mathcal{D} &= 4\pi\rho^{(\mathrm{e})} \\ \nabla\mathcal{H} &= 0 \end{aligned} \tag{1.3.3}$$

The third equation is obtained using the continuity equation

$$\frac{\partial \rho}{\partial t} = -\nabla\mathbf{j} = -\nabla\frac{\partial \mathcal{P}}{\partial t}$$

whence

$$\rho = -\nabla\mathcal{P}$$

as we have it in electrostatics.

Equations (1.3.3) are similar to the ordinary set of equations of the optics of dielectrics; however, with the definition chosen above, the induction $\mathcal{D}$ includes both the effects of motion of free charges and the magnetization currents. Thus there is no need to introduce magnetic induction. The vector $\mathcal{P}$ implies magnetic phenomena as well (this will be shown later in the text).

Both approaches (equations (1.3.2) and (1.3.3)) are completely equivalent. We can therefore use either of the approaches: to construct the theory of "metallic" type using matter equations of the type

$$\mathbf{j} = \sigma \mathcal{E}$$

or the theory of "dielectric" type, using the relation between the total polarization and the field,

$$\mathcal{P} = \chi \mathcal{E}$$

Since $\mathcal{P}$ includes all the currents, only one susceptibility χ is introduced; there is no need in introducing magnetic susceptibility.

1.4. Expansion of Current in Powers of Field

Our main objective now is to calculate the current induced in a system by an external field. In the general case, the problem of calculating the mean current in an arbitrary field and for an arbitrary system of a macroscopic number of particles has no solution.

Only one regular approach is known which allows an analysis of the solution in the general form. It involves the expansion of the response in powers of field:

$$\mathbf{j}(\mathbf{r}, t) = \sum_{k=0}^{\infty} \mathbf{j}^{(\mathbf{k})}(\mathbf{r}, t) \qquad j^{(k)}(\mathbf{r}, t) \propto \mathcal{E}^k$$

The ordinary electrodynamics is linear:

$$\mathbf{j} = \sigma \mathcal{E} \qquad \mathcal{P} = \chi \mathcal{E}$$

Formerly, optics was also limited to linear terms. In the most general terms, the linear relation between coordinates- and time-dependent current vectors and the field has the form

$$j_{\alpha}^{(1)}(\mathbf{r}, t) = \int \sigma_{\alpha\beta}^{(1)}(\mathbf{r}, \mathbf{r}'; t, t') \mathcal{E}_{\beta}(\mathbf{r}', t') \mathrm{d}^3 r' \mathrm{d}t' \tag{1.4.1}$$

This formula is a linear relation between the current and the electric field $\mathcal{E}$. The magnetic field is not involved here since it can be expressed in terms of $\mathcal{E}$ using the second Maxwell's equation in (1.3.1) or (1.3.2). Furthermore, its direct effect on charges is smaller by a factor v/c than that of the electric field.

With the advent of lasers, many nonlinear effects were discovered in optics. They can be described only by taking into account higher-order terms in the expansion of current in powers of field. In the general case, these nonlinear terms look as follows:

$$j_\alpha^{(2)}(\mathbf{r},t) = \int \sigma_{\alpha\beta\gamma}^{(2)}(\mathbf{r},\mathbf{r}',\mathbf{r}'';t,t',t'')\mathcal{E}_\beta(\mathbf{r}',t')\mathcal{E}_\gamma(\mathbf{r}'',t'')\,\mathrm{d}^3r'\,\mathrm{d}^3r''\,\mathrm{d}t'\,\mathrm{d}t'' \tag{1.4.2}$$

$$j_\alpha^{(3)}(\mathbf{r},t) = \int \sigma_{\alpha\beta\gamma\delta}^{(3)}(\mathbf{r},\mathbf{r}',\mathbf{r}'',\mathbf{r}''';t,t',t'',t''')\,\mathcal{E}_\beta(\mathbf{r}',t') \times \mathcal{E}_\gamma(\mathbf{r}'',t'')\mathcal{E}_\delta(\mathbf{r}''',t''')\mathrm{d}^3r'\,\mathrm{d}^3r''\,\mathrm{d}^3r'''\,\mathrm{d}t'\,\mathrm{d}t''\,\mathrm{d}t''' \tag{1.4.3}$$

Terms of fourth and higher orders in field have similar form.

We can make use of the expansion in powers of field as long as the external fields are much lower than the internal Coulomb fields, so that perturbations cannot greatly change the system and its properties:

$$\mathcal{E}/\mathcal{E}_{\mathrm{at}} \ll 1$$

Here $\mathcal{E}_{\mathrm{at}} \sim e/a_0^2 \sim 10^9\,\mathrm{V\,cm^{-1}}$ is a typical interatomic field. A field of the order of $\mathcal{E}_{\mathrm{at}}$ destroys atoms and molecules over times of the order of atomic time scale, that is, $\sim 10^{-16}$ s. Fields of this strength are hardly ever produced in laboratories; matter can only exist in such fields as a plasma, with a different nonlinearity parameter. The nonlinearity parameter $\mathcal{E}/\mathcal{E}_{\mathrm{at}}$ can be used for gases, liquids, and nonconducting solids. In the case of plasmas, metals, and semiconductors, on the other hand, all of which contain free charges, the possibility of expanding the current in a series in powers of field requires additional analysis. We will discuss it later. If the expansion of current in a series in powers of field is found not to be valid, for example, for the resonance interaction, one resorts to various simplified models which make it possible to solve the system of equations exactly, without power expansion.

By analogy to (1.4.1)–(1.4.3), we can write a series expansion of polarization in powers of field:

$$\mathcal{P}_\alpha(\mathbf{r},t) = \int \chi_{\alpha\beta}^{(1)}(\mathbf{r},\mathbf{r}';t,t')\mathcal{E}_\beta(\mathbf{r}',t')\mathrm{d}^3r'\,\mathrm{d}t' + \int \chi_{\alpha\beta\gamma}^{(2)}(\mathbf{r},\mathbf{r}'\mathbf{r}'';t,t',t'')\mathcal{E}_\beta(\mathbf{r}',t')\mathcal{E}_\gamma(\mathbf{r}'',t'')\mathrm{d}^3r'\,\mathrm{d}^3r''\,\mathrm{d}t'\,\mathrm{d}t'' + \ldots \tag{1.4.4}$$

It is necessary to comment on the limits of integration in (1.4.1)–(1.4.4). By virtue of the causality principle, the current and polarization at a time t depend only on the values of fields at earlier moments of time. Consequently, integration in time — in t', t'', etc. — is made from $-\infty$ to t. Owing to the constraints due to relativity, the integration in space is carried over the region in which $|\mathbf{r}-\mathbf{r}'| \leq c|t-t'|$ since the values of fields at points which do not satisfy this condition cannot influence the current at a point $\mathbf{r}$ at a time moment t (no interaction can propagate at a speed above the speed of light). As a rule, however, the finiteness of the limits of integration over $\mathbf{r}'$ and $\mathbf{r}''$, etc. is unimportant. It is acceptable to integrate in $\mathbf{r}'$ with infinite limits because a substantially smaller volume of space really contributes to the integral: the response functions σ and χ fall off to almost zero over distances $|\mathbf{r}-\mathbf{r}'|$ that are much smaller than the indicated relativistic limits.

As we see from expressions (1.4.1)–(1.4.4), the current or polarization at a point $\mathbf{r}$ at a time moment t are determined in the general case by the values of the field $\mathcal{E}$ at other spatial points $\mathbf{r}'$ — one then speaks of the nonlocality of the response, and by the values at preceding moments of time t' — this is known as retardation.

The nature of the retardation effect is sufficiently obvious. A particle "remembers" for some time the field at the preceding moments of time. For example, a free charge remembers the acceleration, and hence the field, of the preceding moments during the time τ of velocity relaxation. For a bound electron, this time interval is of the order of the atomic time scale

$$T_{\mathrm{at}} \sim \hbar/I_0 \sim 10^{-16}\,\mathrm{s}$$

that is, of the time during which the velocity of an atomic electron remains unchanged. This time is of the order of the period of optical radiation waves, so that retardation is significant in the optical and still shorter wavelength ranges. Figure 1.1 shows a typical dependence of $\sigma^{(1)}$ or $\chi^{(1)}$ on $t-t'$.

Nonlocality is a less familiar phenomenon. It arises because a particle arrives at a point $\mathbf{r}$ from $\mathbf{r}'$ and carries the memory of the action exerted on it at the point $\mathbf{r}'$. Consequently, the nonlocality radius for free charges, that is, the distance $|\mathbf{r}-\mathbf{r}'|$ on which the response functions σ and χ fall off almost to zero, is found to be of the order of the free path length $l = v\tau$. For bound charges, on the other hand, it is of the order of atomic or molecular size a_0 (Figure 1.2). The significance of the nonlocality effect depends on the ratio of the nonlocality radius to the characteristic distance over which the field changes significantly. Typically, phenomenological electrodynamics of continuous media uses as this distance the wavelength λ.

For the optical and the ultraviolet wavelength ranges, and all the more so for the infrared and radio ranges, we have $\lambda \gg a_0 \sim 10^{-8}\,\mathrm{cm}$. Consequently, if $|\mathbf{r}-\mathbf{r}'| \leq a_0$, the field at a point $\mathbf{r}'$ at such frequencies is almost

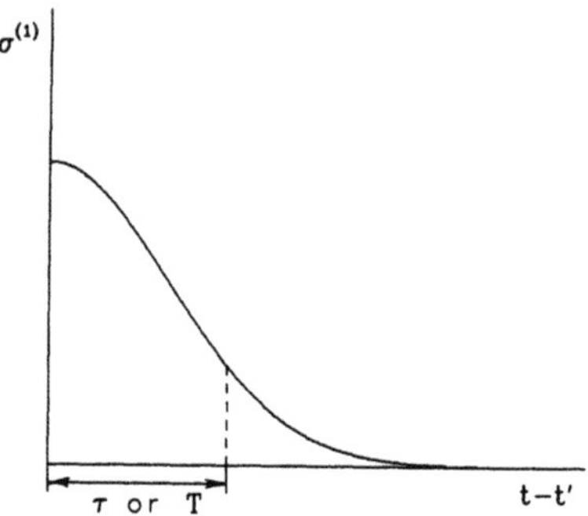

Figure 1.1. Response as a function of time.

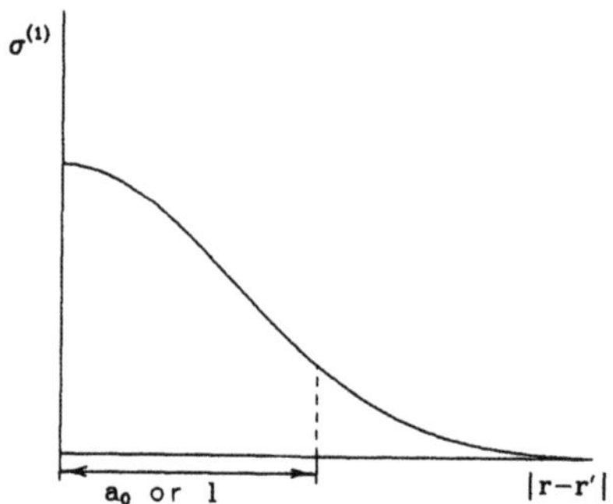

Figure 1.2. Response as a function of distance.

identical to the field at a point **r**. This is the reason why the nonlocality effect is usually only secondary while retardation is almost always significant. There are cases, however, when nonlocality is decisive (long free path lengths in metals or plasma, polaritons in semiconductors and dielectrics).

The nonlocality radius for a free charge is determined by the smaller of two lengths: the free path length $l = v\tau$ or the displacement of a charge over one field oscillation period $v\lambda/c$. If the nonlocality period is dictated by the second of the two lengths, the ratio of the nonlocality radius to the wavelength is of the order of $v/c \ll 1$. Nevertheless, exceptional cases are possible here even if the speed of light c/n is low (near its absorption band of the matter) and comparable with that of a particle v; nonlocality is then strong.

Sometimes nonlocality must be taken into account even if it is small. Thus gyrotropy (the effect of rotation of the plane of polarization in matter) is a corollary of nonlocality. In the framework of the approach we have chosen, it will be shown that magnetic susceptibilities are manifestations of nonlocality.

All nonlocality effects vanish after averaging over a physically infinitely small volume of a size much larger than a_0 or $v\tau$. This is an additional argument in favor of rejecting this type of averaging.

As for the microscopic approach to the description of the electromagnetic field in a medium, the nonlocality effects are always significant. It will become clear later, especially when crystals are considered, that the propagation of even long-wavelength field in condensed media is always accompanied with a small-scale field structure whose amplitude is not low and which originates from the fields of individual particles; these last fields are induced by the main field. The wavelengths characteristic of this structure are precisely of the order of interparticle distance, a_0, so that nonlocality effects in it are definitely not small. This structure determines the familiar phenomenon: the field acting on each particle in the medium differs from the mean field, and this must always be taken into account in a microscopic analysis of response.

1.5. Linear Susceptibilities and Their Properties

In the general case, the material equations (1.4.1)–(1.4.4) for stationary and spatially uniform media are integral and can be transformed to algebraic form using the Fourier transform.

A medium is said to be stationary if its properties are invariant in time in the absence of strong time-dependent fields. The choice of the origin $t = 0$ of the frame of reference is arbitrary; time enters the response of the medium only as the difference $t - t'$,

$$\sigma^{(1)}_{\alpha\beta}(\mathbf{r}, \mathbf{r}'; t - t') \qquad \chi^{(1)}_{\alpha\beta}(\mathbf{r}, \mathbf{r}'; t - t')$$

By definition, the spatial uniformity implies that the properties of the medium are identical at all its points. If this is so, $\sigma^{(1)}(\mathbf{r}, \mathbf{r}'; t - t')$ is not altered by the displacement $\mathbf{r} \to \mathbf{r} + \boldsymbol{\rho}$, $\mathbf{r}' \to \mathbf{r}' + \boldsymbol{\rho}$:

$$\sigma^{(1)}_{\alpha\beta}(\mathbf{r}, \mathbf{r}'; t - t') = \sigma^{(1)}_{\alpha\beta}(\mathbf{r} + \boldsymbol{\rho}, \mathbf{r}' + \boldsymbol{\rho}; t - t')$$

This requirement is equivalent to stating that $\sigma^{(1)}_{\alpha\beta}$ is a function only of the difference in coordinates:

$$\sigma^{(1)}_{\alpha\beta}(\mathbf{r}, \mathbf{r}'; t - t') = \sigma^{(1)}_{\alpha\beta}(\mathbf{r} - \mathbf{r}'; t - t')$$

The spatial uniformity is a property characterizing liquids, gases and amorphous solids, but not crystals that we will discuss separately. If the field and all other quantities are expanded in Fourier integrals,

$$\mathcal{E}_\alpha(\mathbf{r},t) = \int E_{\mathbf{k}\omega\alpha} e^{i\mathbf{kr}-i\omega t} d^3k\, d\omega/(2\pi)^4 \tag{1.5.1}$$

and inversely,

$$E_{\mathbf{k}\omega\alpha} = \int \mathcal{E}_\alpha(\mathbf{r},t) e^{-i\mathbf{kr}+i\omega t}\, d^3r\, dt \tag{1.5.2}$$

we obtain

$$j^{(1)}_{\mathbf{k}\omega\alpha} = \sigma^{(1)}_{\alpha\beta}(\mathbf{k},\omega) E_{\mathbf{k}\omega\beta} \tag{1.5.3}$$

where $\sigma^{(1)}_{\alpha\beta}(\mathbf{k},\omega)$ is the Fourier transform of the function $\sigma^{(1)}_{\alpha\beta}(\mathbf{r}-\mathbf{r}',t-t')$:

$$\sigma^{(1)}_{\alpha\beta}(\mathbf{k},\omega) = \int_0^\infty\!\!\int \sigma^{(1)}_{\alpha\beta}(\mathbf{R},\tau) e^{-i\mathbf{kR}+i\omega t}\, d\tau\, d^3R$$

and $\mathbf{R} = \mathbf{r} - \mathbf{r}'$; also, $\tau = t - t' > 0$, which reflects the causality principle: the action must precede the response.

We have mentioned already that instead of the current density, it is often more convenient to operate with polarization $\mathcal{P}$, and instead of electric conduction $\sigma^{(1)}$, the susceptibility $\chi^{(1)}$. Since

$$\mathbf{j}(\mathbf{r},t) = \frac{\partial \mathcal{P}(\mathbf{r},t)}{\partial t}$$

or, in terms of Fourier transforms,

$$j_{\mathbf{k}\omega\alpha} = -i\omega P_{\mathbf{k}\omega\alpha}$$

then

$$P^{(1)}_{\mathbf{k}\omega\alpha} = \chi^{(1)}_{\alpha\beta}(\mathbf{k},\omega) E_{\mathbf{k}\omega\beta} \tag{1.5.4}$$

where

$$\chi^{(1)}_{\alpha\beta}(\mathbf{k},\omega) = \frac{i}{\omega}\sigma^{(1)}_{\alpha\beta}(\mathbf{k},\omega) \tag{1.5.5}$$

Relation (1.5.5) between χ and σ is not limited to the case of linearity because it has been obtained from the general relation between current and polarization.

Now we can also consider the dielectric permittivity

$$\varepsilon_{\alpha\beta}(\mathbf{k},\omega) = \delta_{\alpha\beta} + 4\pi\chi^{(1)}_{\alpha\beta}(\mathbf{k},\omega)$$

which relates induction $\mathcal{D}$ to the electric field strength $\mathcal{E}$ when we use Maxwell's equation (1.3.3) written in the "dielectric" form.

For nonuniform media, we can consider functions of the type $\chi^{(1)}_{\alpha\beta}(\mathbf{r},\mathbf{r}',\omega)$, which is the Fourier transform of $\chi^{(1)}_{\alpha\beta}(\mathbf{r},\mathbf{r}',\tau)$ in time only. One refers to the dependence of σ, χ and ε on frequency as their dispersion, and to their dependence on $\mathbf{k}$, as to their spatial dispersion.

The dispersion is a corollary of the retardation effect, with the spatial dispersion following from nonlocality.

In optics, and in electrodynamics in general, the frequency of radiation is $\omega = ck/n$, where n is the refractive index of a medium; it may seem at the first glance, therefore, that the number of independent parameters of the Fourier transform decreases. The actual situation is, however, that ω and $\mathbf{k}$ play here the role of formal parameters of the Fourier transform and are thus independent variables. Relations of the type of (1.5.3) must be essentially valid not only for a free field which is a superposition of plane waves with $\omega = ck/n$ but for any fields (e.g., static fields) produced by an arbitrary distribution of charges and currents.

Fields, currents, and polarizations are real, which imposes constraints on the Fourier transforms. Since $\mathcal{E}(\mathbf{r},t)$ and $\mathbf{j}(\mathbf{r},t)$ are real, definitions (1.5.2) and (1.5.4) imply that

$$E_{\mathbf{k}\omega\alpha} = E^*_{-\mathbf{k},-\omega,\alpha} \qquad j_{\mathbf{k}\omega\alpha} = j^*_{-\mathbf{k},-\omega,\alpha} \qquad P_{\mathbf{k}\omega\alpha} = P^*_{-\mathbf{k},-\omega,\alpha}$$

Hence,

$$\begin{aligned} \sigma^{(1)}_{\alpha\beta}(\mathbf{k},\omega) &= \sigma^{(1)*}_{\alpha\beta}(-\mathbf{k},-\omega) \\ \chi^{(1)}_{\alpha\beta}(\mathbf{k},\omega) &= \chi^{(1)*}_{\alpha\beta}(-\mathbf{k},-\omega) \end{aligned} \tag{1.5.6}$$

When we later derive the general microscopic expression for susceptibility and conduction, we will be able to prove a less trivial proposition

$$\chi^{(1)}_{\alpha\beta}(\mathbf{k},\omega) = \chi^{(1)}_{\beta\alpha}(-\mathbf{k},\omega) \tag{1.5.7}$$

which is equivalent to the symmetry principle for kinetic coefficients.

In any specific medium, other symmetry properties are found which impose additional constraints on the tensors χ and σ. In the general case of stationary and spatially uniform medium, however, it is impossible to derive any other symmetry properties except those given above, unless one looks into the specific structure of the medium.

Now we can check whether anything has been lost (magnetic effects, etc.) when, in contrast to the traditional approach, only the induction is introduced. To do this, we look at the low-frequency range where conduction currents, displacement currents, and magnetization currents are clearly separated, that is, we make use of the quasistationary currents approximation.

First we write the expression for the total current neglecting magnetization currents:

$$j_\alpha(t) = \tilde{\sigma}_{\alpha\beta}\mathcal{E}_\beta + \frac{\partial}{\partial t}(\tilde{\chi}_{\alpha\beta}\mathcal{E}_\beta) = \frac{\partial \mathcal{P}_\alpha}{\partial t}$$

We also neglect here both retardation and nonlocality, since the wavelength is in the low-frequency range $\lambda \gg l$, where l is the sample size; $\tilde{\sigma}_{\alpha\beta}$ is the conductivity at low frequencies due to free charges, and $\tilde{\chi}_{\alpha\beta}$ is the dielectric susceptibility due to bound charges. The tensors $\tilde{\sigma}$ and $\tilde{\chi}$ are real.

For the total polarization we obtain

$$\mathcal{P}_\alpha(t) = \tilde{\chi}_{\alpha\beta}\mathcal{E}_\beta(t) + \int_{-\infty}^{t} \tilde{\sigma}_{\alpha\beta}\mathcal{E}_\beta(t')\,\mathrm{d}t'$$

and in terms of the Fourier transforms,

$$P_{\omega\alpha} = \tilde{\chi}_{\alpha\beta}E_{\omega\beta} + \frac{\mathrm{i}}{\omega}\tilde{\sigma}_{\alpha\beta}E_{\omega\beta} = \left(\tilde{\chi}_{\alpha\beta} + \frac{\mathrm{i}}{\omega}\tilde{\sigma}_{\alpha\beta}\right)E_{\omega\beta}$$

Comparing it with the general relation in the approach chosen,

$$P_{\omega\alpha} = \chi^{(1)}_{\alpha\beta}(\omega)E_{\omega\beta}$$

we obtain

$$\chi^{(1)}_{\alpha\beta}(\omega) = \tilde{\chi}_{\alpha\beta} + \frac{\mathrm{i}}{\omega}\tilde{\sigma}_{\alpha\beta} \tag{1.5.8}$$

In what follows, it is shown that $\tilde{\chi}$ corresponds to the nondissipative component of the total susceptibility χ, and $\mathrm{i}\tilde{\sigma}/\omega$ corresponds to the dissipative component.

Therefore, the contribution of the free and bound charges is not lost in the approach described. Let us look now at how magnetic currents are described when only induction and only susceptibility are used. It may seem at the first glance that the magnetization currents are lost since only the electric field is found in expressions for current, such as (1.4.1). In actual fact, however, the electric and magnetic fields are related through Maxwell's equations.

Let us take standard expressions for the magnetization current,

$$\mathbf{j}_m(t) = c\,\mathrm{curl}\,\mathcal{M} = c\,\mathrm{curl}\,\tilde{\chi}_m(\mathcal{H} - 4\pi\mathcal{M}) \simeq c\,\mathrm{curl}\,\tilde{\chi}_m\mathcal{H}$$

For the sake of simplicity, we will consider the case of isotropic medium, which is characterized by scalar magnetic susceptibility $\tilde{\chi}_m$ at low frequencies.

Let us make use of Maxwell's equation

$$\operatorname{curl}\boldsymbol{\mathcal{E}} = -\frac{1}{c}\frac{\partial \boldsymbol{\mathcal{H}}}{\partial t}$$

Integrating it in time and assuming that all fields are zero at $t = -\infty$, we obtain

$$\boldsymbol{\mathcal{H}}(\mathbf{r}, t) = -c \int_{-\infty}^{t} \operatorname{curl}\boldsymbol{\mathcal{E}}(\mathbf{r}, t')\, dt'$$

Therefore, the magnetization current can be written in the form

$$\mathbf{j}_m(\mathbf{r}, t) = -c^2\tilde{\chi}_m \int_{-\infty}^{t} \operatorname{curl}\, \operatorname{curl}\boldsymbol{\mathcal{E}}(\mathbf{r}, t')\, dt'$$

$$= -c^2\tilde{\chi}_m \int_{-\infty}^{t} (\operatorname{grad} \nabla\boldsymbol{\mathcal{E}} - \Delta\boldsymbol{\mathcal{E}})\, dt'$$

In Cartesian coordinates, this vector relation is rewritten as

$$j_{m\alpha}(\mathbf{r}, t) = -c^2\tilde{\chi}_m \int_{-\infty}^{t} \left(\frac{\partial^2}{\partial x_\alpha \partial x_\beta} - \frac{\partial^2}{\partial {x_\gamma}^2}\, \delta_{\alpha\beta}\right) \mathcal{E}_\beta dt' \tag{1.5.9}$$

We will show now that a contribution of the form of (1.5.9) can be singled out from the general expression (1.4.1). To do this, we separate from the total current its component which contains second derivatives of electric field with respect to spatial coordinates. This can be done only if the nonlocality effect, that is, spatial dispersion, is taken into account.

Let us assume that nonlocality effects are small, that is, that the characteristic length of nonuniformity of the field (this is typically the wavelength λ) is much greater than nonlocality radius (as a rule, it is of the order of a_0). In this case, the response $\sigma^{(1)}_{\alpha\beta}(\mathbf{r} - \mathbf{r}', t - t')$ vanishes only where $|\mathbf{r} - \mathbf{r}'| \lesssim a_0$.

Let us expand $\mathcal{E}_\beta(\mathbf{r}', t')$ in the integrand of formula (1.4.1) into a Taylor series in the neighborhood of a point $\mathbf{r}$:

$$\mathcal{E}_\beta(\mathbf{r}', t') = \mathcal{E}_\beta(\mathbf{r}, t') + \frac{\partial \mathcal{E}_\beta(\mathbf{r}, t')}{\partial x_\gamma}(x'_\gamma - x_\gamma)$$

$$+ \frac{1}{2}\frac{\partial^2 \mathcal{E}_\beta(\mathbf{r}, t)}{\partial x_\gamma \partial x_\delta}(x'_\gamma - x_\gamma)(x'_\delta - x_\delta) + \dots$$

If this expansion is substituted into (1.4.1), the field $\mathcal{E}_\beta(\mathbf{r}, t')$ and its derivatives can be factored out of the integral in spatial variables. Let us introduce the following notation:

$$\sigma^{(1)}_{\alpha\beta}(t-t') = \int \sigma^{(1)}_{\alpha\beta}(\mathbf{r}-\mathbf{r}', t-t')\mathrm{d}^3r'$$

$$a_{\alpha\beta\gamma}(t-t') = \int (x'_\gamma - x_\gamma)\sigma^{(1)}_{\alpha\beta}(\mathbf{r}-\mathbf{r}', t-t')\mathrm{d}^3r'$$

$$b_{\alpha\beta\gamma\delta}(t-t') = \int (x'_\gamma - x_\gamma)(x'_\delta - x_\delta)\sigma^{(1)}_{\alpha\beta}(\mathbf{r}-\mathbf{r}', t-t')\mathrm{d}^3r'$$

This gives

$$j_\alpha(\mathbf{r}, t) = \int_{-\infty}^{t} \Big\{ \sigma^{(1)}_{\alpha\beta}(t-t')\mathcal{E}_\beta(\mathbf{r}, t') + a_{\alpha\beta\gamma}(t-t')\frac{\partial \mathcal{E}_\beta(\mathbf{r}, t')}{\partial x_\gamma} + \frac{1}{2} b_{\alpha\beta\gamma\delta}(t-t')\frac{\partial^2 \mathcal{E}_\beta(\mathbf{r}, t')}{\partial x_\gamma \partial x_\beta} + \ldots \Big\} \mathrm{d}t' \tag{1.5.10}$$

The first term in braces is the response of the system to the field if the spatial dispersion is neglected. We have already discussed it.

For media that are symmetric with respect to inversion, the second term containing $\partial\mathcal{E}_\beta/\partial c_\gamma$ vanishes since $a_{\alpha\beta\gamma} = 0$.

Note that the third term has a contribution corresponding to magnetic currents. If we assume

$$\tilde{b}_{\alpha\beta\gamma\delta} = -2\tilde{\chi}_m c^2(\delta_{\alpha\gamma}\delta_{\beta\delta} - \delta_{\alpha\beta}\delta_{\gamma\delta})$$

we thereby single out from (1.5.10) a contribution which coincides with the expression for the magnetic current, (1.5.9).

In the approach chosen here, the magnetization current is a corollary of weak spatial dispersion. It is the induction current due to the second derivatives of electric field with respect to coordinates. The ratio of the magnitude of this current to the main contribution (the first term in (1.5.10)) is of the order of $(a_0/\lambda)^2 \sim (v_{\mathrm{at}}/c)^2 \ll 1$. This estimate explains why the magnetic susceptibility (for paramagnetic materials, $\tilde{\chi}_m \sim 10^{-6}$) is small in comparison with the susceptibility of dielectrics ($\chi^{(1)}_{\alpha\beta} \sim 1$). We were thus able to show that expressions of the type of (1.4.1) include all currents: conduction, displacement and magnetization ones.

Let us switch to the Fourier representation in the expression (1.5.9) for the magnetization current. We obtain

$$j_{m\mathbf{k}\omega\alpha} = -c^2\tilde{\chi}_m \frac{\mathrm{i}}{\omega} k^2\Big(\delta_{\alpha\beta} - \frac{k_\alpha k_\beta}{k^2}\Big)\mathcal{E}_{\mathbf{k}\omega\beta} \tag{1.5.11}$$

The expression

$$\Pi_{\perp\alpha\beta} = (\delta_{\alpha\beta} - k_\alpha k_\beta / k^2)$$

is the operation of projecting on the directions perpendicular to $\mathbf{k}$; it singles out only the transverse fields. This corresponds to the fact that the magnetization current is determined only by the transverse (vortex) component of the electric field. Formula (1.5.11) implies an expression for the contribution to $\chi^{(1)}(\mathbf{k}, \omega)$ due to magnetization currents. Taking into account (1.5.3) and (1.5.6), we derive from (1.5.11) that

$$\chi^{(1)}_{m\alpha\beta}(\mathbf{k}, \omega) = \frac{c^2 k^2 \tilde{\chi}_m}{\omega^2}\left(\delta_{\alpha\beta} - \frac{k_\alpha k_\beta}{k^2}\right) \tag{1.5.12}$$

Note that this contribution to the susceptibility $\chi^{(1)}$ at low frequencies has a singularity ω^{-2}. The origin of this singularity is purely formal. The corresponding vortex electric field $\mathcal{E} \sim \partial H/\partial t$ is very low and is proportional to the frequency so that the formula for susceptibility includes the factor ω^{-1}. The second factor ω^{-1} originates in definition (1.5.5).

1.6. Relation of Dissipation of Energy in a Medium to the Anti-Hermitian Component of Linear Susceptibility

We have seen already that the imaginary component of susceptibility (1.5.8) is determined at low frequencies in the range of quasistationary currents by the static electric conductivity of the medium, that is, is related to the dissipation of the field's energy. We can now show that this meaning of the imaginary component χ is retained at all frequencies.

The dissipation of the energy of the field is the work that the field does over the particles of the medium. The energy dissipated per unit time in unit volume of a region V whose dimensions are much larger than the nonlocality radius is

$$Q = \frac{1}{V}\int_V j_\alpha(\mathbf{r}, t)\mathcal{E}_\alpha(\mathbf{r}, t)\,\mathrm{d}^3 r \tag{1.6.1}$$

It is not easy to speak of losses at a point when nonlocality is taken into account; hence, integration over the volume V is carried out. The dissipation, as well as the response of the medium, becomes locally meaningless. For harmonic fields and currents, for which

$$\mathcal{E}_\alpha(\mathbf{r}, t) = 1/2(E_{\omega\alpha}(\mathbf{r})\mathrm{e}^{-\mathrm{i}\omega t} + E^*_{\omega\alpha}(\mathbf{r})\mathrm{e}^{\mathrm{i}\omega t})$$
$$j_\alpha(\mathbf{r}, t) = 1/2(j_{\omega\alpha}(\mathbf{r})\mathrm{e}^{-\mathrm{i}\omega t} + j^*_{\omega\alpha}(\mathbf{r})\mathrm{e}^{\mathrm{i}\omega t})$$

we obtain from (1.6.1) after averaging over time

$$\begin{aligned}\overline{\dot{Q}} &= \frac{1}{4V}\int_V \Big(j_{\omega\alpha}(\mathbf{r})E^*_{\omega\alpha}(\mathbf{r}) + j^*_{\omega\alpha}(\mathbf{r})E_{\omega\alpha}(\mathbf{r})\Big)\,\mathrm{d}^3r \\ &= \frac{\omega}{4\mathrm{i}V}\int_V \mathrm{d}^3r \int_V \mathrm{d}^3r' \Big(\chi^{(1)}_{\alpha\beta}(\mathbf{r},\mathbf{r}';\omega)E^*_{\omega\alpha}(\mathbf{r})E_{\omega\beta}(\mathbf{r}') \\ &\quad - \chi^{(1)*}_{\alpha\beta}(\mathbf{r},\mathbf{r}';\omega)E_{\omega\alpha}(\mathbf{r})E^*_{\omega\beta}(\mathbf{r}')\Big) \\ &= \frac{\omega}{2V}\int_V \mathrm{d}^3r \int_V \mathrm{d}^3r'\, E^*_{\omega\alpha}(\mathbf{r})\frac{\chi^{(1)}_{\alpha\beta}(\mathbf{r},\mathbf{r}';\omega) - \chi^{(1)*}_{\beta\alpha}(\mathbf{r}',\mathbf{r};\omega)}{2\mathrm{i}}E_{\omega\beta}(\mathbf{r}')\end{aligned} \tag{1.6.2}$$

We have used here the expression (1.4.1) for the current and a relation of the type of (1.5.5) between σ and χ; $\chi^{(1)}_{\alpha\beta}(\mathbf{r},\mathbf{r}';\omega)$ stand for the Fourier transforms $\chi^{(1)}_{\alpha\beta}(\mathbf{r},\mathbf{r}';t-t')$ in time. The integration in $\mathbf{r}'$ is carried out both over the volume V and over $\mathbf{r}$ because it is assumed that if $|\mathbf{r}-\mathbf{r}'|$ is greater than the nonlocality radius, then $\chi^{(1)}_{\alpha\beta}(\mathbf{r},\mathbf{r}';\omega) \simeq 0$. At the last step, summation and integration variable in the second term in (1.6.2) have been exchanged: $\alpha \leftrightarrow \beta$ and $\mathbf{r} \leftrightarrow \mathbf{r}'$.

For plane waves, that is,

$$E_{\omega\alpha}(\mathbf{r}) = E_{\mathbf{k}\omega\alpha}\mathrm{e}^{\mathrm{i}\mathbf{k}\mathbf{r}}$$

we obtain for spatially uniform medium, instead of (1.6.2)

$$\overline{\dot{Q}} = \frac{\omega}{2}\frac{1}{2\mathrm{i}}\Big(\chi^{(1)}_{\alpha\beta}(\mathbf{k},\omega) - \chi^{(1)*}_{\beta\alpha}(\mathbf{k},\omega)\Big)E^*_{\mathbf{k}\omega\alpha}E_{\mathbf{k}\omega\beta} \tag{1.6.3}$$

If the field $\boldsymbol{\mathcal{E}}$ is taken as a sum of plane waves with different values of $\mathbf{k}$ (a Fourier integral or Fourier series), then integration over a large volume V yields a sum, or integral, of expressions of the type of (1.6.3) over $\mathbf{k}$.

Expressions (1.6.2) and (1.6.3) show that losses are determined by the anti-Hermitian component of the linear susceptibility tensor

$$\frac{1}{2\mathrm{i}}(\chi^{(1)}_{\alpha\beta}(\mathbf{r},\mathbf{r}';\omega) - \chi^{(1)*}_{\beta\alpha}(\mathbf{r}',\mathbf{r};\omega))$$

or, for a spatially uniform medium, by

$$\frac{1}{2\mathrm{i}}(\chi^{(1)}_{\alpha\beta}(\mathbf{k},\omega) - \chi^{(1)*}_{\beta\alpha}(\mathbf{k},\omega))$$

If spatial dispersion is absent, the principle of symmetry of kinetic coefficients (1.5.8) implies that

$$\chi^{(1)}_{\alpha\beta}(\mathbf{k},\omega) = \chi^{(1)}_{\alpha\beta}(\omega) = \chi^{(1)}_{\beta\alpha}(\omega)$$

In this case the anti-Hermitian part of the tensor $\chi^{(1)}$ is its imaginary part.

In many cases, we have

$$\chi^{(1)}_{\alpha\beta}(-\mathbf{k},\omega) = \chi^{(1)}_{\alpha\beta}(\mathbf{k},\omega)$$

even if the spatial dispersion is nonzero (nongyrotropic cases). In this situation, losses are determined, again by virtue of (1.5.7), by the imaginary part of the susceptibility tensor,

$$\overline{\dot{Q}} = \frac{\omega}{2}\mathrm{Im}\chi^{(1)}_{\alpha\beta}E^{*}_{\mathbf{k}\omega\alpha}E_{\mathbf{k}\omega\beta}$$

The susceptibilities which we began discussing in section 1.4, are the true susceptibilities that relate polarization and the statistically average field in the medium. These are the relations which complement the system of Maxwell's equations (1.3.1) or (1.3.3) and are introduced in the phenomenological theory.

Another approach is to begin with external fields (in the absence of the medium) or sources $\mathbf{j}^{(e)}$ and $\rho^{(e)}$ of external fields, as in equations (1.3.2) and (1.3.3). These sources would produce, in the absence of the medium, external fields $\mathcal{E}^{(e)}$, and we can pose the problem of finding the response of the system, that is, the current and charge densities $\mathbf{j}$ and ρ, to the external field. The expansion of the current in powers of external field yields expressions of the same type as (1.4.1), (1.4.3) and (1.4.4), but with $\mathcal{E}$ replaced by $\mathcal{E}^{(e)}$, and σ and χ by the responses to the external field, $\sigma^{(e)}$ and $\chi^{(e)}$.

The point is that it is $\chi^{(e)}$ (or $\sigma^{(e)}$) and not χ that the microscopic theory calculates using quantum mechanics. We will show this to be true in the next chapter when deriving the general expressions for susceptibilities.

In applications, however, one needs the response to the true field, that is, the susceptibility χ. The problem then is to establish a relation between χ and $\chi^{(e)}$, in order to find the response in the medium to the true field from the response to the external field calculated in terms of quantum mechanics.

We begin with Maxwell's equations with external currents, (1.3.2) or (1.3.3). As usual, we can exclude the magnetic field and write equations for the electric field:

$$\mathrm{curl}\,\mathrm{curl}\,\mathcal{E} + \frac{1}{c^2}\frac{\partial^2\mathcal{E}}{\partial t^2} + \frac{4\pi}{c^2}\frac{\partial^2\mathcal{P}}{\partial t^2} = -\frac{4\pi}{c^2}\frac{\partial j^{(e)}}{\partial t} \tag{1.6.4}$$

After Fourier transforms of all fields of the type (1.5.1) or (1.5.2) are found, we arrive at Fourier transforms

$$\mathcal{D}_{0\alpha\beta}^{-1}(\mathbf{k},\omega)E_{\mathbf{k}\omega\beta} - \frac{4\pi\omega^2}{c^2}P_{\mathbf{k}\omega\alpha} = \frac{4\pi \mathrm{i}\omega}{c^2}j^{(\mathrm{e})}_{\mathbf{k}\omega\alpha} \tag{1.6.5}$$

We have introduced here a tensor

$$\mathcal{D}_{0\alpha\beta}^{-1}(\mathbf{k},\omega) = \left(k^2 - \frac{\omega^2}{c^2}\right)\delta_{\alpha\beta} - k_\alpha k_\beta \tag{1.6.6}$$

Equation (1.6.5) is obtained from (1.6.4) if we recall that the field

$$\mathcal{E} = \mathbf{E}\mathrm{e}^{\mathrm{i}\mathbf{kr}} + \mathrm{C.C.}$$

(C.C. stands for complex-conjugate expressions) implies that

$$\operatorname{curl}\mathcal{E} = \mathrm{i}[\mathbf{kE}]\mathrm{e}^{\mathrm{i}\mathbf{kr}} + \mathrm{C.C.}$$
$$\operatorname{curl}\operatorname{curl}\mathcal{E} = -[\mathbf{k}[\mathbf{kE}]]\mathrm{e}^{\mathrm{i}\mathbf{kr}} + \mathrm{C.C.} = (-\mathbf{k}(\mathbf{kE}) + k^2\mathbf{E})\mathrm{e}^{\mathrm{i}\mathbf{kr}} + \mathrm{C.C.}$$

We will need the tensor inverse to $\mathcal{D}_0^{-1}$. It is easy to find if we single out in $\mathcal{D}_0^{-1}$ its longitudinal and transverse parts. First we recast $\mathcal{D}_0^{-1}$ to the form

$$\mathcal{D}_{0\alpha\beta}^{-1}(\mathbf{k},\omega) = \left(k^2 - \frac{\omega^2}{c^2}\right)\Pi_{\perp\alpha\beta}(\mathbf{k}) - \frac{\omega^2}{c^2}\Pi_{\parallel\alpha\beta}(\mathbf{k}) \tag{1.6.7}$$

where

$$\Pi_{\parallel\alpha\beta}(\mathbf{k}) = \frac{k_\alpha k_\beta}{k^2} \qquad \text{and} \qquad \Pi_{\perp\alpha\beta}(\mathbf{k}) = \delta_{\alpha\beta} - \frac{k_\alpha k_\beta}{k^2}$$

are the operators of projection to the directions which are parallel and perpendicular, respectively, to $\mathbf{k}$.

The projection operators $\Pi_\parallel$ and $\Pi_\perp$ have the following properties:

$$\Pi_\perp\Pi_\perp = \Pi_\perp \qquad \Pi_\parallel\Pi_\parallel = \Pi_\parallel \qquad \Pi_\perp\Pi_\parallel = \Pi_\parallel\Pi_\perp = 0 \qquad \Pi_\perp + \Pi_\parallel = 1$$

If an arbitrary operator Γ can be written as

$$\Gamma = \gamma_\parallel\Pi_\parallel + \gamma_\perp\Pi_\perp$$

where $\gamma_\parallel$ and $\gamma_\perp$ are numerical coefficients, then it is not difficult to verify that

$$\Gamma^{-1} = \frac{1}{\gamma_\parallel}\Pi_\parallel + \frac{1}{\gamma_\perp}\Pi_\perp$$

Indeed,

$$\Gamma\Gamma^{-1} = (\gamma_\parallel\Pi_\parallel + \gamma_\perp\Pi_\perp)\left(\frac{1}{\gamma_\parallel}\Pi_\parallel + \frac{1}{\gamma_\perp}\Pi_\perp\right) = \Pi_\parallel + \Pi_\perp = 1$$

Equation (1.6.7) implies, therefore, that

$$\mathcal{D}_0 \equiv (\mathcal{D}_0^{-1})^{-1} = \frac{1}{k^2 - \omega^2/c^2} \Pi_\perp - \frac{c^2}{\omega^2} \Pi_\parallel \tag{1.6.8}$$

By multiplying equation (1.6.5) by $\mathcal{D}_0$, we obtain

$$\left(1 - \frac{4\pi\omega^2}{c^2} \mathcal{D}_0 \chi^{(1)}\right) E = \frac{4\pi i\omega}{c^2} \mathcal{D}_0 j^{(e)} \tag{1.6.9}$$

For the sake of brevity, we have dropped the tensor suffices α, β and made use of the matrix notation.

Assuming $\chi^{(1)} = 0$ in (1.6.9), we obtain an expression for the external field (the field produced by the current $j^{(e)}$ in the absence of matter):

$$E^{(e)} = \frac{4\pi i\omega}{c^2} \mathcal{D}_0 j^{(e)} \tag{1.6.10}$$

Equation (1.6.9) gives

$$\left(1 - \frac{4\pi\omega^2}{c^2} \mathcal{D}_0 \chi^{(1)}\right) E = E^{(e)} \tag{1.6.11}$$

that is,

$$E = \left(1 - \frac{4\pi\omega^2}{c^2} \mathcal{D}_0 \chi^{(1)}\right)^{-1} E^{(e)} \tag{1.6.12}$$

Formulas (1.6.11) and (1.6.12) express the external field in terms of the true field, and vice versa.

The definition

$$P^{(1)} = \chi^{(1)} E = \chi^{(1e)} E^{(e)}$$

and equation (1.6.12) imply that

$$\chi^{(1e)} = \chi^{(1)} \left(1 - \frac{4\pi\omega^2}{c^2} \mathcal{D}_0 \chi^{(1)}\right)^{-1} \tag{1.6.13}$$

This formula expresses $\chi^{(1e)}$ in terms of $\chi^{(1)}$, although the reverse relation is more interesting. Equation (1.6.13) is readily solved for $\chi^{(1)}$:

$$\chi^{(1)} = \chi^{(1e)} \left(1 - \frac{4\pi\omega^2}{c^2} \mathcal{D}_0 \chi^{(1)}\right) = \chi^{(1e)} - \frac{4\pi\omega^2}{c^2} \chi^{(1e)} \mathcal{D}_0 \chi^{(1)}$$

Transferring the last term of the righthand side to the left, we obtain

$$\left(1 + \frac{4\pi\omega^2}{c^2} \chi^{(1e)} \mathcal{D}_0\right) \chi^{(1)} = \chi^{(1e)}$$

that is,

$$\chi^{(1)} = \left(1 + \frac{4\pi\omega^2}{c^2}\chi^{(1e)}\mathcal{D}_0\right)^{-1}\chi^{(1e)} \tag{1.6.14}$$

The identity

$$\left(1 + \frac{4\pi\omega^2}{c^2}\chi^{(1e)}\mathcal{D}_0\right)\chi^{(1e)} = \chi^{(1e)}\left(1 + \frac{4\pi\omega^2}{c^2}\mathcal{D}_0\chi^{(1e)}\right)$$

which implies

$$\left(1 + \frac{4\pi\omega^2}{c^2}\chi^{(1e)}\mathcal{D}_0\right)^{-1}\chi^{(1e)} = \chi^{(1e)}\left(1 + \frac{4\pi\omega^2}{c^2}\mathcal{D}_0\chi^{(1e)}\right)^{-1}$$

enables us to rewrite (1.6.14) in an equivalent form:

$$\chi^{(1)} = \chi^{(1e)}\left(1 + \frac{4\pi\omega^2}{c^2}\mathcal{D}_0\chi^{(1e)}\right)^{-1} \tag{1.6.15}$$

Formulas (1.6.14) and (1.6.15) make it possible to find expressions for $\chi^{(1)}$ once the susceptibilities for the external field $\chi^{(1e)}$ have been found from the microscopic theory.

Using expressions (1.6.11)–(1.6.15), we can express the losses in terms of the external fields and the susceptibilities $\chi^{(1e)}$. Equation (1.6.11) implies that

$$\begin{aligned}
\frac{1}{2i}E^{(e)+}(\chi^{(1e)} - \chi^{(1e)+})E^{(e)} &= \frac{1}{2i}(E^{(e)+}\chi^{(1)}E - E^{+}\chi^{(1)+}E^{(e)}) \\
&= \frac{1}{2i}\Bigg(E^{+}\left(1 - \frac{4\pi\omega^2}{c^2}\chi^{(1)+}\mathcal{D}_0\right)\chi^{(1)}E \\
&\quad - E^{+}\chi^{(1)+}\left(1 - \frac{4\pi\omega^2}{c^2}D_0\chi^{(1)}\right)E\Bigg) \\
&= \frac{1}{2i}E^{+}(\chi^{(1)} - \chi^{(1)+})E
\end{aligned}$$

(the superscript + denotes Hermitian conjugation, that is, transposition of the matrix, or the column-to-row conversion with the subsequent operation of complex conjugation).

Correspondingly, expression (1.6.3) for losses can be rewritten in terms of susceptibility $\chi^{(1e)}$ and the external field $E^{(e)}$:

$$\overline{\dot{Q}} = \frac{\omega}{2}\frac{1}{2i}E^{(e)*}_{\mathbf{k}\omega\alpha}\left(\chi^{(1e)}_{\alpha\beta}(\mathbf{k},\omega) - \chi^{(1e)}_{\beta\alpha}(\mathbf{k},\omega)\right)E^{(e)}_{\mathbf{k}\omega\beta} \tag{1.6.16}$$

The dissipation of the energy of the field in a medium thus can be given by two equivalent formulas: either in terms of the mean field and the imaginary part of susceptibility, or in terms of the external field and the imaginary part of the response to the external field. The physical meaning expressing the losses via the external field is that the fields E and $E^{(e)}$ differ by fields that are generated by the induced currents. The induced field then describes the internal interaction between particles of a medium and does not lead to dissipation. What really dissipates is the energy of the external field.

1.7. Nonlinear Susceptibilities

The retardation effect and nonlocality (and, correspondingly, dispersion and spatial dispersion) characterize nonlinear susceptibilities as well as linear ones.

The general relation of the current density $\mathbf{j}$ and the field $\mathbf{E}$ in the second order in field is given by (1.4.2). The integration in time is carried out from $-\infty$ to t, and that in coordinates, over the entire space.

In a stationary and spatially uniform medium we have

$$\sigma^{(2)}_{\alpha\beta\gamma} = \sigma^{(2)}_{\alpha\beta\gamma}(\mathbf{r}-\mathbf{r}', \mathbf{r}-\mathbf{r}''; t-t', t-t'')$$

It will again be more convenient to transfer to the Fourier representation (1.5.1) and (1.5.2). Let us introduce the notation

$$\mathbf{R}' = \mathbf{r}-\mathbf{r}' \qquad \mathbf{R}'' = \mathbf{r}-\mathbf{r}'' \qquad \tau' = t-t' \qquad \tau'' = t-t''$$

The quantities τ' and τ'' are the response delay times at the moment of time t relative to the fields acting at the moments t' and t''. By virtue of the causality principle, $\tau' \geq 0$ and $\tau'' \geq 0$.

If (1.4.2) is multiplied by $\exp(\mathrm{i}\omega t - \mathrm{i}\mathbf{k}\mathbf{r})$ and integrated over t and r, we obtain

$$j^{(2)}_{\mathbf{k}\omega\alpha} = \int \mathrm{d}t\, \mathrm{d}^3 r\, \mathrm{e}^{-\mathrm{i}\mathbf{k}\mathbf{r}+\mathrm{i}\omega t} \int_0^\infty \mathrm{d}\tau'\, \mathrm{d}\tau'' \int \mathrm{d}^3 R'\, \mathrm{d}^3 R''\, \sigma^{(2)}_{\alpha\beta\gamma}(\mathbf{R}', \mathbf{R}''; \tau', \tau'')$$
$$\times \int \frac{\mathrm{d}^3 k'\, \mathrm{d}^3 k''\, \mathrm{d}\omega'\, \mathrm{d}\omega''}{(2\pi)^8}\, \mathrm{e}^{\mathrm{i}\mathbf{k}(\mathbf{r}-\mathbf{R}')} \exp[\mathrm{i}\mathbf{k}''(\mathbf{r}-\mathbf{R}'') - \mathrm{i}\omega'(t-\tau')$$
$$- \mathrm{i}\omega''(t-\tau'')] E_{\mathbf{k}'\omega'\beta} E_{\mathbf{k}''\omega''\gamma}$$

Integration in t and r can now be carried out; using then the integral representation of the δ function

$$(2\pi)^4 \delta(\omega-\omega'-\omega'')\delta^3(\mathbf{k}-\mathbf{k}'-\mathbf{k}'') = \int \mathrm{e}^{-\mathrm{i}(\mathbf{k}-\mathbf{k}'-\mathbf{k}'')\mathbf{r}+\mathrm{i}(\omega-\omega'-\omega'')t}\, \mathrm{d}t\, \mathrm{d}^3 r$$

we find

$$j^{(2)}_{\mathbf{k}\omega\alpha} = \int \frac{d^3k'\, d^3k''\, d\omega'\, d\omega''}{(2\pi)^4} \delta^3(\mathbf{k} - \mathbf{k}' - \mathbf{k}'')\, \delta(\omega - \omega' - \omega'') \times \sigma^{(2)}_{\alpha\beta\gamma}(\mathbf{k}', \mathbf{k}''; \omega', \omega'') E_{\mathbf{k}'\omega'\beta} E_{\mathbf{k}''\omega''\gamma} \qquad (1.7.1)$$

where

$$\sigma^{(2)}_{\alpha\beta\gamma}(\mathbf{k}', \mathbf{k}''; \omega', \omega'') = \int_0^\infty d\tau'\, d\tau'' \int d^3R'\, d^3R'' \sigma^{(2)}_{\alpha\beta\gamma}(\mathbf{R}', \mathbf{R}''; \tau', \tau'') \times \exp(-i\mathbf{k}'\mathbf{R}' - i\mathbf{k}''\mathbf{R}'' + i\omega'\tau' + i\omega''\tau'')$$

is the Fourier transform of the function $\sigma^{(2)}_{\alpha\beta\gamma}$.

The relation of the current $j^{(2)}_{\mathbf{k}\omega\alpha}$ to the fields is not algebraic here, but integral. However, if only two plane harmonic modes of the field, $\mathbf{k}'\omega'$ and $\mathbf{k}''\omega''$, are nonzero, it produces a current

$$j^{(2)}_{\mathbf{k}'+\mathbf{k}'', \omega'+\omega'', \alpha} = \sigma^{(2)}_{\alpha\beta\gamma} E_{\mathbf{k}'\omega'\beta} E_{\mathbf{k}''\omega''\gamma}$$

As in the linear case, not the conductivity $\sigma^{(2)}$ but susceptibility $\chi^{(2)}$ is often used. Their relationship is similar to (1.5.5):

$$\chi^{(2)}_{\alpha\beta\gamma}(\mathbf{k} = \mathbf{k}' + \mathbf{k}''; \omega = \omega' + \omega'') = \frac{i}{\omega} \sigma^{(2)}_{\alpha\beta\gamma}(\mathbf{k} = \mathbf{k}' + \mathbf{k}''; \omega = \omega' + \omega'') \qquad (1.7.2)$$

Owing to δ functions, (1.7.1) can be integrated, for example, in $\mathbf{k}''$ and ω'' but it is more convenient to retain the form of (1.7.1) in which δ functions indicate the condition of summation of frequencies and the synchronism condition:

$$\omega = \omega' + \omega'' \qquad \mathbf{k} = \mathbf{k}' + \mathbf{k}'' \qquad (1.7.3)$$

The situation with cubic susceptibility and higher-order susceptibilities is quite similar.

Let us see what form the loss expressions will take if the quadratic contribution to current is taken into account. We assume that there are three waves which satisfy conditions (1.7.3). For simplicity, we neglect spatial dispersion. By analogy to (1.6.3), we obtain the following expression for the losses at a frequency ω, caused by the quadratic current $j^{(2)}$:

$$\overline{\dot{Q}_\omega} = \overline{j_\alpha(\mathbf{r}, t)\mathcal{E}_\alpha(\mathbf{r}, t)} = \frac{\omega}{2}\frac{1}{2i}(\chi^{(2)}_{\alpha\beta\gamma} E^*_{\omega\alpha} E_{\omega'\beta} E_{\omega''\gamma} - \chi^{(2)*}_{\alpha\beta\gamma} E_{\omega\alpha} E^*_{\omega'\beta} E^*_{\omega''\gamma}) = \frac{\omega}{2} \operatorname{Im}(\chi^{(2)}_{\alpha\beta\gamma}(\omega = \omega' + \omega'') E^*_{\omega\alpha} E_{\omega'\beta} E_{\omega''\gamma}) \qquad (1.7.4)$$

The averaging is carried out over time, so that rapidly oscillating terms in $\dot{Q}$ are eliminated.

In the linear case, the losses were expressed in terms of the imaginary part of susceptibility; here, they are found in terms of the imaginary part of the product of susceptibility $\chi^{(2)}$ by the amplitudes of the fields. These amplitudes are complex so that the losses depend on the relative phases of the fields at the frequencies ω, ω' and ω''.

The physical meaning of this result is that only one wave at the frequency ω is produced in the linear case, and the polarization arises at the same frequency. Losses depend on the relative phases of current and field (or polarization and field), and this relation between phases is determined by the phase of susceptibility. In the nonlinear case, however, there are three fields, and the polarization at the frequency ω is produced by the fields at the frequencies ω' and ω'', so that the losses depend on the relative phases of all three fields.

In some cases, losses may be negative, that is, there may be amplification at the frequency ω. Negative losses may simply indicate a transfer of the field energy from some modes to others. Therefore, formula (1.7.4) covers both the true losses and energy redistribution.

If the medium is transparent at all three frequencies (there is no real dissipation of energy), then energy changes in each of the modes can only be caused by energy transfer to other modes. Therefore,

$$\overline{\dot{Q}_\omega} + \overline{\dot{Q}'_\omega} + \overline{\dot{Q}''_\omega} = 0$$

This leads to certain constraints on the tensor $\chi^{(2)}$. To find them, we use (1.7.4) and similar expressions for $\overline{\dot{Q}'_\omega}$ and $\overline{\dot{Q}''_\omega}$ and arrive at

$$\frac{\omega}{2}\mathrm{Im}(\chi^{(2)}_{\alpha\beta\gamma}(\omega = \omega' + \omega'')E^*_{\omega\alpha}E_{\omega'\beta}E_{\omega''\gamma}) + \frac{\omega'}{2}\mathrm{Im}(\chi^{(2)}_{\alpha\beta\gamma}(\omega' = \omega - \omega'') \times E^*_{\omega'\alpha}E_{\omega\beta}E^*_{\omega''\gamma}) + \frac{\omega''}{2}\mathrm{Im}(\chi^{(2)}_{\alpha\beta\gamma}(\omega'' = \omega - \omega')E^*_{\omega''\alpha}E_{\omega\beta}E^*_{\omega'\gamma}) = 0 \tag{1.7.5}$$

The summation indices α, β and γ in this expression can be rearranged so that each term will contain an identical product of fields, $E^*_{\omega\alpha}E_{\omega'\beta}E_{\omega''\gamma}$. Equality (1.7.5) must hold for any value of amplitudes of the fields, which requires that

$$\omega\chi^{(2)}_{\alpha\beta\gamma}(\omega = \omega' + \omega'') - \omega'\chi^{(2)\,*}_{\beta\alpha\gamma}(\omega' = \omega - \omega'') - \omega''\chi^{(2)\,*}_{\gamma\alpha\beta}(\omega'' = \omega - \omega') = 0 \tag{1.7.6}$$

It will be shown later that if there is no true dissipation, that is, if we are far from absorption bands, the susceptibility is a real quantity. We can assume, therefore, that instead of imposing the limitation (1.7.6), we can choose stronger symmetry constraints:

$$\chi^{(2)}_{\alpha\beta\gamma}(\omega=\omega'+\omega'')=\chi^{(2)}_{\beta\alpha\gamma}(\omega'=\omega-\omega'')=\chi^{(2)}_{\gamma\alpha\beta}(\omega''=\omega-\omega') \quad (1.7.7)$$

These symmetry relations can be formulated into a rule: when rearranging frequencies in $\chi^{(2)}$, also rearrange the corresponding polarization indices. This can be shown by the diagram in Figure 2.2 (see Chapter 2).

The significance of this result lies in the fact that having measured $\chi^{(2)}$ for one of the processes, say, addition of frequencies $\omega=\omega'+\omega''$, we can at the same time determine $\chi^{(2)}$ for the processes of frequency subtraction, that is, $\omega'=\omega-\omega''$ and $\omega''=\omega-\omega'$.

Relation (1.7.7) will be given rigorous proof in the next chapter, using an explicit expression for quadratic susceptibilities.

Similar arguments hold also for the cubic susceptibility and higher-order susceptibilities. The difference lies only in that in the case of the cubic susceptibility and odd-order susceptibilities in general, degenerate cases are possible, when losses are expressed in terms of the imaginary component of a susceptibility.

Among all other processes described by $\chi^{(3)}_{\alpha\beta\gamma\delta}(\omega=\omega'+\omega''+\omega''')$, there are special cases when, for instance, $\omega''=-\omega'$. Since the field is real, a field component at a frequency $-\omega'$ is always present in addition to the component at a frequency ω'. This implies

$$E_{\omega''\gamma}=E^*_{\omega'\gamma}$$

By analogy to (1.7.4), losses are found from the expression

$$\overline{\dot{Q}}_\omega=\frac{\omega}{2}\,\mathrm{Im}(\chi^{(3)}_{\alpha\beta\gamma\delta}(\omega=\omega'-\omega'+\omega)E^*_{\omega\alpha}E_{\omega'\beta}E^*_{\omega'\gamma}E_{\omega\delta})$$

and are thus essentially independent of the phase of the field at the frequencies ω and ω'.

In this case, losses are determined by the imaginary part of the cubic susceptibility and correspond to the two-photon absorption or two-photon scattering.

1.8. Susceptibilities of a Crystal

The results presented in Sections 1.5–1.7 mostly treated spatially uniform media. We have already mentioned that crystals must be treated as a special case. Classical optics of crystals uses averaging over a physically infinitely small volume of the medium, which contains a large number of particles; for crystals, this means a volume whose dimensions are much greater than a_0 or the interatomic distance. After this averaging, a crystal can be treated as a spatially uniform medium, which substantially simplifies the solution of problems in crystal optics.

We have indicated, however, that the procedure of averaging over a physically infinitely small volume (small in comparison with the wavelength) cannot be used for the short-wavelength electromagnetic field (the x-ray range) and that even in the optical range this averaging eliminates a number of principally important effects.

If, however, we forgo the averaging over volume and if we restrict the averaging to statistical averaging over the Gibbs distribution, we come across certain mathematical complications caused by inherent nonuniformity of crystals. By definition, a crystal is a periodic nonuniform medium. The periodicity is imposed by the Bravais lattice of the crystal,

$$\mathbf{R_n} = n_1\mathbf{a}_1 + n_2\mathbf{a}_2 + n_3\mathbf{a}_3$$

where $\mathbf{R_n}$ is an arbitrary vector of the Bravais lattice, $\mathbf{a}_1$, $\mathbf{a}_2$, $\mathbf{a}_3$ are the basis vectors of the lattice and n_1, n_2, and n_3 are integers composing an integral-valued vector $\mathbf{n}$.

An arbitrary scalar function $\rho(\mathbf{r})$, which is periodical on the Bravais lattice, has the property

$$\rho(\mathbf{r} + \mathbf{R_n}) = \rho(\mathbf{r}) \tag{1.8.1}$$

One example of such a function can be the mean electron density in the crystal. If $\rho(\mathbf{r})$ is expanded in a Fourier series,

$$\rho(\mathbf{r}) = \sum_{\mathbf{k}} \rho_{\mathbf{k}} \mathrm{e}^{\mathrm{i}\mathbf{kr}}$$

then condition (1.8.1) implies that the equality

$$\mathrm{e}^{\mathrm{i}\mathbf{kR_n}} = 1$$

holds for an arbitrary vector of the Bravais lattice, that is,

$$\mathbf{kR_n} = 2\pi M \tag{1.8.2}$$

where $M = 0, \pm1, \pm2 \ldots$ is an integer. If $\mathbf{k}_1$ and $\mathbf{k}_2$ satisfy condition (1.8.2), their linear combination with integral coefficients also satisfies (1.8.2). Hence, all vectors $\mathbf{k}$ satisfying (1.8.2) make up a lattice known as the reciprocal lattice of the crystal.

The basis vectors of the reciprocal lattice can be written in the form

$$\mathbf{g}_1 = \frac{2\pi[\mathbf{a}_2\mathbf{a}_3]}{(\mathbf{a}_1\mathbf{a}_2\mathbf{a}_3)} \qquad \mathbf{g}_2 = \frac{2\pi[\mathbf{a}_3\mathbf{a}_1]}{(\mathbf{a}_1\mathbf{a}_2\mathbf{a}_3)} \qquad \mathbf{g}_3 = \frac{2\pi[\mathbf{a}_1\mathbf{a}_2]}{(\mathbf{a}_1\mathbf{a}_2\mathbf{a}_3)}$$

where the mixed vector product

$$(\mathbf{a}_1\mathbf{a}_2\mathbf{a}_3) = \mathbf{a}_1[\mathbf{a}_2\mathbf{a}_3] = [\mathbf{a}_1\mathbf{a}_2]\mathbf{a}_3$$

equals the volume Ω of a unit cell of the Bravais lattice. An arbitrary vector of the reciprocal lattice can then be rewritten as

$$\mathbf{k} = \mathbf{g}_\mathbf{l} = l_1\mathbf{g}_1 + l_2\mathbf{g}_2 + l_3\mathbf{g}_3 \tag{1.8.3}$$

where l_1, l_2, and l_3 are integers.

The vectors $\mathbf{g}_i$ possess the property

$$\mathbf{g}_i\mathbf{a}_k = 2\pi\delta_{ik}$$

which guarantees that (1.8.2) is satisfied for any vector (1.8.3). No other vector exists which satisfies condition (1.8.2). Therefore, an arbitrary function $\rho(\mathbf{r})$, which is periodic on the lattice, can be written as a Fourier series

$$\rho(\mathbf{r}) = \sum_{\mathbf{l}} \rho^{\mathbf{l}} \mathrm{e}^{\mathrm{i}\mathbf{g}_\mathbf{l}\mathbf{r}} \tag{1.8.4}$$

The linear susceptibility $\chi^{(1)}$ of a stationary and uniform medium is $\chi^{(1)}_{\alpha\beta}(\mathbf{r}-\mathbf{r}'; t-t')$. For crystals, stationarity does take place but the uniformity, expressed by the condition

$$\chi^{(1)}_{\alpha\beta}(\mathbf{r}+\mathbf{R}, \mathbf{r}'+\mathbf{R}; t-t') = \chi^{(1)}_{\alpha\beta}(\mathbf{r}, \mathbf{r}'; t-t')$$

does not hold for an arbitrary translation $\mathbf{R}$.

Instead, the condition that holds is the periodicity

$$\chi^{(1)}_{\alpha\beta}(\mathbf{r}+\mathbf{R}_\mathbf{n}, \mathbf{r}'+\mathbf{R}_\mathbf{n}; t-t') = \chi^{(1)}_{\alpha\beta}(\mathbf{r}, \mathbf{r}'; t-t')$$

for any vector $\mathbf{R}_\mathbf{n}$ of the Bravais lattice, since this translation brings both points $\mathbf{r}$ and $\mathbf{r}'$ to positions that are equivalent to their original positions.

This means that the function $\chi^{(1)}_{\alpha\beta}$ has the form

$$\chi^{(1)}_{\alpha\beta} = \chi^{(1)}_{\alpha\beta}(\mathbf{r}, \mathbf{r}-\mathbf{r}'; t-t')$$

which depends periodically on the first argument:

$$\chi^{(1)}_{\alpha\beta}(\mathbf{r}+\mathbf{R}_\mathbf{n}, \mathbf{r}-\mathbf{r}'; t-t') = \chi^{(1)}_{\alpha\beta}(\mathbf{r}, \mathbf{r}-\mathbf{r}'; t-t')$$

Such a function can be expanded in a Fourier series in the first argument by analogy to (1.8.4) and into a Fourier integral in the remaining variables:

$$\chi^{(1)}_{\alpha\beta}(\mathbf{r}, \mathbf{R}, \tau) = \sum_{\mathbf{l}} \int \frac{\mathrm{d}^3k\,\mathrm{d}\omega}{(2\pi)^4}\, \mathrm{e}^{\mathrm{i}\mathbf{k}\mathbf{R}+\mathrm{i}\mathbf{g}_\mathbf{l}\mathbf{r}-\mathrm{i}\omega\tau} \chi^{\mathbf{l}}_{\alpha\beta}(\mathbf{k}, \omega) \tag{1.8.5}$$

where $\tau = t - t'$ and $\mathbf{R} = \mathbf{r} - \mathbf{r}'$.

According to (1.4.3), therefore,

$$\mathcal{P}^{(1)}_{\alpha}(\mathbf{r},t) = \int_{-\infty}^{t} \mathrm{d}t' \int \mathrm{d}^3 r' \chi^{(1)}_{\alpha\beta}(\mathbf{r},\mathbf{r}-\mathbf{r}';t-t')\mathcal{E}_{\beta}(\mathbf{r}',t')$$

The Fourier transform yields

$$P^{(1)}_{\mathbf{k}\omega\alpha} = \int \mathcal{P}^{(1)}_{\alpha}(\mathbf{r},t)\mathrm{e}^{-\mathrm{i}\mathbf{k}\mathbf{r}+\mathrm{i}\omega t}\mathrm{d}t\,\mathrm{d}^3 r = \int \mathrm{d}t\,\mathrm{d}^3 r\,\mathrm{d}\tau\,\mathrm{d}^3 R\,\mathrm{e}^{-\mathrm{i}\mathbf{k}\mathbf{r}+\mathrm{i}\omega t}$$
$$\times \sum_{\mathbf{l}} \int \frac{\mathrm{d}^3 k'\,\mathrm{d}\omega'}{(2\pi)^4} \exp(\mathrm{i}\mathbf{k}\mathbf{R} + \mathrm{i}\mathbf{g}_{\mathbf{l}}\mathbf{r} - \mathrm{i}\omega'\tau)$$
$$\times \chi^{\mathbf{l}}_{\alpha\beta}(\mathbf{k}',\omega')\mathcal{E}_{\beta}(\mathbf{r}-\mathbf{R},t-\tau)$$

Integration in t and $\mathbf{r}$ gives

$$P^{(1)}_{\mathbf{k}\omega\alpha} = \int \mathrm{d}\tau\,\mathrm{d}^3 R \sum_{\mathbf{l}} \int \frac{\mathrm{d}^3 k'\,\mathrm{d}\omega'}{(2\pi)^4} \exp[-\mathrm{i}(\mathbf{k}-\mathbf{g}_{\mathbf{l}}-\mathbf{k}')\mathbf{R} + \mathrm{i}(\omega-\omega')\tau]$$
$$\times \chi^{\mathbf{l}}_{\alpha\beta}(\mathbf{k}',\omega')E_{\mathbf{k}-\mathbf{g}_{\mathbf{l}},\omega,\beta}$$

Integrating now over τ and $\mathbf{R}$ and making use of

$$\int \exp[-\mathrm{i}(\mathbf{k}-\mathbf{g}_{\mathbf{l}}-\mathbf{k}')\mathbf{R} + \mathrm{i}(\omega-\omega')\tau]\mathrm{d}\tau\,\mathrm{d}^3 R = (2\pi)^4\delta^3(\mathbf{k}-\mathbf{k}'+\mathbf{g}_{\mathbf{l}})\delta(\omega-\omega')$$

we find that

$$P^{(1)}_{\mathbf{k}\omega\alpha} = \sum_{\mathbf{l}} \chi^{\mathbf{l}}_{\alpha\beta}(\mathbf{k}-\mathbf{g}_{\mathbf{l}},\omega)E_{\mathbf{k}-\mathbf{g}_{\mathbf{l}},\omega\beta} \tag{1.8.6}$$

Formula (1.8.6) replaces the relation

$$P^{(1)}_{\mathbf{k}\omega\alpha} = \chi^{(1)}_{\alpha\beta}(\mathbf{k},\omega)E_{\mathbf{k}\omega\beta}$$

which is valid for the uniform medium. Equation (1.8.6) involves a set of susceptibilities $\chi^{\mathbf{l}}_{\alpha\beta}$ numbered by an integral-valued vector $\mathbf{l}$.

Let a field of frequency ω_0, with wave vector $\mathbf{q}$, be excited in the crystal by an external source:

$$\mathcal{E}_{\beta}(\mathbf{r}',t') = E_{\beta}\mathrm{e}^{\mathrm{i}\mathbf{q}\mathbf{r}'-\mathrm{i}\omega_0 t'}$$

This gives

$$E_{\mathbf{k}\omega\beta} = \int E_{\beta}\mathrm{e}^{\mathrm{i}\mathbf{q}\mathbf{r}'-\mathrm{i}\omega_0 t'}\,\mathrm{e}^{-\mathrm{i}\mathbf{k}\mathbf{r}'+\mathrm{i}\omega t'}\,\mathrm{d}t'\,\mathrm{d}^3 r' = (2\pi)^4 E_{\beta}\delta(\omega-\omega_0)\delta^3(\mathbf{k}-\mathbf{q})$$

According to (1.8.6), the Fourier transform of polarization is

$$P^{(l)}_{\mathbf{k}\omega\alpha} = (2\pi)^4 \sum_{\mathrm{l}} \chi^{\mathrm{l}}_{\alpha\beta}(\mathbf{k}-\mathbf{g}_{\mathrm{l}},\omega) E_\beta \delta(\omega-\omega_0)\delta^3(\mathbf{k}-\mathbf{g}_{\mathrm{l}}-\mathbf{q})$$

Converting now to polarization $\mathcal{P}^{(1)}_\alpha(\mathbf{r},t)$, we obtain

$$\begin{aligned}\mathcal{P}^{(1)}_\alpha(\mathbf{r},t) &= \int \frac{\mathrm{d}^3k\,\mathrm{d}\omega}{(2\pi)^4} P^{(1)}_{\mathbf{k}\omega\alpha} \mathrm{e}^{\mathrm{i}\mathbf{k}\mathbf{r}-\mathrm{i}\omega t} = \sum_{\mathrm{l}} \chi^{\mathrm{l}}_{\alpha\beta}(\mathbf{q},\omega_0) E_\beta \mathrm{e}^{\mathrm{i}(\mathbf{q}+\mathbf{g}_{\mathrm{l}})\mathbf{r}-\mathrm{i}\omega_0 t} \\ &= \Big(\sum_{\mathrm{l}} \chi^{\mathrm{l}}_{\alpha\beta}(\mathbf{q},\omega_0) E_\beta \mathrm{e}^{\mathrm{i}\mathbf{g}_{\mathrm{l}}\mathbf{r}}\Big) \mathrm{e}^{\mathrm{i}\mathbf{q}\mathbf{r}-\mathrm{i}\omega_0 t} \qquad (1.8.7)\end{aligned}$$

Expression (1.8.7) shows that $\mathcal{P}^{(1)}_\alpha$ is a sum of plane waves at frequency ω_0, with wave vectors $\mathbf{q}+\mathbf{g}_{\mathrm{l}}$ which differ from the wave vector $\mathbf{q}$ of the field by the reciprocal lattice vector $\mathbf{g}_{\mathrm{l}}$. This last representation of $\mathcal{P}^{(1)}_\alpha$ in (1.8.7) shows that the set of these waves can also be interpreted as a wave with wave vector $\mathbf{q}$ but modulated by a function which is periodic on the lattice. This is implied by the fact that the function in parentheses in (1.8.7) is a periodic function of the type of (1.8.4). Therefore, polarization appears in the medium at a frequency ω_0 but its wave vectors are $\mathbf{q}+\mathbf{g}_{\mathrm{l}}$. This polarization generates a field at a frequency ω_0 and with spatial frequencies which differ from $\mathbf{q}$ by an arbitrary vector $\mathbf{g}_{\mathrm{l}}$ of the reciprocal lattice of the crystal. The field can thus be written in the form

$$\mathcal{E}_\alpha(\mathbf{r},t) = \sum_{\mathrm{l}} E^{\mathrm{l}}_\alpha \mathrm{e}^{\mathrm{i}(\mathbf{q}+\mathbf{g}_{\mathrm{l}})\mathbf{r}-\mathrm{i}\omega_0 t} = \left(\sum_{\mathrm{l}} E^{\mathrm{l}}_\alpha \mathrm{e}^{\mathrm{i}\mathbf{g}_{\mathrm{l}}\mathbf{r}}\right) \mathrm{e}^{\mathrm{i}\mathbf{q}\mathbf{r}-\mathrm{i}\omega_0 t}$$

and interpreted either as an ensemble of plane waves or as a plane wave modulated by a periodic function. The wave vector of the envelope of this wave is $\mathbf{q}$; the periodic function describes how the field changes within one unit cell.

The pattern of field distribution in the crystal is shown in Figure 1.3. The formation of an entire set of waves in a crystal manifests how the true "effective" field differs from the mean field; classic optics of crystals neglects this effect.

Let a field with wave vector $\mathbf{q}$ be incident on a crystalline plate from the vacuum. Will the crystal then emit waves generated in it with wave vectors that differ from $\mathbf{q}$? For these waves to escape into vacuum, it is necessary that for them, as well as for the incident wave, the law of dispersion for a free wave in vacuum be satisfied:

$$\omega_0^2 = c^2(\mathbf{q}+\mathbf{g}_{\mathrm{l}})^2$$

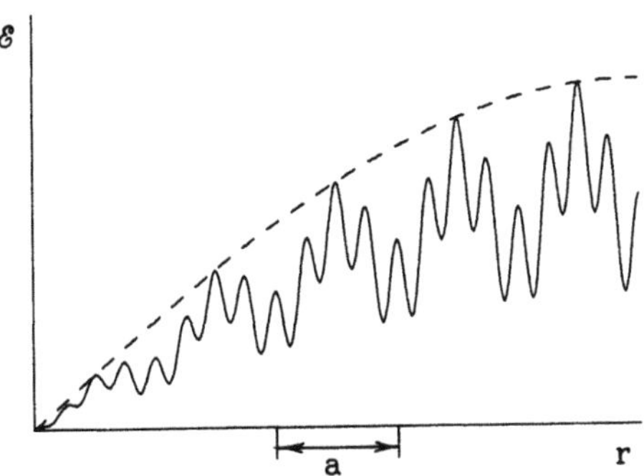

Figure 1.3. Schematic pattern of field distribution in the crystal (a is the lattice period). Broken line traces the envelope $e^{i\mathbf{qr}}$.

Since we always have

$$\omega_0^2 = c^2\mathbf{q}^2$$

this immediately implies that

$$\mathbf{g}_l^2 = -2\mathbf{q}\mathbf{g}_l = 2qg_l \cos\Theta$$

where Θ is the angle between the vectors $\mathbf{q}$ and $-\mathbf{g}_l$. In the optical range, this last condition is never satisfied for $\mathbf{l} \neq 0$. Indeed, in the optical range we have $\lambda \sim 10^{-4}\,\text{cm}$ so that $q \sim 2\pi/\lambda \sim 10^{-4}\,\text{cm}^{-1}$ while $g_l \sim 2\pi/a_0 \sim 10^8\,\text{cm}^{-1}$; hence we always have $g_l \gg q$. Waves with $\mathbf{l} \neq 0$ undergo total internal reflection at the crystal interfaces, do not escape from the crystal, and thus are not observed outside crystals.

In the x-ray region, $q \sim g_l$ and the condition

$$g_l = 2q\cos\Theta \tag{1.8.8}$$

can be satisfied.

Equation (1.8.2) written for the vector $\mathbf{g}_l$,

$$\mathbf{g}_l\mathbf{R}_\mathbf{n} = 2\pi M$$

implies that for a fixed M, lattice vectors $\mathbf{R}_\mathbf{n}$ lie in the plane for which $\mathbf{g}_l$ is a vector of the normal. A set of planes for $M = 0, \pm 1, \pm 2$ etc. are the set of crystallographic planes fixed by the reciprocal lattice vector $\mathbf{g}_l$ and the spacing between the neighboring planes,

$$d = 2\pi/g_l$$

Condition (1.8.8) thus reduces to

$$\frac{2\pi}{d} = 2\,\frac{2\pi}{\lambda}\cos\Theta$$

or

$$\lambda = 2d\cos\Theta$$

This equality describes the minimal reciprocal lattice vector perpendicular to the set of atomic planes of the crystal. For other reciprocal lattice vectors perpendicular to atomic planes, we have

$$d = 2\pi n / g_l$$

so that in the general case condition (1.8.8) is equivalent to the familiar Bragg condition for the diffraction of x-ray radiation in crystals

$$n\lambda = 2d\cos\Theta$$

If this condition is satisfied, the diffracted wave emitted by the crystal will correspond to the reflection from the atomic planes determined by the vector $\mathbf{g}_l$ and perpendicular to this vector.

Regardless of whether additional waves escape from the crystal or not, these waves are always present inside the crystal. Consequently, the wave vector of the field in the crystal is therefore determined only up to an arbitrary vector of reciprocal lattice. If $\mathbf{q}$ is the wave vector of the components of the field, then there is inevitably a component with the wave vector $\mathbf{q}+\mathbf{g}_l$, where $\mathbf{g}_l$ is an arbitrary reciprocal lattice vector. For the basis vector, it is expedient to choose the vector of minimal length from the countable set of these vectors. This requirement is equivalent to a convention that the wave vector is to be chosen within the (first) Brillouin zone.

By definition, the Brillouin zone in the reciprocal lattice of a crystal is the minimal polyhedron bounded by planes drawn through the midpoints of various vectors of the reciprocal lattice normally to them.

It can be shown that this is a particular case of choosing a primitive unit cell of the reciprocal lattice. The entire reciprocal lattice can be filled by translating Brillouin zones by reciprocal lattice vectors. Choosing a primitive unit cell is not an unambiguous operation. Another possible way is to choose as a primitive unit cell a parallelepiped constructed on the basis vectors $\mathbf{g}_1$, $\mathbf{g}_2$ and $\mathbf{g}_3$. All primitive unit cells have identical volume but the advantage of the Brillouin zone is that it possesses the same point symmetry characterizing the reciprocal lattice. The primitive cell, chosen in a similar manner for the original Bravais lattice, is known as the Wigner–Seitz unit cell.

It is logical to normalize all wave vectors to the Brillouin zone. If $\mathbf{k}$ and $\mathbf{k}'$ are arbitrary wave vectors that differ, as in (1.8.6), by one of the reciprocal lattice vectors, then there exist reciprocal lattice vectors $\mathbf{g}_l$ and $\mathbf{g}_{l'}$, such that

$$\mathbf{k} = \mathbf{q} + \mathbf{g}_l \qquad \mathbf{k}' = \mathbf{q} + \mathbf{g}_{l'}$$

and $\mathbf{q}$ lies in the first Brillouin zone. It is now possible to introduce convenient symmetric notation. A wave component in a crystal is now characterized by a pair of vectors $\mathbf{q}$ and $\mathbf{g_l}$.

In this notation, one writes $P^{\mathbf{l}}_{\mathbf{q}\omega\alpha}$ instead of $P_{\mathbf{k}\omega\alpha}$ and, accordingly, $E^{\mathbf{l}}_{\mathbf{q}\omega\beta}$ instead of $E_{\mathbf{k}'\omega\beta}$. The relation

$$P_{\mathbf{k}\omega\alpha} = \sum_{\mathbf{l}''} \chi^{\mathbf{l}''}_{\alpha\beta}(\mathbf{k} - \mathbf{g}_{\mathbf{l}''}, \omega) E_{\mathbf{k}-\mathbf{g}_{\mathbf{l}''},\omega,\beta}$$

is now rewritten as

$$\begin{aligned} P^{\mathbf{l}}_{\mathbf{q}\omega\alpha} &\equiv P_{\mathbf{q}+\mathbf{g}_{\mathbf{l}},\omega,\alpha} \sum_{\mathbf{l}''} \chi^{\mathbf{l}''}_{\alpha\beta}(\mathbf{q} + \mathbf{g}_{\mathbf{l}} - \mathbf{g}_{\mathbf{l}''}, \omega) E_{\mathbf{q}+\mathbf{g}_{\mathbf{l}}-\mathbf{g}_{\mathbf{l}''}\omega\beta} \\ &= \sum_{\mathbf{l}'} \chi^{\mathbf{l}-\mathbf{l}'}_{\alpha\beta}(\mathbf{q} + \mathbf{g}_{\mathbf{l}'}, \omega) E_{\mathbf{q}+\mathbf{g}_{\mathbf{l}'}\omega\beta} = \sum_{\mathbf{l}'} \chi^{\mathbf{l}\mathbf{l}'}_{\alpha\beta}(\mathbf{q}, \omega) E^{\mathbf{l}'}_{\mathbf{q}\omega\beta} \end{aligned}$$

where we have denoted

$$\chi^{\mathbf{l}\mathbf{l}'}_{\alpha\beta}(\mathbf{q}, \omega) \equiv \chi^{\mathbf{l}-\mathbf{l}'}_{\alpha\beta}(\mathbf{q} + \mathbf{g}_{\mathbf{l}'}, \omega)$$

The rows and columns of the matrix $\chi^{\mathbf{l}\mathbf{l}'}_{\alpha\beta}$ are enumerated by a pair of indices (one of them a vector): α, $\mathbf{l}$ and β, $\mathbf{l}'$. Although this matrix is of infinite rank, it is virtually always sufficient to truncate it to a finite number of values of $\mathbf{l}$: the field amplitudes and polarizations with high $\mathbf{l}$ are small. In what follows we use the symbolic matrix notation. Instead of

$$P^{\mathbf{l}}_{\mathbf{q}\omega\alpha} = \sum_{\beta,\mathbf{l}'} \chi^{\mathbf{l}\mathbf{l}'}_{\alpha\beta}(\mathbf{q}, \omega) E^{\mathbf{l}'}_{\mathbf{q}\omega\beta}$$

we write

$$P_{\mathbf{q}\omega} = \chi(\mathbf{q}, \omega) E_{\mathbf{q}\omega}$$

or even

$$P = \chi E$$

1.9. Mean Susceptibility of Crystals

In order to describe completely the microstructure of the electromagnetic field in a crystal, describing also the difference between the actual field and the mean field, it was necessary to introduce the susceptibility tensor $\chi^{\mathbf{l}\mathbf{l}'}_{\alpha\beta}(\mathbf{q}, \omega)$ instead of $\chi_{\alpha\beta}(\mathbf{q}, \omega)$. In the usual optics of crystals, however, where the field wavelength in the crystal is large in comparison with interatomic distance ($q \ll g_{\mathbf{l}}$), the averaged tensor $\overline{\chi}_{\alpha\beta}(\mathbf{q},\omega)$ is quite sufficient to consider practically all optical phenomena at the phenomenological level.

How is the tensor $\chi^{\mathbf{l}}_{\alpha\beta}(\mathbf{q},\omega)$ related to the tensor $\overline{\chi}_{\alpha\beta}$ which is used in the optics of crystals to describe averaged fields? Clearly, the mean field is the component $\overline{E}_\beta = E^0_\beta$ that corresponds to $\mathbf{l} = 0$, since the averaging over the crystal unit cell causes all rapidly oscillating components which include the factor $\exp(i\mathbf{g}_{\mathbf{l}}\mathbf{r})$ to vanish. Likewise, $\overline{\mathbf{P}}_\alpha = \mathbf{P}^0_\alpha$. It could be assumed that $\overline{\chi}_{\alpha\beta} = \chi^{00}_{\alpha\beta}$, but this is wrong. The point is that χ^{00} gives the amplitude of a spatially slowly varying (long-wavelength) polarization which is generated by the long-wavelength field component. However, high-frequency field harmonics also generate polarization harmonics, including a long-wavelength field component $\mathbf{P}^0_\alpha$.

In order to find the relation we are seeking, it is necessary to express the amplitudes of short-wavelength field harmonics in terms of the long-wavelength field E^0_β; using then $\chi^{0\mathbf{l}}_{\alpha\beta}$, we can find the long-wavelength (mean) polarization P^0_α in terms of the long-wavelength (mean) field E^0_β.

In order to find the relation sought, we write Maxwell's equations including external currents which possess only the long-wavelength component, omit all harmonics with $\mathbf{l} \neq 0$ and arrive at an equation for the field component E^0_β.

Equation for the electric field in symbolic form is written, by analogy to (1.6.9), as

$$\left(1 - \frac{4\pi\omega^2}{c^2}\mathcal{D}_0(\mathbf{q},\omega)\chi(\mathbf{q},\omega)\right)E_{\mathbf{q}\omega} = \frac{4\pi i\omega}{c^2}\mathcal{D}_0(\mathbf{q},\omega)\, j^{(e)}_{\mathbf{q}\omega} \tag{1.9.1}$$

The function $\mathcal{D}_0(\mathbf{q},\omega)$ of field propagation in vacuum looks, in complete form, similarly to (1.6.8):

$$\begin{aligned}\mathcal{D}^{\mathbf{ll'}}_{0\alpha\beta}(\mathbf{q},\omega) = \Bigg\{&\frac{1}{(\mathbf{q}+\mathbf{g}_{\mathbf{l}})^2 - \omega/c^2}\left(\delta_{\alpha\beta} - \frac{(\mathbf{q}+\mathbf{g}_{\mathbf{l}})_\alpha(\mathbf{q}+\mathbf{g}_{\mathbf{l}})_\beta}{(\mathbf{q}+\mathbf{g}_{\mathbf{l}})^2}\right)\\ &- \frac{c^2}{\omega^2}\frac{(\mathbf{q}+\mathbf{g}_{\mathbf{l}})_\alpha(\mathbf{q}+\mathbf{g}_{\mathbf{l}})_\beta}{(\mathbf{q}+\mathbf{g}_{\mathbf{l}})^2}\Bigg\}\delta_{\mathbf{ll'}} = \mathcal{D}^{\mathbf{ll'}}_{0\perp\alpha\beta}(\mathbf{q},\omega) + \mathcal{D}^{\mathbf{ll'}}_{0\|\alpha\beta}(\mathbf{q},\omega)\end{aligned} \tag{1.9.2}$$

The matrix $\mathcal{D}_0$ is diagonal in $\mathbf{l}$ and has a small transverse component for $\mathbf{l} \neq 0$,

$$\frac{\mathcal{D}_{0\|}}{\mathcal{D}_{0\perp}} \sim \frac{(\mathbf{q}+\mathbf{g}_{\mathbf{l}})^2 - \omega^2/c^2}{\omega^2/c^2} \gg 1$$

since in the optical range we have $\omega/c \approx q$ and $q/g_{\mathbf{l}} \ll 1$. This means that the short-wavelength fields generated in the crystal are almost purely longitudinal Coulomb fields. The functions $\mathcal{D}_0^{-1}$ and $\mathcal{D}_0$ can be considered not only in the Fourier transform space but also in the coordinate form. In this last case, we only have to replace $\mathrm{i}k_\alpha$ by $\partial/\partial x_\alpha$ and $-\mathrm{i}\omega$ by $\partial/\partial t$, after which $\boldsymbol{\mathcal{D}}_0^{-1}$ is transformed into a differential operator and $\boldsymbol{\mathcal{D}}_0$ transforms into an integral linear operator inverse to this differential operator.

The current $j^{(\mathrm{e})}$ has only the long-wavelength component

$$j^{(\mathrm{e})}_{\mathbf{q}\omega\alpha} = 0 \qquad \text{for } \mathbf{l} \neq 0$$

so that the righthand side of equation (1.9.1) vanishes if $\mathbf{l} \neq 0$. Let us split (1.9.1) into two equations, one of which corresponds to $\mathbf{l} = 0$ and the other to $\mathbf{l} \neq 0$. To do this, we introduce

$$\begin{aligned} \overline{E}^{\mathbf{l}}_\alpha &= E^{\mathbf{l}}_\alpha \delta_{\mathbf{l}0} \\ \widetilde{E}^{\mathbf{l}}_\alpha &= E^{\mathbf{l}}_\alpha (1 - \delta_{\mathbf{l}0}) \end{aligned}$$

that is, we split the field into long-wavelength and short-wavelength parts,

$$E = \overline{E} + \widetilde{E}$$

Similarly, by using the diagonality of $\boldsymbol{\mathcal{D}}_0$ in $\mathbf{l}$, we split it into long- and short-wavelength matrices:

$$\begin{aligned} \boldsymbol{\mathcal{D}}_0 &= \overline{\boldsymbol{\mathcal{D}}}_0 + \widetilde{\boldsymbol{\mathcal{D}}}_0 \\ \overline{\boldsymbol{\mathcal{D}}}^{\mathbf{l}\mathbf{l}'}_{0\alpha\beta} &= \boldsymbol{\mathcal{D}}^{\mathbf{l}\mathbf{l}'}_{0\alpha\beta} \delta_{\mathbf{l}0} \\ \widetilde{\boldsymbol{\mathcal{D}}}^{\mathbf{l}\mathbf{l}'}_{0\alpha\beta} &= \boldsymbol{\mathcal{D}}^{\mathbf{l}\mathbf{l}'}_{0\alpha\beta} (1 - \delta_{\mathbf{l}0}) \end{aligned}$$

Here, as before, $\delta_{\mathbf{l}0}$ is the Kronecker delta:

$$\delta_{\mathbf{l}\mathbf{l}'} = \begin{cases} 1, & \mathbf{l} = \mathbf{l}' \\ 0, & \mathbf{l} \neq \mathbf{l}' \end{cases}$$

Obviously, $\tilde{j}^{(\mathrm{e})} = 0$, $\bar{j}^{(\mathrm{e})} = j^{(\mathrm{e})}$ and equation (1.9.1) separates into a pair of equations

$$\left(1 - \frac{4\pi\omega^2}{c^2} \overline{\boldsymbol{\mathcal{D}}}_0 \chi\right) \overline{E} = \frac{4\pi\omega^2}{c^2} \overline{\boldsymbol{\mathcal{D}}}_0 \chi \widetilde{E} - \frac{4\pi\mathrm{i}\omega^2}{c^2} \overline{\boldsymbol{\mathcal{D}}}_0 j^{(\mathrm{e})} \tag{1.9.3}$$

$$\left(1 - \frac{4\pi\omega^2}{c^2} \widetilde{\boldsymbol{\mathcal{D}}}_0 \chi\right) \widetilde{E} = \frac{4\pi\omega^2}{c^2} \widetilde{\boldsymbol{\mathcal{D}}}_0 \chi \overline{E} \tag{1.9.4}$$

The short-wavelength field $\widetilde{E}$ can be found from equation (1.9.4) in terms of long-wavelength field $\bar{E}$:

$$\widetilde{E} = \frac{4\pi\omega^2}{c^2} \left[1 - \frac{4\pi\omega^2}{c^2} \widetilde{\boldsymbol{\mathcal{D}}}_0 \chi\right]^{-1} \widetilde{\boldsymbol{\mathcal{D}}}_0 \chi \overline{E} \tag{1.9.5}$$

Substituting (1.9.5) into (1.9.3), we now obtain the equation for the long-wavelength field component only, that is, for the mean field

$$\left(1 - \frac{4\pi\omega^2}{c^2}\overline{\mathcal{D}}_0\chi\right)\overline{E} - \left(\frac{4\pi\omega^2}{c^2}\right)^2 \overline{\mathcal{D}}_0\chi\left[1 - \frac{4\pi\omega^2}{c^2}\widetilde{\mathcal{D}}_0\chi\right]^{-1}\widetilde{\mathcal{D}}_0\chi\overline{E}$$
$$= \frac{4\pi i\omega}{c^2}\overline{\mathcal{D}}_0 j^{(e)}$$

By comparing this equation with a phenomenological equation

$$\left(1 - \frac{4\pi\omega^2}{c^2}\overline{\mathcal{D}}_0\overline{\chi}\right)\overline{E} = \frac{4\pi i\omega}{c^2}\overline{\mathcal{D}}_0 j^{(e)}$$

we obtain the expression for the mean susceptibility:

$$\overline{\chi} = \left\{\chi + \frac{4\pi\omega^2}{c^2}\chi\left[1 - \frac{4\pi\omega^2}{c^2}\widetilde{\mathcal{D}}_0\chi\right]^{-1}\widetilde{\mathcal{D}}_0\chi\right\}^{00} \tag{1.9.6}$$

Making use of the identity

$$\chi = \chi\left[1 - \frac{4\pi\omega^2}{c^2}\widetilde{\mathcal{D}}_0\chi\right]^{-1}\left(1 - \frac{4\pi\omega^2}{c^2}\widetilde{\mathcal{D}}_0\chi\right)$$

we rewrite the above expression in a shorter form:

$$\overline{\chi} = \left\{\chi\left[1 - \frac{4\pi\omega^2}{c^2}\widetilde{\mathcal{D}}_0\chi\right]^{-1}\right\}^{00} \tag{1.9.7}$$

In a more expanded form, (1.9.7) yields the tensor $\overline{\chi}_{\alpha\beta}(\mathbf{q},\omega)$:

$$\overline{\chi}_{\alpha\beta} = \left\{\chi\left[1 - \frac{4\pi\omega^2}{c^2}\widetilde{\mathcal{D}}_0\chi\right]^{-1}\right\}^{00}_{\alpha\beta}$$

By analogy to how we obtained (1.6.15) out of (1.6.14), we can recast (1.9.7) in another, equivalent form:

$$\overline{\chi} = \left\{\left[1 - \frac{4\pi\omega^2}{c^2}\chi\widetilde{\mathcal{D}}_0\right]^{-1}\chi\right\}^{00} \tag{1.9.8}$$

If the operator in (1.9.7) or (1.9.8) is expanded in a series,

$$\left[1 - \frac{4\pi\omega^2}{c^2}\chi\widetilde{\mathcal{D}}_0\right]^{-1} = 1 + \frac{4\pi\omega^2}{c^2}\chi\widetilde{\mathcal{D}}_0 + \left(\frac{4\pi\omega^2}{c^2}\right)^2\chi\widetilde{\mathcal{D}}_0\chi\widetilde{\mathcal{D}}_0 + \ldots$$

we obtain the series

$$\overline{\chi} = \left\{\chi + \chi\frac{4\pi\omega^2}{c^2}\widetilde{\mathcal{D}}_0\chi + \chi\frac{4\pi\omega^2}{c^2}\widetilde{\mathcal{D}}_0\chi\frac{4\pi\omega^2}{c^2}\widetilde{\mathcal{D}}_0\chi + \ldots\right\}^{00}$$

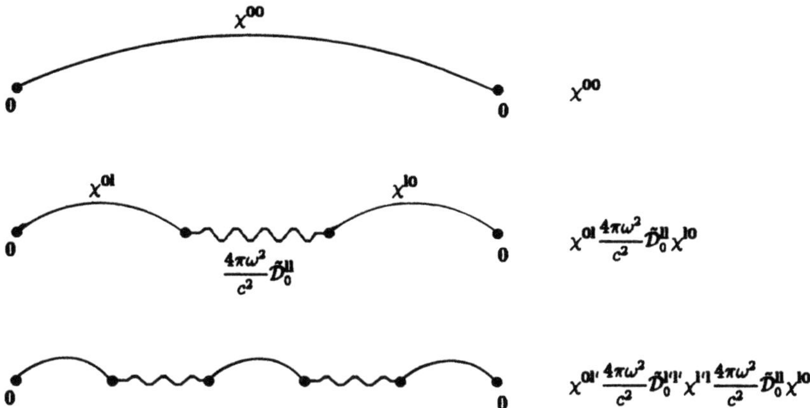

Figure 1.4. Diagrams for the successive terms of the mean susceptibility $\overline{\chi}$ expanded in powers of χ. The operator $4\pi\omega^2\widetilde{\mathcal{D}}_0/c^2$ describes the propagation of the short-wavelength field.

The terms of this series can be put in correspondence with the diagrams shown in Figure 1.4. The meaning of this expansion corresponds to what we said above: $\overline{\chi}$ differs from χ_{00} because the long-wavelength field generates short-wavelength components $\widetilde{E}$, which, propagating through the crystal, again generate $\overline{E}$.

Expressions (1.9.7) and (1.9.8) allow us to find the general expression for the mean susceptibility of the crystal. Since $\widetilde{\mathcal{D}}_{0\|} \gg \widetilde{\mathcal{D}}_{0\perp}$, we can in fact replace $4\pi\omega^2\widetilde{\mathcal{D}}_0/c^2$ in (1.9.7) and (1.9.8) by $4\pi\omega^2\widetilde{\mathcal{D}}_{0\|}/c^2 = -4\widetilde{\Pi}_{\|}$.

The representation of $\overline{\chi}$ in the form (1.9.6) can be written approximately as

$$\overline{\chi}_{\alpha\beta} = \chi^{00}_{\alpha\beta} + \chi^{01}_{\alpha\|}[1 + 4\pi\widetilde{\chi}_{\|}]^{-1,\mathbf{l}\mathbf{l}'}\chi^{\mathbf{l}'0}_{\|\beta}$$

where $\widetilde{\chi}_{\|}$ is the longitudinal susceptibility for the short-wavelength longitudinal field, or

$$\widetilde{\chi}_{\alpha\beta} = \chi^{00}_{\alpha\beta} + \chi^{01}_{\alpha\|}(\widetilde{\varepsilon}_{\|}^{-1})^{\mathbf{l}\mathbf{l}'}\chi^{\mathbf{l}'0}_{\|\beta}$$

where $epsilon_{\|} = 1 + 4\pi\chi_{\|} = k_\alpha\varepsilon_{\alpha\beta}k_\beta k^{-2}$ and $\tilde{\varepsilon}^{\mathbf{l}\mathbf{l}'}$ contains no components with $\mathbf{l} = 0$ or $\mathbf{l}' = 0$.

If not only the short-wavelength but also the long-wavelength field is purely longitudinal (when we are interested in the response of the medium to the longitudinal field), then formulas (1.9.6)–(1.9.8) allow substantial simplification. Fairly cumbersome algebra ultimately gives

$$\widetilde{\varepsilon} = \frac{1}{(\varepsilon_{\|}^{-1})^{00}}$$

This formula is obvious for longitudinal fields and for the longitudinal response. In this case ($\mathcal{D}$ and $\mathcal{E}$ are longitudinal)

$$D^{\mathbf{l}} = \varepsilon_{\|}^{\mathbf{l}\mathbf{l}'} E^{\mathbf{l}'}$$

or

$$E^{\mathbf{l}} = (\varepsilon_{\|}^{-1})^{\mathbf{l}\mathbf{l}'} D^{\mathbf{l}'}$$

For this reason we have, among other things, that

$$E^{0} = (\varepsilon_{\|}^{-1})^{00} D^{0} \tag{1.9.9}$$

since equation $\nabla \mathcal{D} = 4\pi\rho^{(e)}$ implies that $D^{\mathbf{l}} = 0$ for $\mathbf{l} \neq 0$.

It follows from (1.9.9) that

$$\frac{1}{\tilde{\varepsilon}_{\|}} = \frac{E^{0}}{D_{0}} = (\varepsilon_{\|}^{-1})^{00}$$

Now we can at last find the relation of the averaged susceptibility to the response to the external field. It is easy to see that $\chi^{(e)}(\mathbf{r}, \mathbf{r}', \omega)$ has the same translation properties as $\chi(\mathbf{r}, \mathbf{r}', \omega)$, so that in the Fourier representation it becomes $\chi_{\alpha\beta}^{(e)\mathbf{l}\mathbf{l}'}(\mathbf{q}, \omega)$. General relations (1.6.13)–(1.6.15) between $\chi^{(e)}$ and χ are still valid, with the only difference that these relations connect matrices via indices α, β and $\mathbf{l}$, $\mathbf{l}'$, and all products must be interpreted in the matrix sense.

By combining (1.9.7) and (1.6.15) and taking into account that $\mathcal{D}_0 = \overline{\mathcal{D}}_0 + \tilde{\mathcal{D}}_0$, it is not difficult to show that

$$\overline{\chi}_{\alpha\beta}(\mathbf{q}, \omega) = \chi_{\alpha\gamma}^{(e)00} \left(\left[1 + \frac{4\pi\omega^2}{c^2} \overline{\mathcal{D}}_0 \chi^{(e)00} \right]^{-1} \right)_{\gamma\beta} \tag{1.9.10}$$

This formula does not contain summation over $\mathbf{l} \neq 0$, that is, has no contribution of short-wavelength spatial harmonics. This only signifies, in fact, that the harmonics which are the fields generated by the particles of the medium are already included in the value of $\chi_{\alpha\beta}^{(e)00}$ when the response to the external field is calculated.

1.10. Field Acting in the Crystal

The entire optics of crystals can be phenomenologically constructed, with the exception of the x-ray range, if we know the tensor $\overline{\chi}_{\alpha\beta}$ which is expressed, as we have demonstrated in the preceding paragraph, via $\chi_{\alpha\beta}^{\mathbf{l}\mathbf{l}'}$. What is the additional information provided by the tensor $\chi_{\alpha\beta}^{\mathbf{l}\mathbf{l}'}$? This tensor relates $\overline{\chi}_{\alpha\beta}$ to the crystal structure at the microscopic level, that is, to calculate $\overline{\chi}_{\alpha\beta}$. The reason for this is that the tensor $\chi^{\mathbf{l}\mathbf{l}'}_{\alpha\beta}$ allows the calculation of the field at any point if the mean field is known.

Let us illustrate this situation using as an example the simplest model of a cubic crystal with pointlike polarizable atoms located at lattice sites. Let $\varkappa$ be the polarizability of one atom; the dipole moment of the atom in the field $\mathcal{E}$ then equals $\varkappa\mathcal{E}$, while the crystal polarization is

$$\mathcal{P}_\alpha(\mathbf{r},t) = \sum_{\mathbf{n}} \varkappa\mathcal{E}_\alpha(\mathbf{r},t)\,\delta(\mathbf{r}-\mathbf{R}_\mathbf{n})$$

where, as is usually done, $\mathbf{n}$ enumerates the lattice sites.

Taking into account the retardation effect for the Fourier transforms in time, we find

$$P_{\omega\alpha}(\mathbf{r}) = \sum_{\mathbf{n}} \varkappa(\omega)E_{\omega\alpha}(\mathbf{r})\,\delta(\mathbf{r}-\mathbf{R}_\mathbf{n})$$

so that

$$\chi(\mathbf{r},\mathbf{r}-\mathbf{r}';\omega) = \varkappa(\omega)\sum_{\mathbf{n}}\delta(\mathbf{r}-\mathbf{R}_\mathbf{n})\,\delta(\mathbf{r}-\mathbf{r}') \tag{1.10.1}$$

Correspondingly, we have

$$\chi^{\mathbf{l}\mathbf{l}'}_{\alpha\beta}(\mathbf{q},\omega) = \varkappa(\omega)n\delta_{\alpha\beta} \tag{1.10.2}$$

where n is the number of atoms per unit volume ($n = \Omega^{-1}$). It is important that $\chi^{\mathbf{l}\mathbf{l}'}_{\alpha\beta}$ is independent of $\mathbf{l}$ and $\mathbf{l}'$. This is a corollary of the fact that the model assumes the atomic size to be negligibly small and, therefore, neglects the nonlocality of the response, as follows from the δ function in (1.10.1).

Making use of (1.9.7), we find the tensor $\overline{\chi}_{\alpha\beta}$ through which to express the optical properties of crystals.

We begin with determining the matrix

$$A^{\mathbf{l}\mathbf{l}'}_{\alpha\beta} = \left\{\left[1 - \frac{4\pi\omega^2}{c^2}\widetilde{\mathcal{D}}_0\chi\right]^{-1}\right\}^{\mathbf{l}\mathbf{l}'}_{\alpha\beta}$$

We take into account in the propagation function $\widetilde{\mathcal{D}}_0$ only the longitudinal part, as we have already justified above. For the matrix A we have the equation

$$\left(1 - \frac{4\pi\omega^2}{c^2}\widetilde{\mathcal{D}}_0\chi\right)A = 1$$

or, in more detailed form,

$$A^{\mathbf{l}\mathbf{l}'}_{\alpha\beta} + 4\pi\varkappa n(1-\delta_{\mathbf{l}0})\frac{(\mathbf{q}+\mathbf{g}_\mathbf{l})_\alpha(\mathbf{q}+\mathbf{g}_\mathbf{l})_\gamma}{|\mathbf{q}+\mathbf{g}_\mathbf{l}|^2}\sum_{\mathbf{l}''} A^{\mathbf{l}''\mathbf{l}'}_{\gamma\beta} = \delta_{\alpha\beta}\delta_{\mathbf{l}\mathbf{l}'} \qquad (1.10.3)$$

We will use the notation

$$a^{\mathbf{l}'}_{\alpha\beta} = \sum_{\mathbf{l}} A^{\mathbf{l}\mathbf{l}'}_{\alpha\beta}$$

Carrying out summation over $\mathbf{l}$ in (1.10.3), we obtain

$$a^{\mathbf{l}'}_{\alpha\beta} = -4\pi\varkappa n\sum_{\mathbf{l}\neq 0}\frac{(\mathbf{q}+\mathbf{g}_\mathbf{l})_\alpha(\mathbf{q}+\mathbf{g}_\mathbf{l})_\gamma}{|\mathbf{q}+\mathbf{g}_\mathbf{l}|^2}a^{\mathbf{l}'}_{\gamma\beta} + \delta_{\alpha\beta} \qquad (1.10.4)$$

Let us calculate the explicit form of the sum

$$\sum_{\mathbf{l}\neq 0}\frac{(\mathbf{q}+\mathbf{g}_\mathbf{l})_\alpha(\mathbf{q}+\mathbf{g}_\mathbf{l})_\gamma}{|\mathbf{q}+\mathbf{g}_\mathbf{l}|^2} = \lim_{\mathbf{r}\to 0}\sum_{\mathbf{l}\neq 0}\frac{(\mathbf{q}+\mathbf{g}_\mathbf{l})_\alpha(\mathbf{q}+\mathbf{g}_\mathbf{l})_\gamma}{|\mathbf{q}+\mathbf{g}_\mathbf{l}|^2}e^{i(\mathbf{q}+\mathbf{g}_\mathbf{l})\mathbf{r}}$$

$$= -\lim_{\mathbf{r}\to 0}\frac{\partial^2}{\partial x_\alpha\partial x_\gamma}\sum_{\mathbf{l}\neq 0}\frac{e^{i(\mathbf{q}+\mathbf{g}_\mathbf{l})\mathbf{r}}}{|\mathbf{q}+\mathbf{g}_\mathbf{l}|^2} = \lim_{\mathbf{r}\to 0}\frac{\partial^2 f(\mathbf{r})}{\partial x_\alpha\partial x_\gamma} \qquad (1.10.5)$$

We have introduced the function

$$f(\mathbf{r}) = \sum_{\mathbf{l}\neq 0}\frac{e^{i(\mathbf{q}+\mathbf{g}_\mathbf{l})\mathbf{r}}}{|\mathbf{q}+\mathbf{g}_\mathbf{l}|^2}$$

which satisfies the equation

$$\Delta f(\mathbf{r})| = \sum_{\mathbf{l}\neq 0} e^{i(\mathbf{q}+\mathbf{g}_\mathbf{l})\mathbf{r}} = e^{i\mathbf{q}\mathbf{r}}\sum_{\mathbf{l}\neq 0} e^{i\mathbf{g}_\mathbf{l}\mathbf{r}} = e^{i\mathbf{q}\mathbf{r}}(\Omega\sum_{\mathbf{n}}\delta(\mathbf{r}-\mathbf{R}_\mathbf{n}) - 1)$$

This last equality is satisfied because the Fourier-series expansion of the periodic function

$$\Omega\sum_n \delta(\mathbf{r}-\mathbf{R}_n)$$

is $\sum_\mathbf{l}\exp(i\mathbf{g}_\mathbf{l}\mathbf{r})$.

In what follows, we assume that $q \ll g_\mathbf{l}$, so that we can set $\mathbf{q}=0$. As we see from the equation for $f(\mathbf{r})$, this function can be interpreted as the potential of negative pointlike charges $\Omega/4\pi$ located at lattice sites, and positive charges which are distributed continuously with density $(4\pi)^{-1}$ and, on average, exactly compensate the negative charges.

By virtue of (1.10.5), the value assumed by this function is essential in the neighborhood of the node $\mathbf{r}=0$. As is typical in electrodynamics,

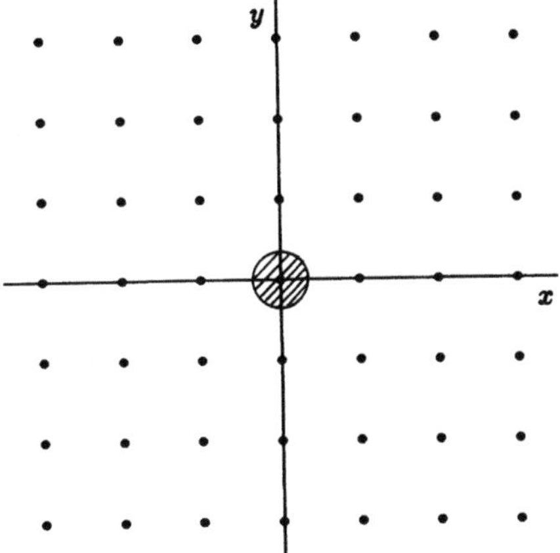

Figure 1.5. The sphere that encloses the charge which must be taken into account in the calculation of the function $\mathbf{f}(\mathbf{r})$ in the neighborhood of the origin of coordinates.

we need to omit the self action of dipoles, that is, we need not take into account the charge at $\mathbf{r} = 0$ when calculating $f(\mathbf{r})$. Let us take a sphere centered at $\mathbf{r} = 0$ with a radius smaller than the cubic lattice period (see Figure 1.5). Hence, $f = f' + f''$, where f' is the potential of the charges inside the sphere and f'' is the potential of the charges outside it.

Obviously, the cubic symmetry of the arrangement of pointlike charges implies that

$$f''(-x, y, 0) = f''(x, y, 0)$$

since f'' must remain unchanged under 180° rotation around the y axis. Hence,

$$\frac{\partial f''(0, y, 0)}{\partial x} = 0$$

and therefore,

$$\frac{\partial^2 f''(0, 0, 0)}{\partial x \, \partial y} = 0$$

Likewise, $\frac{\partial^2 f''(0,0,0)}{\partial x_\alpha \partial x_\gamma} = 0$ for $\alpha \neq \gamma$. By definition, the function $f''(\mathbf{r})$ satisfies inside a sphere the equation

$$\Delta f'' = \left(\frac{\partial^2}{\partial x^2} + \frac{\partial^2}{\partial y^2} + \frac{\partial^2}{\partial z^2}\right) f'' = 0$$

Furthermore, by virtue of the cubic symmetry,

$$\frac{\partial^2 f''}{\partial x^2} = \frac{\partial^2 f''}{\partial y^2} = \frac{\partial^2 f''}{\partial z^2}$$

so that

$$\frac{\partial^2 f''(0,0,0)}{\partial x_\alpha \partial x_\gamma} = 0 \quad \text{for any } \alpha \text{ and } \gamma$$

The function f'' in (1.10.5) can be disregarded and only f' must be found, since the calculation of the sum (1.10.5) requires second derivatives at $\mathbf{r} = 0$. The function f' in the neighborhood of $\mathbf{r} = 0$ satisfies the equation

$$\Delta f'(\mathbf{r}) = 1$$

whose solution in spherical coordinates gives

$$f'(\mathbf{r}) = -\frac{r^2}{6} + \text{const}$$

From equation (1.10.5) we now obtain

$$\sum_{\mathbf{l}\neq 0} \frac{(\mathbf{q}+\mathbf{g}_\mathbf{l})_\alpha (\mathbf{q}+\mathbf{g}_\mathbf{l})_\gamma}{|\mathbf{q}+\mathbf{g}_\mathbf{l}|^2} = -\frac{1}{3}\delta_{\alpha\gamma}$$

By substituting this result into (1.10.4), we arrive at

$$a^{\mathbf{l}'}_{\alpha\beta} = \frac{\delta_{\alpha\beta}}{1 - 4\pi\varkappa n/3}$$

And finally,

$$\overline{\chi}_{\alpha\beta} = (\chi A)^{00}_{\alpha\beta} = \varkappa n \sum_{\mathbf{l}} A^{\mathbf{l}0}_{\alpha\beta} = \varkappa n a^0_{\alpha\beta} = \frac{\varkappa n}{1 - 4\pi\varkappa n/3}\delta_{\alpha\beta}$$

This last expression is precisely the familiar Clausius–Mosotti formula which relates the mean susceptibility of a crystal with the polarizability of an individual atom and where the factor $a^0_{\alpha\beta}$ shows how the field acting on each atom differs from the mean field. This formula implies

$$\overline{\varepsilon} = 1 + 4\pi\overline{\chi} = \frac{1 + 8\pi\varkappa n/3}{1 - 4\pi\varkappa n/3}$$

or, if we solve it for $\varkappa$,

$$\frac{4\pi}{3}\varkappa n = \frac{\overline{\varepsilon} - 1}{\overline{\varepsilon} + 2}$$

This is the familiar Lorentz–Lorenz formula.

In principle, such calculations can be carried out for any structure, provided we know the tensor $\chi^{\mathbf{l}\mathbf{l}'}_{\alpha\beta}$, including also cubic crystals with non-pointlike atoms. In this last case, $\chi^{\mathbf{l}\mathbf{l}'}_{\alpha\beta}$ drop off as $\mathbf{l}$ and $\mathbf{l}'$ increase.

1.11. Nonlinear Susceptibilities of Crystals

As we have mentioned already, the quadratic response of a stationary medium to the field is given by the formula

$$\mathcal{P}_{\alpha}^{(2)}(\mathbf{r},t)=\int dt'\,dt''\int d^3r'\,d^3r''\chi_{\alpha\beta\gamma}^{(2)}(\mathbf{r},\mathbf{r}',\mathbf{r}'',t-t',t-t'')$$
$$\times\,\mathcal{E}_{\beta}(\mathbf{r}',t')\,\mathcal{E}_{\gamma}(\mathbf{r}'',t'')$$

In a crystal, $\chi^{(2)}$ satisfies the periodicity conditions

$$\chi_{\alpha\beta\gamma}^{(2)}(\mathbf{r}+\mathbf{R_n},\mathbf{r}'+\mathbf{R_n},\mathbf{r}''+\mathbf{R_n};t-t',t-t'')$$
$$=\chi_{\alpha\beta\gamma}^{(2)}(\mathbf{r},\mathbf{r}',\mathbf{r}'';t-t',t-t'')$$

Therefore, $\chi^{(2)}$ has the form

$$\chi_{\alpha\beta\gamma}^{(2)}=\chi_{\alpha\beta\gamma}^{(2)}(\mathbf{r},\mathbf{R}',\mathbf{R}'';\tau',\tau'')$$

where $\tau'=t-t'$, $\tau''=t-t''$, $\mathbf{R}'=\mathbf{r}-\mathbf{r}'$, $\mathbf{R}''=\mathbf{r}-\mathbf{r}''$ and the dependence on $\mathbf{r}$ is periodic.

By analogy to (1.8.5), the Fourier transform of this function has the form

$$\chi_{\alpha\beta\gamma}^{(2)}(\mathbf{r},\mathbf{R}',\mathbf{R}'';\tau',\tau'')=\sum_{\mathbf{l}}\int\frac{d^3k'\,d^3k''\,d\omega'd\omega''}{(2\pi)^8}$$
$$\times\exp(i\mathbf{k}'\mathbf{R}'+i\mathbf{k}''\mathbf{R}''+i\mathbf{g}_{\mathbf{l}}\mathbf{r}-i\omega'\tau'-i\omega''\tau'')$$
$$\times\,\chi_{\alpha\beta\gamma}^{\mathbf{l}}(\mathbf{k}',\mathbf{k}'';\omega',\omega'')\qquad(1.11.1)$$

We can again introduce symmetric notation

$$\chi_{\alpha\beta\gamma}^{\mathbf{l}\mathbf{l}'\mathbf{l}''}(\mathbf{q}=\mathbf{q}'+\mathbf{q}'',\omega=\omega'+\omega'')$$

All wave vectors $\mathbf{q}$, $\mathbf{q}'$, $\mathbf{q}''$ belong to the first Brillouin zone. In optical fields, $\mathbf{q}'$ and $\mathbf{q}''$ are small and their vector sum definitely falls into the first Brillouin zone. In the general case,

$$\mathbf{q}'+\mathbf{q}''=\mathbf{q}+\mathbf{g}_{\mathbf{l}}$$

where $\mathbf{g}_{\mathbf{l}}$ is a reciprocal lattice vector. If $\mathbf{l}=0$, one speaks of normal processes, and if $\mathbf{l}\neq 0$, of umklapp processes. In optics, umklapp processes are important only in the x-ray range.

In the same fashion, symmetric notation is introduced for cubic susceptibility $\chi^{(3)}$ and for higher-order susceptibilities:

$$\chi^{(3)\,\mathrm{ll'l''l'''}}_{\alpha\beta\gamma\delta}(\mathbf{q}+\mathbf{g}_\mathrm{l}=\mathbf{q}'+\mathbf{q}''+\mathbf{q}''',\ \omega=\omega'+\omega''+\omega''')$$

It is again necessary to clarify the relation of the mean nonlinear susceptibility, which is widely used in the phenomenological nonlinear optics, to the susceptibilities we discuss here.

Instead of (1.9.1), we obtain equations for the field at a frequency ω, up to terms cubic in the field:

$$\begin{aligned}\left(1-\frac{4\pi\omega_i^2}{c^2}\boldsymbol{\mathcal{D}}_{0i}\chi_i^{(1)}\right)E_i &= \frac{4\pi\omega_i^2}{c^2}\boldsymbol{\mathcal{D}}_{0i}\chi_i^{(2)}e_jE_k\\ &+\frac{4\pi\omega_i^2}{c^2}\boldsymbol{\mathcal{D}}_{0i}\chi_i^{(3)}E_lE_mE_n+\frac{4\pi\omega_i}{c^2}\boldsymbol{\mathcal{D}}_{0i}j_i^{(\mathrm{e})}\\ \omega_i=\omega_j+\omega_k&=\omega_l+\omega_m+\omega_n\end{aligned} \tag{1.11.2}$$

The righthand side contains terms which originate with the nonlinear polarization due to the fields at the frequencies ω_i, ω_k, ω_l, ω_m, ω_n. For the sake of brevity, $\chi_i^{(1,2,3)}$ denotes susceptibilities of different orders, which generate polarization at a frequency ω_i with a wave vector $\mathbf{q}_i$ in response to the fields E_i or E_j, E_k or E_l, E_m and E_n.

Similarly to (1.9.5), equation (1.11.2) makes it possible to express short-wave fields $\widetilde{E}_i$ in terms of long-wave fields $\bar{E}_i$. Here it is more convenient to write the total field E_i in terms of its long-wavelength component.

To achieve this, we first obtain from (1.11.2)

$$\widetilde{E}_i=\frac{4\pi\omega_i^2}{c^2}\widetilde{\boldsymbol{\mathcal{D}}}_{0i}\chi_i^{(1)}E_i+\frac{4\pi\omega_i^2}{c^2}\widetilde{\boldsymbol{\mathcal{D}}}_{0i}\chi_i^{(2)}E_jE_k+\frac{4\pi\omega_i^2}{c^2}\widetilde{\boldsymbol{\mathcal{D}}}_{0i}\chi_i^{(3)}E_lE_mE_n$$

whence

$$\begin{aligned}\overline{E}_i=E_i-\widetilde{E}_i&=\left(1-\frac{4\pi\omega_i^2}{c^2}\widetilde{\boldsymbol{\mathcal{D}}}_{0i}\chi_i^{(1)}\right)E_i\\ &-\frac{4\pi\omega_i^2}{c^2}\widetilde{\boldsymbol{\mathcal{D}}}_{0i}\chi_i^{(2)}E_jE_k-\frac{4\pi\omega_i^2}{c^2}\widetilde{\boldsymbol{\mathcal{D}}}_{0i}\chi_i^{(3)}E_lE_mE_n\end{aligned}$$

By multiplying this equation by

$$A_i=\left[1-\frac{4\pi\omega_i^2}{c^2}\widetilde{\boldsymbol{\mathcal{D}}}_{0i}\chi_i^{(1)}\right]^{-1}$$

we obtain

$$E_i=A_i\overline{E}_i+\frac{4\pi\omega_i^2}{c^2}A_i\widetilde{\boldsymbol{\mathcal{D}}}_{0i}\chi_i^{(2)}E_jE_k+\frac{4\pi\omega_i^2}{c^2}A_i\widetilde{\boldsymbol{\mathcal{D}}}_{0i}\chi_i^{(3)}E_lE_mE_n \tag{1.11.3}$$

Correspondence Rules

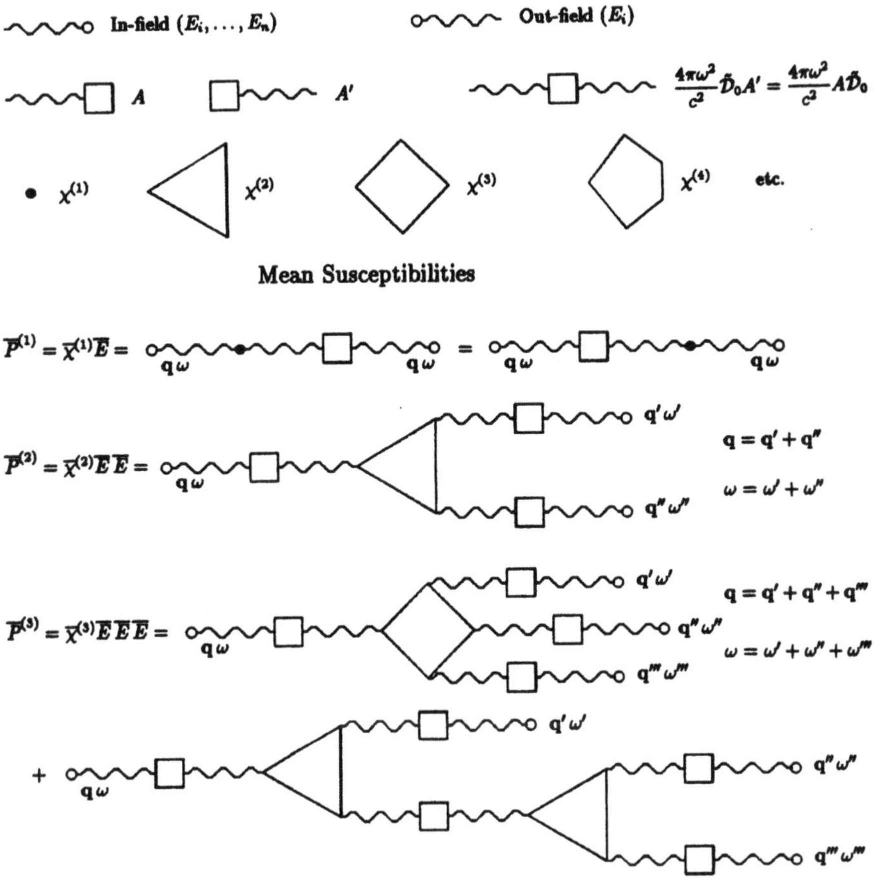

Figure 1.6. Diagrams for the calculation of the mean susceptibilities.

Similar formulas are obviously valid for all fields (E_j, E_k, $E_l, \ldots$). An expression for E_i in terms of the long-wavelength (mean) field $\overline{E}$ can be found by iterations (i.e., assuming the second term on the righthand side of (1.11.3) to be much smaller than the first and much greater than the third, and substituting for E_s, $s = j, k, l, m, n$, the expressions obtained in terms of $\overline{E}$ and powers of E via (1.11.3)). We also need to truncate the expressions to terms cubic in $\bar{E}$.

The resulting expression must be substituted into

$$P_i = \chi_i^{(1)} E_i + \chi_i^{(2)} E_j E_k + \chi_i^{(3)} E_l F_{\cdot n} E_n$$

again dropping all terms that are higher than cubic in $\overline{E}$.

The polarization P_i is then obtained in terms of the long-wavelength fields. Its long-wavelength component $\overline{P}_i = P_i^0$ now determines the mean linear and nonlinear susceptibilities.

We obtain

$$\overline{P}_i = \overline{\chi}_i^{(1)}\overline{E}_i + \overline{\chi}_i^{(2)}\overline{E}_j\overline{E}_k + \overline{\chi}_i^{(3)}\overline{E}_l\overline{E}_m\overline{E}_n$$

where

$$\begin{aligned}
\overline{\chi}_i^{(1)} &= (\chi_i^{(1)}A_i)^{00} = (A_i'\chi_i^{(1)})^{00} \\
\overline{\chi}_i^{(2)} &= (A_i'\chi_i^{(2)}A_jA_k)^{000} \\
\overline{\chi}_i^{(3)} &= (A_i'\chi_i^{(3)}A_kA_lA_m)^{0000} + \left(A_i'\chi_i^{(2)}A_p\frac{4\pi\omega_p^2}{c^2}\widetilde{\mathcal{D}}_{0p}\chi_p^{(2)}A_iA_mA_n\right)^{0000}
\end{aligned} \tag{1.11.4}$$

$$A_i' = \left[1 - \frac{4\pi\omega_i^2}{c^2}\chi_i^{(1)}\widetilde{\mathcal{D}}_{0i}\right]^{-1} \qquad \omega_p = \omega_l + \omega_m \qquad \omega_i = \omega_p + \omega_k$$

The physical meaning of the matrices A and A' is that they take into account the processes of creation and propagation of spatial field harmonics, which in their turn produce the long-wavelength polarization. Expressions (1.11.4) for the mean susceptibilities can be put in correspondence with diagrams shown in Figure 1.6; the correspondence rules are shown in the figure. All terms of the expressions for $\overline{\chi}^{(4)}$, $\overline{\chi}^{(5)}$, etc. can be obtained by putting the higher-order susceptibility tensors in correspondence with polygons that have more vertices than $\chi^{(3)}$, and then composing all possible diagrams which include these polygons and the elements found in the diagrams in Figure 1.6. This is a typical example of the diagram technique, which produces various new formulas without additional calculations, by applying the correspondence rules to the set of all possible diagrams.

2

Quantum-Mechanical Theory of Linear and Nonlinear Susceptibility

2.1. Density Matrix of a System

The wave function $\psi(q,t)$, or the vector of state, corresponds to the most complete description of a system in quantum mechanics. It is quite often, however, especially for large systems, that a statistical description is necessary: we can only speak of a probability of the system occupying a particular pure state which is described by the wave function. The state of the system is then characterized by the density matrix (density operator).

Let w_j be the probability for the system to be in a state ψ_j. The state ψ_j is some finite or infinite set of states; in general, this set of functions is not necessarily orthogonal or complete, and can even be overdetermined.

For a pure state $\psi_j(q)$, the mean values of an arbitrary physical quantity to which corresponds an operator L are

$$\overline{L}_j = \int \psi_j^*(q) L \psi_j(q)\, dq$$

where q is defined as the complete set of the generalized coordinates of the system.

For a mixed state, the mean value is

$$\overline{L} = \sum_j w_j \overline{L}_j = \sum_j w_j \int \psi_j^* L \psi_j\, dq$$

Let us introduce the density matrix (in the coordinate representation)

$$\rho(q,q',t) = \sum_j w_j \psi_j(q,t)\psi_j^*(q',t) \tag{2.1.1}$$

The mean value of the operator L then becomes

$$\overline{L} = \mathrm{Sp}\{L\rho\} \equiv \int \lim_{q'\to q} L_q \rho(q,q',t)\, dq \tag{2.1.2}$$

where the index q with the operator L signifies that the operator acts on $\rho(q, q', t)$ via the variable q.

As we see from (2.1.1), the density matrix is an averaged bilinear combination of wave functions; at the same time, the mean value of a physical quantity in a pure state is expressed in a bilinear form in terms of the wave function. The averaging of bilinear expressions cannot be reduced to using a bilinear expression of a certain averaged wave function; hence, the density matrix describes a more general statistical ensemble than the wave function. If in (2.1.1) $w_j = 1$ for a certain j and $w_j = 0$ for all other j, we have the case of a pure state.

The condition

$$\sum_j w_j = 1$$

implies the normalization for $\rho(q, q', t)$:

$$\mathrm{Sp}\{\rho\} \equiv \int \lim_{q' \to q} \rho(q, q', t)\, dq = 1$$

In any other noncoordinate representation,

$$\psi_j(q) = \sum_n c_{nj} \varphi_n(q)$$

where $\varphi_n(q)$ is the complete set of normalized orthogonal wave functions. This gives

$$\overline{L}_j = \sum_{n'n} c^*_{n'j} L_{n'n} c_{nj}$$

where $L_{n'n} = \int \varphi^*_{n'} L \varphi_n \, dq$ are the matrix elements of the operator L and

$$\overline{L} = \sum_j w_j \overline{L}_j = \sum_{n'n} \rho_{nn'} L_{n'n} \equiv \sum_n (\rho L)_{nn} = \mathrm{Sp}\{\rho L\} = \mathrm{Sp}\{L\rho\} \quad (2.1.3)$$

In (2.1.3), $\rho_{nn'}$ are the matrix elements of the density matrix in the φ_n representation,

$$\rho_{nn'} = \sum_j w_j c^*_{n'j} c_{nj} \quad (2.1.4)$$

and the trace Sp denotes the sum of the diagonal elements of the matrix. The normalization condition for the matrix $\rho_{nn'}$ now becomes

$$\mathrm{Sp}\{\rho\} = \sum_j w_j \sum_n c^*_{nj} c_{nj} = 1$$

Expression (2.1.2) is a particular case of formula (2.1.3) for the continuous (coordinate) representation.

The need to introduce the density matrix also arises when we consider a subsystem of a large system, even though the entire system is described by a wave function. Assume again that q are the coordinates of a subsystem while Q are all other coordinates of the larger system. If the Hamiltonian of the larger system is not reducible to the sum of the Hamiltonian $\mathcal{H}_q$ of the subsystem and the Hamiltonian $\mathcal{H}_Q$ of the surrounding system while their interaction V_{qQ} is significant, then $\mathcal{H} = \mathcal{H}_q + \mathcal{H}_Q + V_{qQ}$. In this situation, the wave function of the larger system cannot, in the general case, be written as a product of the wave functions of the subsystem and the system around it (it cannot be factorized) and we are justified to operate with the concept of the wave function of the subsystem.

A large system can always be regarded as isolated, by including into it everything that interacts in any way with the subsystem or with other parts of the larger system. Hence, it is always possible to introduce for it a wave function or a density matrix

$$R(q, Q; q', Q'; t)$$

which is defined by a relation of the type of (2.1.1). As for the mean value of a physical quantity L which refers to the subsystem only, it can be given in terms of a reduced density matrix $\rho(q, q', t)$ which is a function only of the coordinates of the subsystem,

$$\overline{L} = \int \lim_{\substack{q' \to q \\ Q' \to Q}} L_q R(q, Q; q', Q'; t)\, \mathrm{d}q\, \mathrm{d}Q = \int \lim_{q' \to q} L_q \rho(q, q', t)\, \mathrm{d}q$$

where

$$\rho(q, q', t) = \int R(q, Q; q', Q; t)\, \mathrm{d}Q$$

In a particular case, R can describe a pure state:

$$R(q, Q; q', Q'; t) = \psi(q, Q, t)\psi^*(q', Q', t)$$

but $\rho(q, q', t)$ still remains nonreducible to the density matrix of a pure state. On the other hand, the wave function of a subsystem cannot in general be introduced while the density matrix, as follows from the above analysis, can always be introduced.

Let the representation be given by the wave functions $\varphi_n(q)$ which describe the subsystem and $\Phi_N(Q)$ describing the surroundings. This does not contradict the statement that a subsystem does not have its own wave function, since any function $\psi(q, Q)$ can be expanded in a generalized Fourier series over a set of the functions $\varphi_n(q)\Phi_N(Q)$, which is complete if the sets φ_n and Φ_N are complete. By virtue of (2.1.3), then

$$\overline{L} = \sum_{\substack{nn' \\ NN'}} R_{nn'}^{NN'} L_{n'n}^{N'N} = \sum_{\substack{nn' \\ NN'}} R_{nn'}^{NN'} L_{nn'} \delta_{NN'} = \sum_{nn'} \rho_{n'n} L_{nn'} = \mathrm{Sp}\{\rho L\}$$

where $\rho_{nn'} = \sum_N R_{nn'}^{NN}$ is the reduced density matrix of the subsystem in the φ_n representation. We have used the fact that L is an operator of the subsystem:

$$L_{n'n}^{N'N} = \int \Phi_{N'}^* \varphi_{n'}^* L \varphi_n \Phi_N \, \mathrm{d}q \, \mathrm{d}Q$$
$$= \int \varphi_{n'}^* L \varphi_n \, \mathrm{d}q \int \Phi_{N'}^* \Phi_N \, \mathrm{d}Q = L_{nn'} \delta_{NN'}$$

It is essential that the trace Sp is independent of the specific representation, so that the mean value

$$L = \mathrm{Sp}\{\rho L\} = \mathrm{Sp}\{L\rho\}$$

can be calculated in any representation. In an isolated system, the equation of motion for the density matrix is a corollary of the Schrödinger equation and is equivalent to it.

Using the definition of the density matrix (2.1.4), we obtain

$$\mathrm{i}\hbar \frac{\mathrm{d}\rho_{nn'}}{\mathrm{d}t} = \mathrm{i}\hbar \sum_j w_j \left(\frac{\mathrm{d}c_{n'j}^*}{\mathrm{d}t} c_{nj} + c_{n'j}^* \frac{\mathrm{d}c_{nj}}{\mathrm{d}t} \right)$$

The Schrödinger equation in the φ_n representation is

$$\mathrm{i}\hbar \frac{\mathrm{d}c_{nj}}{\mathrm{d}t} = \sum_m \mathscr{H}_{nm} c_{mj}$$
$$-\mathrm{i}\hbar \frac{\mathrm{d}c_{n'j}^*}{\mathrm{d}t} = \sum_m \mathscr{H}_{n'm}^* c_{mj}^*$$

Since the Hamiltonian is Hermitian,

$$\mathscr{H}_{n'm}^* = \mathscr{H}_{mn'}$$

we find

$$\mathrm{i}\hbar \frac{\mathrm{d}\rho_{nn'}}{\mathrm{d}t} = \sum_m \mathscr{H}_{nm} \left(\sum_i w_i c_{mj} c_{n'j}^* \right)$$
$$- \sum_m \left(\sum_i w_j c_{nj} c_{mj}^* \right) \mathscr{H}_{mn'} = (\mathscr{H}\rho - \rho\mathscr{H})_{nn'}$$

In the operator notation, this equation for the density matrix of an isolated system is written as

$$\mathrm{i}\hbar \frac{\mathrm{d}\rho}{\mathrm{d}t} = [\mathscr{H}, \rho] \tag{2.1.5}$$

where $[\mathcal{H}, \rho] \equiv \mathcal{H}\rho - \rho\mathcal{H}$.

In this form, the equation of motion is independent of the choice of a specific representation. In the coordinate representation, we have

$$i\hbar \frac{\partial \rho(q,q',t)}{\partial t} = \mathcal{H}_q \rho(q,q',t) - \mathcal{H}^*_{q'} \rho(q,q',t)$$

In order to find the equation of motion of a subsystem which interacts with its surroundings (referred to henceforth as the thermostat), we write

$$i\hbar \frac{\mathrm{d}R}{\mathrm{d}t} = [\mathcal{H}, R] = [\mathcal{H}_q, R] + [\mathcal{H}_Q, R] + [V_{qQ}, R]$$

In the representation defined by the basis $\psi_n^N(q,Q) = \varphi_n(q)\Phi_N(Q)$, this equation becomes

$$i\hbar \frac{\mathrm{d}R_{nn'}^{NN'}}{\mathrm{d}t} = \sum_{mM} (\mathcal{H}_{nm}^{NM} R_{mn'}^{MN'} - R_{nm}^{NM} \mathcal{H}_{mn'}^{MN'}) \tag{2.1.6}$$

We must take into account that

$$\mathcal{H}_{qnm}^{NM} = \mathcal{H}_{qnm} \delta_{NM} \qquad \mathcal{H}_{Qnm}^{NM} = \mathcal{H}_Q^{NM} \delta_{nm}$$

because the operator $\mathcal{H}_q$ acts only on the variables q while $\mathcal{H}_Q$ acts on the variables Q.

Taking the trace of equation (2.1.6) in the variables Q, that is, multiplying it by $\delta_{NN'}$ and summing up over N, N', we obtain

$$i\hbar \frac{\mathrm{d}\rho_{nn'}}{\mathrm{d}t} = [\mathcal{H}_q, \rho]_{nn'} + \Gamma_{nn'}$$

Note that the term containing $\mathcal{H}_Q$ has vanished,

$$\sum_N [\mathcal{H}_Q, R]_{nm}^{NN} = \sum_{NN'} (\mathcal{H}_Q^{NN'} R_{nm}^{N'N} - R_{nm}^{NN'} \mathcal{H}_Q^{N'N}) = 0$$

because the summation indices N and N' in the second term can be transposed.

By definition, the matrix $\mathbf{\Gamma}_{n'n}$ is

$$\mathbf{\Gamma}_{n'n} = \sum_N [V_{qQ}, R]_{nn'}^{NN}$$

The equation in the operator form for the density matrix is

$$i\hbar \frac{\mathrm{d}\rho}{\mathrm{d}t} = [\mathcal{H}_q, \rho] + \mathbf{\Gamma} \tag{2.1.7}$$

where $\mathbf{\Gamma}$ is the relaxation matrix (relaxation operator).

Equation (2.1.7) indeed becomes an equation for ρ if a way is found to relate $\mathbf{\Gamma}$ to ρ. This is possible in the important particular case of the thermostat being such a large system that the effect of the subsystem on its state is negligible. Assuming then the state of the thermostat to be equilibrium or nonequilibrium but fixed, one can carry out averaging over this state; the influence of the small subsystem on the thermostat must be ignored or taken into account using perturbation theory.

Let us consider one particular model producing results of much greater significance than one would expect: the random collisions model.

The random collisions model is formulated on the basis of an analysis of processes in the gas and uses the following assumptions:

1. The system and the thermostat interact by short pulses; after each pulse, the system evolves as an insulated one.
2. An instantaneous interaction with the thermostat is so great that the system immediately changes to the equilibrium (or stationary) state which is determined by the density matrix ρ_0.

Therefore, the density matrix of the system at $t = t_0$ (immediately after the pulse) is $\rho(t_0) = \rho_0$ and then evolves until the next pulse in accordance with the equation

$$i\hbar \frac{d\rho}{dt} = [\mathcal{H}, \rho].$$

After the pulse, the system completely "forgets" its history. The time moments of collisions are, however, quite random and it is of interest to find the density matrix averaged over time of collision:

$$\overline{\rho}(t) = \int_{-\infty}^{t} w(t, t_0)\rho(t, t_0)\, dt_0 \tag{2.1.8}$$

Here $w(t, t_0)$ is the probability for the last collision before time t to have taken place at time t_0, and $\rho(t, t_0)$ is the density matrix of the isolated system which satisfies the initial condition $\rho(t_0, t_0) = \rho_0$.

Let τ be the time between collisions, that is, dt_0/τ is the probability of a collision during dt_0. Obviously,

$$w(t, t_0)\, dt_0 = \frac{dt_0}{\tau} p(t, t_0)$$

where $p(t, t_0)$ is the probability that there are no collisions between the times t_0 and t. The function $p(t, t_0)$ obeys the equation

$$p(t, t_0 - dt_0) = p(t, t_0)\left(1 - dt_0/\tau\right)$$

because $(1 - dt_0/\tau)$ is the probability that there were no collisions during the interval dt_0. This equation is equivalent to the differential equation

$$\frac{\partial p}{\partial t_0} = \frac{p}{\tau}$$

whose solution

$$p(t, t_0) = \exp\left(-\frac{t - t_0}{\tau}\right)$$

satisfies the initial condition

$$p(t, t) = 1$$

Therefore,

$$w(t, t_0) = \frac{\partial p}{\partial t} = \frac{1}{\tau} \exp\left(-\frac{t - t_0}{\tau}\right)$$

Differentiating (2.1.8) with respect to time t, we obtain

$$\begin{aligned} i\hbar \frac{\partial \overline{\rho}}{\partial t} &= i\hbar \frac{\partial}{\partial t} \int_{-\infty}^{t} w(t, t_0)\rho(t, t_0)\, dt_0 \\ &= i\hbar \frac{\partial}{\partial t} \int_0^{\infty} \frac{1}{\tau} \exp\left(-\frac{\Theta}{\tau}\right) \rho(t, t - \Theta)\, d\Theta \\ &= \frac{i\hbar}{\tau} \int_0^{\infty} \exp\left(-\frac{\Theta}{\tau}\right) \frac{\partial \rho(t, t - \Theta)}{\partial t}\, d\Theta \end{aligned}$$

By virtue of the equation of motion for the density matrix, the differentiation of $\rho(t, t_0)$ with respect to the first argument yields $(i\hbar)^{-1}[\mathcal{H}, \rho]$. The differentiation with respect to the second argument is equivalent to the differentiation with respect to Θ (with reversed sign).

This gives

$$i\hbar \frac{\partial \overline{\rho}}{\partial t} = [\mathcal{H}, \overline{\rho}] - \frac{i\hbar}{\tau} \int_0^{\infty} \exp\left(-\frac{\Theta}{\tau}\right) \frac{\partial \rho(t, t - \Theta)}{\partial \Theta}\, d\Theta$$

The integral in the righthand side is taken by parts, which yields

$$\begin{aligned} i\hbar \frac{\partial \overline{\rho}}{\partial t} = [\mathcal{H}, \overline{\rho}] &- \frac{i\hbar}{\tau} \int_0^{\infty} \frac{1}{\tau} \exp\left(-\frac{\Theta}{\tau}\right) \rho(t, t - \Theta)\, d\Theta \\ &- \frac{i\hbar}{\tau} \exp\left(-\frac{\Theta}{\tau}\right) \rho(t, t - \Theta) \Big|_0^{\infty} \end{aligned}$$

Finally, since $\rho(t,t) = \rho_0$, the equation for the mean density matrix $\overline{\rho}$ takes the form

$$\mathrm{i}\hbar \frac{\mathrm{d}\rho}{\mathrm{d}t} = [\mathscr{H}, \rho] - \mathrm{i}\hbar \frac{\rho - \rho_0}{\tau} \tag{2.1.9}$$

If $\mathscr{H} = 0$, then for $\rho(t_1) = \rho_0 + \Delta\rho$ we have

$$\rho(t) = \rho_0 + \Delta\rho \exp\left(-\frac{t - t_1}{\tau}\right)$$

that is, an increment to the equilibrium density matrix decays in time. Hence, each specific system relaxes jumpwise after a pulse but averaging over the systems results in exponential relaxation of the density matrix.

In fact, equations like (2.1.9) are encountered much more often than assumptions of the random collisions model are met. They hold in all those cases in which deviations from the equilibrium are small. Indeed, Γ in (2.1.7) is a functional of ρ:

$$\mathbf{\Gamma} = \mathbf{\Gamma}\{\rho\}$$

so that if deviations from the equilibrium are small, we have

$$\mathbf{\Gamma} = \mathbf{\Gamma}_0 + \mathbf{\Gamma}_1(\rho - \rho_0) + \ldots$$

However, since $\boldsymbol{\rho}_0$ is an equilibrium density matrix, $\mathbf{\Gamma} = 0$ when $\boldsymbol{\rho} = \boldsymbol{\rho}_0$, that is, $\mathbf{\Gamma}_0 = 0$.

If a specific representation is chosen, equation (2.1.7) becomes

$$\mathrm{i}\hbar \frac{\mathrm{d}\rho_{nn'}}{\mathrm{d}t} = [\mathscr{H}, \rho]_{nn'} + \sum_{mm'} K_{nn',mm'}(\rho - \rho_0)_{mm'} \tag{2.1.10}$$

where K is the generalized relaxation matrix. Generally speaking, relaxation times are different for different matrix elements. In the random collisions model, however, they all equal the same quantity τ.

In view of the delay, $K(\rho - \rho_0)$ is replaced by

$$\int_{-\infty}^{t} K(t - t')(\rho(t') - \rho_0)\,\mathrm{d}t'$$

This is the maximum generalization and the most general expression for the case of small deviations from equilibrium. In the random collisions model, this relation (which is, furthermore, Markov-type with $\mathbf{\Gamma}$ independent of the values of ρ at the preceding times) holds as well if deviations from the equilibrium are large.

2.2. Representation of Interaction

In Section 2.1, we were using the Schrödinger representation. This representation puts physical quantities which do not depend explicitly on time in correspondence with constant operators, while the wave function, or vector of state, changes in time as described by the Schrödinger equation

$$i\hbar \frac{\partial \psi}{\partial t} = \mathscr{H}\psi$$

Since the Hamiltonian $\mathscr{H}$ is Hermitian, the vector ψ has invariant norm and can be rewritten in the form

$$\psi(t) = U(t)\psi(0)$$

where $U(t)$ is a unitary operator (i.e., one conserving the vector norm), which satisfies the equation

$$i\hbar \frac{dU}{dt} = \mathscr{H}U$$

and the initial condition $U(0) = 1$.

Sometimes it is more convenient to transfer the time dependence on the operators. If L is the operator of an arbitrary physical quantity, then the corresponding Heisenberg operator, or the operator in the Heisenberg representation, is

$$L_{\mathrm{H}}(t) = U^{-1}(t)LU(t) \tag{2.2.1}$$

As one can verify by straightforward differentiation of (2.2.1), Heisenberg operators satisfy the equations

$$i\hbar \frac{dL_{\mathrm{H}}}{dt} = [L_{\mathrm{H}}, \mathscr{H}_{\Gamma}]$$

where $\mathscr{H}_{\mathrm{H}} = U^{-1}\mathscr{H}U$.

Since U is a unitary operator, it possesses the property

$$U^{-1} = U^{+}$$

where U^{+} is the operator that is Hermitian-conjugate to U.

The mean value of a physical quantity can now be found as a mean value of the operator L_{H} over the wave function $\psi(0)$, which is now time-independent:

$$\overline{L} = \int \psi^{*}(t)L\psi(t)\,dq = \int \psi^{*}(0)L_{\mathrm{H}}(t)\psi(0)\,dq$$

If, however, $\mathcal{H} = \mathcal{H}_0$ is time-independent, then formally

$$U(t) = \exp\left(-\frac{\mathrm{i}}{\hbar}\mathcal{H}_0 t\right)$$

A function of an operator is defined either by a series

$$\exp\left(-\frac{\mathrm{i}}{\hbar}\mathcal{H}_0 t\right) = \sum_k \frac{1}{k!}\left(-\frac{\mathrm{i}}{\hbar}\mathcal{H}_0 t\right)^k$$

or by a condition that if ψ_n is an eigenfunction of the operator $\mathcal{H}_0$ which corresponds to the eigenvalue E_n, then

$$\exp\left(-\frac{\mathrm{i}}{\hbar}\mathcal{H}_0 t\right)\psi_n = \exp\left(-\frac{\mathrm{i}}{\hbar}E_n t\right)\psi_n$$

In the energy representation which is determined by the eigenfunctions ψ_n of the Hamiltonian $\mathcal{H}_0$, we have

$$U_{n'n} = \exp\left(-\frac{\mathrm{i}}{\hbar}E_n t\right)\delta_{n'n}$$

while

$$(L_{\mathrm{H}}(t))_{n'n} = \exp\left(\frac{\mathrm{i}}{\hbar}(E_{n'} - E_n)t\right)L_{n'n}$$

Both the Schrödinger representation and the Heisenberg representation in which the wave function and the density matrix are independent of time are widely used in nonrelativistic quantum mechanics. The relativistic quantum theory operates more often with the Heisenberg representation.

In addition to these two representations, the so-called interaction representation is also used quite often; in this representation, the time dependence is partly transferred to the operators while the wave function varies slowly.

Let $\mathcal{H} = \mathcal{H}_0 + \mathcal{H}'(t)$, where $\mathcal{H}_0$ is the Hamiltonian of time-independent nonperturbed system and $\mathcal{H}'(t)$ is the perturbation. In the interaction representation,

$$\widetilde{L}(t) = \exp\left(\frac{\mathrm{i}}{\hbar}\mathcal{H}_0 t\right)L\exp\left(-\frac{\mathrm{i}}{\hbar}\mathcal{H}_0 t\right)$$

$$\widetilde{\psi}(t) = \exp\left(\frac{\mathrm{i}}{\hbar}\mathcal{H}_0 t\right)\psi(t)$$

In the energy representation defined by $\mathcal{H}_0$, the matrix elements of an operator $\widetilde{L}$ in the interaction representation are, as in the Heisenberg representation for $\mathcal{H}' = 0$,

$$(\widetilde{L}(t))_{n'n} = \exp\left(\frac{\mathrm{i}}{\hbar}(E_{n'} - E_n)t\right)L_{n'n}$$

and the wave function satisfies the equation

$$\mathrm{i}\hbar\frac{\partial\widetilde{\psi}}{\partial t} = -\exp\Big(\frac{\mathrm{i}}{\hbar}\mathcal{H}_0 t\Big)\mathcal{H}_0\psi(t) + \exp\Big(\frac{\mathrm{i}}{\hbar}\mathcal{H}_0 t\Big)\mathcal{H}\psi(t)$$
$$= \exp\Big(\frac{\mathrm{i}}{\hbar}\mathcal{H}_0 t\Big)\mathcal{H}'(t)\exp\Big(\frac{\mathrm{i}}{\hbar}\mathcal{H}_0 t\Big)\widetilde{\psi}(t)$$

This equation can be written as

$$\mathrm{i}\hbar\frac{\partial\widetilde{\psi}}{\partial t} = \widetilde{\mathcal{H}}'(t)\widetilde{\psi} \tag{2.2.2}$$

which coincides with the Schrödinger equation but includes, instead of the Hamiltonian, the perturbation in the interaction representation. The matrix elements of the operator $\widetilde{\mathcal{H}}'(t)$ are

$$(\widetilde{\mathcal{H}}'(t))_{n'n} = \Big[\exp\Big(\frac{\mathrm{i}}{\hbar}\mathcal{H}_0 t\Big)\mathcal{H}'(t)\exp\Big(-\frac{\mathrm{i}}{\hbar}\mathcal{H}_0 t\Big)\Big]_{n'n}$$
$$= \exp\Big(\frac{\mathrm{i}}{\hbar}(E_{n'} - E_n)t\Big)(\mathcal{H}'(t))_{n'n}$$

Note that the perturbation $\mathcal{H}'(t)$ — in the Schrödinger representation — is, generally, time-dependent, since it may include external fields.

The mean value of a physical quantity can be calculated in any representation and is independent of the choice of representation; it can, for instance, be calculated in terms of the interaction representation:

$$\overline{L} = \int \widetilde{\psi}^*(t)\widetilde{L}(t)\widetilde{\psi}(t)\,\mathrm{d}q$$

When going from the Schrödinger to the interaction representation, the density matrix transforms as operators of physical quantities do:

$$\tilde{\rho}(t) = \exp\Big(\frac{\mathrm{i}}{\hbar}\mathcal{H}_0 t\Big)\rho(t)\exp\Big(-\frac{\mathrm{i}}{\hbar}\mathcal{H}_0 t\Big)$$

It can be shown directly that the mean values are not affected. Indeed,

$$\overline{L} = \mathrm{Sp}\{\widetilde{L}\tilde{\rho}\} = \mathrm{Sp}\Big\{\exp\Big(\frac{\mathrm{i}}{\hbar}\mathcal{H}_0 t\Big)L\rho\exp\Big(-\frac{\mathrm{i}}{\hbar}\mathcal{H}_0 t\Big)\Big\} = \mathrm{Sp}\{L\rho\}$$

The proof makes use of the possibility of cyclic permutation in the trace:

$$\mathrm{Sp}\{AB\} = \sum_{nm} A_{nm}B_{mn} = \sum_{nm} B_{mn}A_{nm} = \mathrm{Sp}\{BA\}$$
$$\mathrm{Sp}\{ABC\} = \mathrm{Sp}\{CAB\}$$

We can find by straightforward differentiation of $\overline{\rho}(t)$, using (2.1.5), that

$$\begin{aligned} i\hbar \frac{\partial \widetilde{\rho}}{\partial t} &= -\exp\left(\frac{i}{\hbar}\mathscr{H}_0 t\right)\mathscr{H}_0(t)\exp\left(-\frac{i}{\hbar}\mathscr{H}_0 t\right) \\ &\quad + \exp\left(\frac{i}{\hbar}\mathscr{H}_0 t\right)[\mathscr{H}, \rho]\exp\left(-\frac{i}{\hbar}\mathscr{H}_0 t\right) \\ &\quad + \exp\left(\frac{i}{\hbar}\mathscr{H}_0 t\right)\rho(t)\mathscr{H}_0\exp\left(-\frac{i}{\hbar}\mathscr{H}_0 t\right) \\ &= \exp\left(\frac{i}{\hbar}\mathscr{H}_0 t\right)[\mathscr{H}', \rho]\exp\left(-\frac{i}{\hbar}\mathscr{H}_0 t\right) = [\tilde{\mathscr{H}}', \tilde{\rho}] \end{aligned}$$

Therefore, the density matrix in the interaction representation satisfies an equation of the same type as in the Schrödinger representation but with $\mathscr{H}$ replaced by $\tilde{\mathscr{H}}'$. If $\mathscr{H}'$ and, hence, $\tilde{\mathscr{H}}'$ are small, then $\tilde{\rho}$ varies slowly.

Hereafter, we constantly use the interaction representation but for the sake of brevity, omit the tilde over operators and density matrix.

2.3. Perturbation Theory for Density Matrix

We thus see that the density matrix in the interaction representation satisfies the equation

$$i\hbar \frac{\partial \rho}{\partial t} = [\mathscr{H}', \rho] \tag{2.3.1}$$

So far we did not take relaxation into account since the external field, and hence the perturbation $\mathscr{H}'$, was assumed to be small and the "heating" of the system was negligible. We will cover this last effect in higher-order approximations, but for the time being we will use for large macroscopic systems equation (2.3.1) without relaxation, assuming that the system functions as a thermostat for itself and thus can be treated as isolated.

No general methods for solving equation (2.3.1) were found so far. The only regular procedure is the expansion in powers of perturbation which is proportional to the external field. We assume

$$\rho(t) = \sum_{k=0}^{\infty} \rho^{(k)}(t) \tag{2.3.2}$$

where $\rho^{(k)} \propto (\mathscr{H}')^k \propto (\boldsymbol{\mathcal{E}}^{(e)})^k$ is proportional to the external field to power k.

If, however, $\mathscr{H}' = 0$, that is, if $\boldsymbol{\mathcal{E}}^{(e)} = 0$, then we find that in the interaction representation

$$\partial\rho/\partial t = 0 \qquad \rho = \rho_0$$

where ρ_0 is the density matrix in the absence of external fields.

Most often, ρ_0 is the equilibrium (Gibbs) density matrix

$$\rho^{(0)} = \rho_0 = \exp\left(\frac{\mathfrak{F} - \mathcal{H}_0}{kT}\right)$$

where $\mathfrak{F}$ is the free energy and T is the temperature. This is not mandatory, however; it is sufficient to assume that ρ_0 is an equilibrium density matrix:

$$[\mathcal{H}_0, \rho_0] = 0$$

This equality guarantees that the matrix ρ_0 is time-independent both in the Schrödinger representation and in the interaction representation at $\mathcal{H}' = 0$. If we write it in the energy representation, we obtain

$$(E_{n'} - E_n)\rho^{(0)}_{n'n} = 0$$

This means that ρ_0 is diagonal in energy.

By substituting expansion (2.3.2) into equation (2.3.1) and equating the terms of equal order in $\mathcal{H}'$, we obtain

$$\mathrm{i}\hbar \mathrm{d}\rho^{(k)}/\mathrm{d}t = [\mathcal{H}', \rho^{(k-1)}]$$

Integrating this equation in time, we find $\rho^{(k)}$ expressed in terms of $\rho^{(k-1)}$:

$$\rho^{(k)}(t) = \frac{1}{\mathrm{i}\hbar} \int_{-\infty}^{t} [\mathcal{H}'(t'), \rho^{(k-1)}(t')]\, \mathrm{d}t' \tag{2.3.3}$$

For the lower limit we took $-\infty$ since it is assumed that in a distant past, $\rho(-\infty) = \rho_0$ and $\rho^{(k)}(-\infty) = 0$ if $k \neq 0$. It is also assumed that the field $\mathcal{E}^{(e)}(t)$ is switched on at a time $t \to -\infty$. If the external field is turned on at $t_0 \neq -\infty$, then $\mathcal{H}'(t') = 0$ at $t' < t_0$ and expression (2.3.3) remains valid.

Using (2.3.3), we find

$$\rho^{(1)}(t) = \frac{1}{\mathrm{i}\hbar} \int_{-\infty}^{t} [\mathcal{H}'(t'), \rho_0]\, \mathrm{d}t'$$

$$\rho^{(2)}(t) = \frac{1}{(\mathrm{i}\hbar)^2} \int_{-\infty}^{t} \mathrm{d}t' \int_{-\infty}^{t'} \mathrm{d}t''[\mathcal{H}'(t'), [\mathcal{H}'(t''), \rho_0]]$$

and generally

$$\rho^{(k)}(t) = \frac{1}{(\mathrm{i}\hbar)^k} \int_{-\infty}^{t} \mathrm{d}t_1 \int_{-\infty}^{t_1} \mathrm{d}t_2 \cdots \int_{-\infty}^{t_{k-1}} \mathrm{d}t_k [\mathcal{H}'(t_1), [\mathcal{H}'(t_2), \ldots$$
$$\ldots [\mathcal{H}'(t_k), \rho_0] \ldots] \tag{2.3.4}$$

The moments of time $t_1, \ldots, t_k$ are ordered:

$$t_k \leq t_{k-1} \leq \cdots \leq t_1 \leq t$$

With the correction of order k to the density matrix known, we can calculate the correction to the mean value of an arbitrary physical quantity which corresponds to an operator L in the interaction representation:

$$\overline{L}(t) = \sum_{k=0}^{\infty} \overline{L}^{(k)}(t)$$

$$\overline{L}^{(k)}(t) = \mathrm{Sp}\{L(t)\rho^{(k)}(t)\} = \frac{1}{(\mathrm{i}\hbar)^k} \int_{-\infty}^{t} \mathrm{d}t_1 \int_{-\infty}^{t_1} \mathrm{d}t_2 \cdots \int_{-\infty}^{t_{k-1}} \mathrm{d}t_k$$
$$\times \mathrm{Sp}\{L(t)[\mathscr{H}'(t_1), [\mathscr{H}'(t_2), \ldots \mathscr{H}'(t_k), \rho_0] \ldots]\} \tag{2.3.5}$$

The averaging in (2.3.5) is carried out with the density matrix ρ_0 placed inside each commutator.

We will now transform (2.3.5) in such a way that integrands include the trace of the product of ρ_0 by an operator.

To do this, we make use of the identity

$$\mathrm{Sp}\{A[A_1, [A_2, \ldots [A_k, B] \ldots]\} = \mathrm{Sp}\{[\ldots [A, A_1], A_2], \ldots A_k]B\} \tag{2.3.6}$$

which holds for arbitrary operators A_1, A_2, ..., A_k, B.

We will prove (2.3.6) using mathematical induction. The proof for $k = 1$ is straightforward, using the linearity of the trace and its invariance under a cyclic permutation of the operators:

$$\begin{aligned}\mathrm{Sp}\{A[A_1, B]\} &= \mathrm{Sp}\{AA_1B - ABA_1\} = \mathrm{Sp}\{AA_1B\} - \mathrm{Sp}\{A_1AB\} \\ &= \mathrm{Sp}\{[A, A_1]B\}\end{aligned}$$

Proving (2.3.6) for $k = 2$ is done similarly, by directly using the properties of traces. Assuming that (2.3.6) holds for $k = l - 1$, we will prove that the identity holds also for $k = l$. Let us denote $C = [A_l, B]$; then we have

$$\begin{aligned}&\mathrm{Sp}\{A[A_1, [A_2, \ldots [A_{l-1}, C] \ldots]\} = \mathrm{Sp}\{[\ldots [A, A_1], A_2], \ldots A_{l-1}] \\ &\quad \times (A_lB - BA_l)\} = \mathrm{Sp}\{[\ldots [A, A_1], A_2], \ldots A_{l-1}]A_lB \\ &\quad - A_l[\ldots [A, A_1], A_2], \ldots A_{l-1}]B\} = \mathrm{Sp}\{[\ldots [A, A_1], A_2], \ldots A_l]B\}\end{aligned}$$

which was to be proved.

Using identity (2.3.6), we can rewrite (2.3.5) as

$$\overline{L}^{(k)}(t) = \frac{1}{(\mathrm{i}\hbar)^k} \int_{-\infty}^{t} \mathrm{d}t_1 \cdots \int_{-\infty}^{t_{k-1}} \mathrm{d}t_k \mathrm{Sp}\{[\ldots [L(t), \mathscr{H}'(t_1)], \ldots \mathscr{H}'(t_k)]\rho_0\} \tag{2.3.7}$$

This last equality can be easily interpreted as

$$\overline{L}(t) = \mathrm{Sp}\{L_{\mathrm{H}}(t)\rho_0\}$$

where

$$L_{\mathrm{H}}(t) = L(t) + \sum_{k=1}^{\infty} \frac{1}{(\mathrm{i}\hbar)^k} \int_{-\infty}^{t} \mathrm{d}t_1 \cdots \int_{-\infty}^{t_{k-1}} \mathrm{d}t_k [\ldots [L(t), \mathscr{H}'(t_1)], \ldots \mathscr{H}'(t_k)] \tag{2.3.8}$$

is a Heisenberg operator, corresponding to $L(t)$, which is expanded in powers of perturbation.

The final result in the form of (2.3.8) could be obtained using the equation of motion for a Heisenberg operator. This method is technically even simpler, but it is physically clearer to begin with the equation of motion for the density matrix in the interaction representation; we have chosen the latter approach.

2.4. Perturbation Operator for Problems of Field–Matter Interaction

In the preceding section we derived the general expressions for the mean value of an arbitrary physical quantity. Our task now is to find the response of the medium to electromagnetic field, that is, the mean polarization or the mean current density, which is the derivative of polarization with respect to time, as a function of field. The linear part of this dependence gives the linear susceptibility $\chi^{(1)}$ while the nonlinear part gives the susceptibilities $\chi^{(2)}$, $\chi^{(3)}$, etc.

First of all, we will need an expression for the current density operator. The classical expression for the current density is (as we have already pointed out in Chapter 1)

$$j_\alpha(\mathbf{r}, t) = \sum_i e_i v_{i\alpha}(t) \delta(\mathbf{r} - \mathbf{r}_i(t)) \tag{2.4.1}$$

In the quantum case, the velocity $v_{i\alpha}$ corresponds to the operator

$$v_{i\alpha} = \frac{1}{m_i}\Big(p_{i\alpha} - \frac{e_i}{c} A_\alpha(\mathbf{r}_i, t)\Big) \tag{2.4.2}$$

where $p_{i\alpha} = -\mathrm{i}\hbar\partial/\partial x_{i\alpha}$.

If we use (2.4.1) and (2.4.2) to try to construct an operator that corresponds to the current density j_α, a question arises about the order in which to put the multiplication terms: they do not commute in the quantum case.

There is a general approach here: the order must be such that one obtains a Hermitian operator, since physical quantities in quantum mechanics are put in correspondence with Hermitian operators. Let operators L and M be Hermitian:

$$L^+ = L \qquad M^+ = M$$

If

$$[L, M] \neq 0$$

then LM and ML are not Hermitian. Indeed,

$$(LM)^+ = M^+L^+ = ML \neq LM$$

and hence, the expression that transforms to the classical product of physical quantities must be taken in the form

$$N = \frac{1}{2}(LM + ML) = N^+$$

(the combination

$$N' = \frac{1}{2i}(LM - ML) = N'^+$$

is also Hermitian; however, this quantity vanishes in the classical limit when we can neglect the commutator of L and M).

Finally, we obtain

$$j_\alpha(\mathbf{r}, t) = \frac{1}{2}\sum_i \frac{e_i}{m_i}\Big(\Big(p_{i\alpha} - \frac{e_i}{c}A_\alpha(\mathbf{r}_i, t)\Big)\delta(\mathbf{r} - \mathbf{r}_i) + \delta(\mathbf{r} - \mathbf{r}_i)\Big(p_{i\alpha} - \frac{e_i}{c}A_\alpha(\mathbf{r}_i, t)\Big)\Big)$$

As in Chapter 1, we single out of A_α the external field, that is, we replace A_α by $A_\alpha + A_\alpha^{(e)}$. The field A_α is the field of the particles of the medium; it is a function of the coordinates and velocities of the particles. Only the external field $A_\alpha^{(e)}$ can be the independent variable in the Hamiltonian of the system, so that the problem is to find the response of the system to the external field. This means that one actually starts with finding the susceptibilities $\chi^{(e)}$ for the external field and then follows with calculating the conventional susceptibilities χ using familiar formulas. In the Coulomb field gauge, A_α takes into account the magnetic interaction between the particles of the medium, which in general is small compared with the Coulombic interaction, by a factor of v^2/c^2. As a rule, it can be neglected in specific calculations.

There exist important qualitative effects in the electromagnetic response, for instance, the response to the low-frequency magnetic field, which is also proportional to v^2/c^2. Therefore, for reasons of greater generality, we carry out the further analysis in an arbitrary gauge and retain the general form of the current operator.

In view of this, we can write

$$
\begin{aligned}
j_\alpha(\mathbf{r},t) &= j_{0\alpha}(\mathbf{r}) - \sum_i \frac{e_i^2}{m_i c} A_\alpha^{(e)}(\mathbf{r}_i,t)\delta(\mathbf{r}-\mathbf{r}_i) \\
j_{0\alpha}(\mathbf{r}) &= \frac{1}{2}\sum_i \frac{e_i}{m_i}\Big[\Big(p_{i\alpha} - \frac{e_i}{c}A_\alpha(\mathbf{r}_i)\delta(\mathbf{r}-\mathbf{r}_i)\Big) \\
&\quad + \delta(\mathbf{r}-\mathbf{r}_i)\Big(p_{i\alpha} - \frac{e_i}{c}A_\alpha(\mathbf{r}_i)\Big)\Big]
\end{aligned}
\tag{2.4.3}
$$

(We retain hats over operators only where confusion could be produced.)

Despite a large number of particles in the medium, the number of different species of particles — electrons and nuclei of a specific isotope — is relatively low. We denote the parameters of particles of a specific species by the subscript s. Then e_s and m_s are the charge and mass of particles of the sth species and ρ_s is the charge density created by these particles. If we also introduce the density n_s of particles of the sth species, then obviously

$$\rho_s = e_s n_s(\mathbf{r})$$

Furthermore,

$$
\begin{aligned}
&\sum_i \frac{e_i^2}{m_i c}\delta(\mathbf{r}-\mathbf{r}_i) = \sum_s \frac{e_s^2}{m_s c}\sum_{i\in s}\delta(\mathbf{r}-\mathbf{r}_i) = \sum_s \frac{e_s^2}{m_s c} n_s(\mathbf{r}) \\
&\sum_i \frac{e_i^2}{m_i c} A_\alpha^{(e)}(\mathbf{r}_i,t)\delta(\mathbf{r}-\mathbf{r}_i) = A_\alpha^{(e)}(\mathbf{r},t)\sum_i \frac{e_i^2}{m_i c}\delta(\mathbf{r}-\mathbf{r}_i) \\
&\quad = \sum_s \frac{e_s^2}{m_s c} n_s(\mathbf{r}) A_\alpha^{(e)}(\mathbf{r},t)
\end{aligned}
$$

Hereafter, we use $j_\alpha(\mathbf{r},t)$ in the form

$$j_\alpha(\mathbf{r},t) = j_{0\alpha}(\mathbf{r}) - \sum_s \frac{e_s^2}{m_s c} n_s(\mathbf{r})\, A_\alpha^{(e)}(\mathbf{r},t) \tag{2.4.4}$$

In order to use the general perturbation theory for the density matrix as presented in Section 2.3, we also need to have an expression for the perturbation operator $\mathscr{H}'$. We have pointed out already that the perturbation is connected with the external field $\mathbf{A}^{(e)}$, so that we need to write the general expression for the nonrelativistic Hamiltonian $\mathscr{H}$ (see expression (1.2.7)), single out the terms that depend on the external field $\mathbf{A}^{(e)}$, and regroup the expression into $\mathscr{H} = \mathscr{H}_0 + \mathscr{H}'$.

The nonperturbed Hamiltonian $\mathscr{H}_0$ must not contain $\mathbf{A}^{(e)}$; it is obtained from $\mathscr{H}$ by setting $\mathbf{A}^{(e)} = 0$, and $\mathscr{H}'$ is formed by the terms of $\mathscr{H}$ which contain no $\mathbf{A}^{(e)}$. Hence,

$$\mathscr{H}_0 = \sum_i \frac{1}{2m_i}\Big(\mathbf{p}_i - \frac{e_i}{c}\,\mathbf{A}(\mathbf{r}_i)\Big)^2 + \frac{1}{2}\sum_{i\neq i'} \frac{e_i e_{i'}}{|\mathbf{r}_i - \mathbf{r}_{i'}|}$$

$$\begin{aligned}\mathscr{H}' = &-\sum_i \frac{e_i}{2m_i c}\Big[\Big(p_{i\alpha} - \frac{e_i}{c}\,A_\alpha(\mathbf{r}_i)\Big)A_\alpha^{(e)}(\mathbf{r}_i,t) \\ &+ A_\alpha^{(e)}(\mathbf{r}_i,t)\Big(p_{i\alpha} - \frac{e_i}{c}\,A_\alpha(\mathbf{r}_i)\Big)\Big] + \sum_i e_i\varphi^{(e)}(\mathbf{r}_i,t) \\ &+ \sum_i \frac{e_i^2}{2m_i c^2}\,A^{(e)2}(\mathbf{r}_i,t)\end{aligned} \tag{2.4.5}$$

For reasons of greater clarity, $\mathscr{H}_0$ is given in the Coulomb gauge; in fact, this does not violate the gauge invariance of the derivation to follow since the explicit form of $\mathscr{H}_0$ is not used at any point. As for $\mathscr{H}'$, it retains its form (2.4.5) under any gauge. However, the final result (2.5.6)–(2.5.7) will be in an explicitly gauge-invariant form and will contain no potentials.

Using δ functions, we can rewrite $\mathscr{H}'$ as a volume integral:

$$\begin{aligned}\mathscr{H}' = \int \mathrm{d}^3 r \sum_i \Big[&-\frac{e_i}{2m_i c}\Big(p_{i\alpha} - \frac{e_i}{c}\,A_\alpha(\mathbf{r})\Big)\delta(\mathbf{r}-\mathbf{r}_i)A_\alpha^{(e)}(\mathbf{r},t) \\ &- \frac{e_i}{2m_i c}\,A_\alpha^{(e)}(\mathbf{r},t)\delta(\mathbf{r}-\mathbf{r}_i)\Big(p_{i\alpha} - \frac{e_i}{c}\,A_\alpha(\mathbf{r})\Big) \\ &+ e_i\delta(\mathbf{r}-\mathbf{r}_i)\varphi^{(e)}(\mathbf{r},t) + \frac{e_i^2}{2m_i c^2}\,\delta(\mathbf{r}-\mathbf{r}_i)A^{(e)2}(\mathbf{r},t)\Big]\end{aligned}$$

In this notation the order of the functions $A_\alpha^{(e)}$ and the operators $p_{i\alpha}$ can be reversed since $p_{i\alpha}$ act on the variables $\mathbf{r}_i$. Transferring $A_\alpha^{(e)}$ in the second term to the righthand side and making use of (2.4.3), we obtain

$$\begin{aligned}\mathscr{H}' = &-\frac{1}{c}\int j_{0\alpha}(\mathbf{r})A_\alpha^{(e)}(\mathbf{r},t)\,\mathrm{d}^3 r + \int \rho(\mathbf{r})\varphi^{(e)}(\mathbf{r},t)\mathrm{d}^3 r \\ &+ \int \sum_{\mathrm{s}} \frac{e_{\mathrm{s}}^2 n_{\mathrm{s}}(\mathbf{r})}{2m_{\mathrm{s}} c^2}\,\mathbf{A}^{(e)2}(\mathbf{r},t)\,\mathrm{d}^3 r\end{aligned} \tag{2.4.6}$$

The charge density is

$$\rho(\mathbf{r}) = \sum_i e_i\delta(\mathbf{r}-\mathbf{r}_i) = \sum_{\mathrm{s}} \rho_{\mathrm{s}}(\mathbf{r}) = \sum_{\mathrm{s}} e_{\mathrm{s}} n_{\mathrm{s}}(\mathbf{r})$$

The formulas (2.4.3)–(2.4.6) are written in the Schrödinger representation. A transition to the interaction representation reduces to replacing the operators $j_{0\alpha}(\mathbf{r})$ by $j_{0\alpha}(\mathbf{r},t)$ and $n_s(\mathbf{r})$ by $n_s(\mathbf{r},t)$ using the formula

$$L(\mathbf{r},t) = \exp\left(\frac{i}{\hbar}\mathscr{H}_0 t\right)L(\mathbf{r})\exp\left(-\frac{i}{\hbar}\mathscr{H}_0 t\right)$$

An important feature of expression (2.4.6) for the perturbation operator is that it contains both terms that are linear in the external field and terms quadratic in it.

We have mentioned already that the magnetic interaction of particles is of the order v^2/c^2 relative to the Coulomb interaction, thus constituting a relativistic effect. A relativistic theory is thus needed for a consistent description of the effect. The same is true for the direct spin–spin interaction between particles. This interaction is absent from the classical Hamilton's function since the spin is a quantum phenomenon and vanishes in the transition to the classical picture.

At the same time, the direct interaction of the spin with the magnetic field can be treated even in the framework of the nonrelativistic quantum mechanics. This interaction also vanishes in the transition to classical mechanics and is absent from classical Hamilton's function; however, it can be found from Dirac's relativistic quantum-mechanical equation for $v/c \ll 1$ by a formal expansion in powers of c^{-1}. Retaining only terms of the order of c^{-1}, we arrive at the Pauli equation, which is actually the Schrödinger equation with the Hamiltonian which includes, in addition to the ordinary nonrelativistic expression for the electron, the spin component

$$\mathscr{H}_s = -\frac{e\hbar}{2mc}\frac{1}{S_e}(\mathbf{S}_e\mathbf{H})$$

where $\mathbf{S}_e$ is the electron spin operator and $S_e = 1/2$ is the electron spin. The physical meaning of this component is to take into account the interaction between the magnetic field $\mathbf{H}$ and the spin magnetic moment of the electron (numerically equal to the Bohr magneton $\mu_0 = |e|\hbar/2mc$); it is directed along the spin vector $\mathbf{S}_e$.

Generalizing this expression to the case of an ensemble of particles with spin $\mathbf{S}_i$ and magnetic moments μ_i, we find

$$\mathscr{H}_s = -\sum_i \frac{\mu_i}{S_i}(\mathbf{S}_i\mathbf{H}(\mathbf{r}_i,t)) = \mu_0\sum_i g_i(\mathbf{S}_i \operatorname{curl}\mathbf{A}(\mathbf{r}_i,t))$$

where g_i is the gyromagnetic factor.

Atomic nuclei also have nonzero spin magnetic moment μ_i which points along the spin vector $\mathbf{S}_i$. By the order of magnitude, these nuclear moments are from 10^{-4} to 10^{-3} of the Bohr magneton.

The spin contribution to the Hamiltonian can be written in the form of a volume integral

$$\mathscr{H}_s = \mu_0 \int \sum_i g_i(\mathbf{S}_i \operatorname{curl} \mathbf{A}(\mathbf{r}, t))\delta(\mathbf{r} - \mathbf{r}_i)\, d^3r$$

Integrating by parts, we transform this integral to the form

$$\mathscr{H}_s = -\frac{1}{c} \int j_s \mathbf{A}(\mathbf{r}, t)\, d^3r$$

where the spin current is

$$j_s(\mathbf{r}) = c\mu_0 \sum_i g_i[\mathbf{S}_i \nabla\delta(\mathbf{r} - \mathbf{r}_i)] \tag{2.4.7}$$

The mean value of the spin current operator $\mathbf{j}_s(\mathbf{r})$ in a state with a wave function ψ is

$$j_s(\mathbf{r}) = c \operatorname{curl} \mathbf{M}(\mathbf{r})$$

where

$$\mathbf{M}(\mathbf{r}) = \sum_i \frac{\mu_i}{S_i} \int (\psi^* \mathbf{S}_i \delta(r - r_i)\psi) \prod_i d^3r_i$$

is the mean density of the spin magnetic moment at a point $\mathbf{r}$.

To take into account the Hamiltonian $\mathscr{H}_s$ and the spin current $\mathbf{j}_s$ is essential for the theories of the spin magnetic resonance phenomena (both electron and nuclear) and for the theory of magnetic phenomena, including that of the magnetic contribution to susceptibility at low frequencies. In optics, it is not usually necessary to take into account the spin current, so that we will as a rule ignore the contribution due to $\mathscr{H}_s$ and $\mathbf{j}_s$. If necessary, we must add the current $\mathbf{j}_s$ to expression (2.4.3) for $\mathbf{j}_0(\mathbf{r})$, add the term $\mathscr{H}_s$ to the expression for $\mathscr{H}_0$, and add the term $\mathscr{H}'_s$ to expression (2.4.5) for the perturbation operator $\mathscr{H}'$, where

$$\mathscr{H}'_s = -\frac{1}{c} \int \mathbf{j}_s(\mathbf{r}) \mathbf{A}^{(e)}(\mathbf{r}, t)\, d^3r$$

2.5. Linear Response of a System to External Field

Using perturbation theory outlined in Section 2.3 and the perturbation operator $\mathscr{H}'$ given by (2.4.6), we can find the expansion of the mean current density in powers of the external field. There is a special point here: current density operator (2.4.4) contains the external field $\mathbf{A}^{(e)}$ and the operator

$\mathcal{H}'$ (2.4.6) contains a term that is quadratic in the external field. Therefore, when calculating the component of the mean current density which is linear in field, one has to retain only field-linear terms in $\mathcal{H}'$ and also average that term in j_α which includes $\mathbf{A}^{(e)}$, with the nonperturbed density matrix ρ_0.

As a result,

$$j_\alpha^{(1)}(\mathbf{r},t) = \mathrm{Sp}\{\rho^{(1)} j_{0\alpha}\} + \mathrm{Sp}\Big\{-\sum_s \frac{e_s^2 n_s(\mathbf{r},t)}{m_s c} A_\alpha^{(e)}(\mathbf{r},t)\rho_0\Big\}$$

The first term in this expression is calculated from (2.3.7) but the part of $\mathcal{H}'$ which is quadratic in field must not be taken into account. The second term is calculated immediately if we notice that

$$\mathrm{Sp}\{n_s(\mathbf{r},t)\rho_0\} = n_s(\mathbf{r})$$

where $n_s(\mathbf{r})$ is the mean, time-independent density of particles of species s for the stationary density matrix ρ_0. We obtain

$$\begin{aligned}\overline{j_\alpha^{(1)}}(\mathbf{r},t) = &-\sum_s \frac{e_s^2 n_s(\mathbf{r})}{m_s c} A_\alpha^{(e)}(\mathbf{r},t)\\ &-\frac{1}{\mathrm{i}\hbar c}\int_{-\infty}^{t} \mathrm{d}t' \int \mathrm{d}^3 r'\\ &\times \mathrm{Sp}\{[j_{0\alpha}(\mathbf{r},t), j_{0\beta}(\mathbf{r}',t)]\rho_0\} A_\beta^{(e)}(\mathbf{r}',t) + \frac{1}{\mathrm{i}\hbar}\int_{-\infty}^{t} \mathrm{d}t' \int \mathrm{d}^3 r'\\ &\times \mathrm{Sp}\{[j_{0\alpha}(\mathbf{r},t), \rho(\mathbf{r}',t')]\rho_0\}\varphi^{(e)}(\mathbf{r}',t')\end{aligned} \tag{2.5.1}$$

It is expedient to find not $\overline{j}_\alpha^{(e)}$ but the derivative of this quantity with respect to time. It is this derivative that we find in Maxwell's equation for the electric field:

$$\mathrm{curl}\,\mathrm{curl}\,\boldsymbol{\mathcal{E}} + \frac{1}{c^2}\frac{\partial^2 \boldsymbol{\mathcal{E}}}{\partial t^2} = -\frac{4\pi}{c}\frac{\partial \mathbf{j}}{\partial t}$$

As we will see later, this derivative can be written in an explicitly invariant form, in terms of the electric field, not of potentials.

Differentiating (2.5.1) with respect to time, we obtain

$$\frac{\partial \bar{j}_\alpha^{(1)}}{\partial t} = -\sum_s \frac{e_s^2 n_s}{m_s c} \frac{\partial A_\alpha^{(e)}(\mathbf{r},t)}{\partial t}$$

$$-\frac{1}{i\hbar c}\int_{-\infty}^{t} dt' \int d^3r' \, \mathrm{Sp}\{[j_{0\alpha}(\mathbf{r},t), j_{0\beta}(\mathbf{r}',t)]\rho_0\} \frac{\partial A_\beta^{(e)}(\mathbf{r}',t')}{\partial t'}$$

$$+\frac{1}{i\hbar}\int d^3r' \, \mathrm{Sp}\{[j_{0\alpha}(\mathbf{r},t), \rho(\mathbf{r}',t)]\rho_0\}\varphi^{(e)}(\mathbf{r}',t)$$

$$-\frac{1}{i\hbar}\int_{-\infty}^{t} dt' \int d^3r' \, \mathrm{Sp}\Big\{\Big[j_{0\alpha}(\mathbf{r},t), \frac{\partial \rho(\mathbf{r}',t')}{\partial t'}\Big]\rho_0\Big\}\varphi^{(e)}(\mathbf{r}',t') \qquad (2.5.2)$$

In obtaining (2.5.2) from (2.5.1), we made use of the equality

$$\mathrm{Sp}\{[j_{0\alpha}(t), j_{0\beta}(t')]\rho_0\} = f_{\alpha\beta}(t-t')$$

because ρ_0 is a stationary density matrix and the mean value of the two-time commutator depends only on the difference between the times t and t'.

In view of this,

$$\frac{\partial}{\partial t}\int_{-\infty}^{t} dt' \, \mathrm{Sp}\{[j_{0\alpha}(t), j_{0\beta}(t')]\rho_0\}A_\beta^{(e)}(t')$$

$$= \frac{\partial}{\partial t}\int_{0}^{\infty} d\tau \, f_{\alpha\beta}(\tau)A_\beta^{(e)}(t-\tau) = \int_{0}^{\infty} d\tau \, f_{\alpha\beta}(\tau)\frac{\partial A_\beta^{(e)}(t-\tau)}{\partial t}$$

$$= \int_{-\infty}^{t} dt' \, \mathrm{Sp}\{[j_{0\alpha}(t), j_{0\beta}(t')]\rho_0\}\frac{\partial A_\beta^{(e)}(t')}{\partial t'}$$

Likewise,

$$\mathrm{Sp}\{[j_{0\alpha}(t), \rho(t')]\rho_0\} = \overline{f}_\alpha(t-t')$$

The differentiation of this function with respect to t can thus be replaced by differentiating it with respect to t' in the integrand.

Now we can use the continuity equation for the operators $\mathbf{j}_0$ and ρ in the interaction representation (when no external field is present):

$$\frac{\partial \rho(\mathbf{r}',t')}{\partial t'} = -\nabla j_0(\mathbf{r}',t') = \frac{\partial}{\partial x'_\beta} j_{0\beta}(\mathbf{r}',t')$$

After integration by parts and recalling that $\varphi^{(e)}$ vanishes at infinity, we transform the last term of (2.5.2) to

$$-\frac{1}{i\hbar}\int\limits_{-\infty}^{t} dt' \int d^3r' \, \mathrm{Sp}\{[j_{0\alpha}(\mathbf{r},t), j_{0\beta}(\mathbf{r}',t')]\rho_0\} \frac{\partial\varphi^{(e)}(\mathbf{r}',t')}{\partial x'_\beta}$$

Combining this term with the second term of (2.5.2) and taking into account that

$$\mathcal{E}^{(e)}_\beta(\mathbf{r},t') = -\frac{1}{c}\frac{\partial A^{(e)}_\beta(\mathbf{r}',t')}{\partial t'} - \frac{\partial\varphi^{(e)}(\mathbf{r}',t')}{\partial x'_\beta}$$

we arrive at the gauge-invariant expression.

Now we only have to consider the next-to-last term in (2.5.2), which resulted from the differentiation of the integral with respect to time at the upper limit. It includes the commutator of the operators for the same moment of time:

$$[j_{0\alpha}(\mathbf{r},t), \rho(\mathbf{r}',t)] = \exp\left(\frac{i}{\hbar}\mathcal{H}_0 t\right)[j_{0\alpha}(\mathbf{r}), \rho(\mathbf{r}')]\exp\left(-\frac{i}{\hbar}\mathcal{H}_0 t\right)$$

To calculate the commutator of time-independent operators, we use the explicit expressions for the operators in the commutator:

$$\begin{aligned}
&\Bigg[\frac{1}{2}\sum_i \frac{e_i}{m_i}\left(p_{i\alpha} - \frac{e_i}{c}A_\alpha(\mathbf{r})\right)\delta(\mathbf{r}-\mathbf{r}_i) \\
&\quad + \frac{1}{2}\sum_i \frac{e_i}{m_i}\delta(\mathbf{r}-\mathbf{r}_i)\left(p_{i\alpha} - \frac{e_i}{c}A_\alpha(\mathbf{r})\right), \sum_{i'} e_{i'}\delta(\mathbf{r}'-\mathbf{r}_{i'})\Bigg] \\
&\quad = \frac{1}{2}\sum_{ii'}\left[\frac{e_i}{m_i}\, p_{i\alpha}\delta(\mathbf{r}-\mathbf{r}_i),\, e_{i'}\delta(\mathbf{r}'-\mathbf{r}_{i'})\right] \\
&\quad + \frac{1}{2}\sum_{ii'}\left[\frac{e_i}{m_i}\,\delta(\mathbf{r}-\mathbf{r}_i)p_{i\alpha},\, e_{i'}\delta(\mathbf{r}'-\mathbf{r}'_i)\right]
\end{aligned} \tag{2.5.3}$$

In double sums, we can retain only the terms with $i = i'$ since the operators referring to a particle with index i commute with the operators referring to the particle i' only if $i \neq i'$.

As a next step, we can make use of the expression for the commutator of the momentum operator p and an arbitrary function $f(x)$ of the coordinate x:

$$[p,\, f(x)] = -i\hbar\frac{\partial f(x)}{\partial x} \tag{2.5.4}$$

This commutation rule can be checked directly, by applying operators in the left and righthand sides of the equality to an arbitrary wave function

$\psi(x)$ and using the expression for the momentum operator in the coordinate representation,

$$p = -\mathrm{i}\hbar \frac{\partial}{\partial x}$$

Using the identity that holds for any operators A, B and C, namely, $[A, B\,C] = A[B,\, C] + [A,\, C]B$ and (2.5.4), we can find the commutators

$$\begin{aligned} [pf_1(x),\, f_2(x)] &= -\mathrm{i}\hbar f_1(x)\,\frac{\partial f_2(x)}{\partial x} \\ [f_1(x)p,\, f_2(x)] &= -\mathrm{i}\hbar f_1(x)\,\frac{\partial f_2(x)}{\partial x} \end{aligned} \tag{2.5.5}$$

for any functions $f_1(x)$ and $f_2(x)$.

Resorting to (2.5.5), we rewrite the commutator in (2.5.3) as

$$\begin{aligned} &-\sum_i \frac{\mathrm{i}\hbar e_i^2}{m_i}\,\delta(\mathbf{r}-\mathbf{r}_i)\,\frac{\partial}{\partial x_{i\alpha}}\,\delta(\mathbf{r}'-\mathbf{r}_i) \\ &= \mathrm{i}\hbar\,\frac{\partial}{\partial x'_\alpha}\sum_i \frac{e_i^2}{m_i}\,\delta(\mathbf{r}-\mathbf{r}_i)\delta(\mathbf{r}'-\mathbf{r}_i) \end{aligned}$$

The thing left for us to do is to integrate by parts in the next to last term in (2.5.2) and recast it to the form

$$-\int \mathrm{d}^3r'\,\mathrm{Sp}\Big\{\sum_i \frac{e_i^2}{m_i}\,\delta(\mathbf{r}-\mathbf{r}_i)\delta(\mathbf{r}'-\mathbf{r}_i)\rho_0\Big\}\frac{\partial\varphi^{(\mathrm{e})}(\mathbf{r}',t)}{\partial x'_\alpha}$$

We have used here the fact that $\exp(-\mathrm{i}/\hbar\mathscr{H}_0 t)\rho_0\exp(\mathrm{i}/\hbar\mathscr{H}_0 t) = \rho_0$. Integrating over r, we obtain

$$-\mathrm{Sp}\Big\{\sum_i \frac{e_i^2}{m_i}\,\delta(\mathbf{r}-\mathbf{r}_i)\rho_0\Big\}\frac{\partial\varphi^{(\mathrm{e})}(\mathbf{r},t)}{\partial x_\alpha} = -\sum_{\mathrm{s}} \frac{e_{\mathrm{s}}^2 n_{\mathrm{s}}(\mathbf{r})}{m_{\mathrm{s}}}\,\frac{\partial\varphi^{(\mathrm{e})}(\mathbf{r},t)}{\partial x_\alpha}$$

Rewriting the next-to-last term of (2.5.2) in this way and combining it with the first term, we find

$$\sum_{\mathrm{s}} \frac{e_{\mathrm{s}}^2 n_{\mathrm{s}}(\mathbf{r})}{m_{\mathrm{s}}}\,\mathcal{E}_\alpha^{(\mathrm{e})}(\mathbf{r},t)$$

This has produced an explicitly gauge-invariant expression for $\partial\bar{j}_\alpha^{(1)}/\partial t$:

$$\begin{aligned} \frac{\partial\bar{j}_\alpha^{(1)}(\mathbf{r},t)}{\partial t} &= \sum_{\mathrm{s}} \frac{e_{\mathrm{s}}^2 n_{\mathrm{s}}(\mathbf{r})}{m_{\mathrm{s}}}\,\mathcal{E}_\alpha^{(\mathrm{e})}(\mathbf{r},t) \\ &+ \frac{1}{\mathrm{i}\hbar}\int_{-\infty}^{t}\mathrm{d}t'\int \mathrm{d}^3r'\,\mathrm{Sp}\{[j_{0\alpha}(\mathbf{r},t), j_{0\beta}(\mathbf{r}',t')]\rho_0\}\mathcal{E}_\beta^{(\mathrm{e})}(\mathbf{r}',t') \end{aligned} \tag{2.5.6}$$

It is not difficult to show that expression (2.5.6) is valid if the spin current j_s (2.4.7) is taken into account in j_0 and the interaction $\mathscr{H}_s$ is taken into account in $\mathscr{H}'$. Taking the Fourier transform of (2.5.6) in time makes it possible to write an expression for the linear susceptibility $\chi^{(1e)}$. Since $\partial j_\alpha/\partial t = \partial^2 \mathcal{P}_\alpha/\partial t^2$, we have

$$-\omega^2 P_{\omega\alpha}(\mathbf{r}) = \sum_s \frac{e_s^2 n_s(\mathbf{r})}{m_s} \delta_{\alpha\beta} \mathcal{E}^{(e)}_{\omega\beta}(\mathbf{r})$$

$$+ \frac{1}{i\hbar} \int_{-\infty}^{+\infty} dt\, e^{i\omega t} \int_0^\infty d\tau \int d^3 r' \operatorname{Sp}\{[j_{0\alpha}(\mathbf{r},t), j_{0\beta}(\mathbf{r}',t')]\rho_0\} \mathcal{E}^{(e)}_\beta(\mathbf{r}', t-\tau)$$

Integration in t' is replaced with integration in $\tau = t - t'$ because

$$\operatorname{Sp}\{[j_{0\alpha}(\mathbf{r},t), j_{0\beta}(\mathbf{r}',t')]\rho_0\}$$

is a function of $t - t'$. Integration in t gives

$$-\omega^2 P_{\omega\alpha}(\mathbf{r}) = \int d^3 r' \Big(\sum_s \frac{e_s^2 n_s(\mathbf{r})}{m_s} \delta_{\alpha\beta} \delta(\mathbf{r} - \mathbf{r}')$$

$$+ \frac{1}{i\hbar} \int_0^\infty d\tau e^{i\omega\tau} \operatorname{Sp}\{[j_{0\alpha}(\mathbf{r},t), j_{0\beta}(\mathbf{r}',t')]\rho_0\} E^{(e)}_{\omega\beta}(\mathbf{r}') \Big)$$

Comparing it with the definition of $\chi^{(1e)}_{\alpha\beta}(\mathbf{r},\mathbf{r}';\omega)$,

$$P_{\omega\alpha}(\mathbf{r}) = \int \chi^{(1e)}_{\alpha\beta}(\mathbf{r},\mathbf{r}';\omega) E^{(e)}_{\omega\beta}(\mathbf{r}') d^3 r'$$

we obtain

$$\chi^{(1e)}_{\alpha\beta}(\mathbf{r},\mathbf{r}';\omega) = -\sum_s \frac{e_s^2 n_s(\mathbf{r})}{m_s} \delta(\mathbf{r} - \mathbf{r}') \delta_{\alpha\beta}$$

$$+ \frac{1}{\hbar\omega^2} \int_0^\infty d\tau e^{i\omega\tau} \operatorname{Sp}\{[j_{0\alpha}(\mathbf{r},t), j_{0\beta}(\mathbf{r}',t')]\rho_0\} \qquad (2.5.7)$$

Note again that the trace in the integrand of the integral over τ is in fact independent of t, being a function only of $\tau = t - t'$.

Formally, this is a corollary of the fact that the stationary density matrix ρ_0 commutes with $\mathscr{H}_0$, so that it also commutes with $\exp(i\mathscr{H}_0 t)$.

Therefore, we find for any arbitrary operators $A(t)$ and $B(t)$ in the interaction representation, which correspond to the Schrödinger operators A and B, that

$$\begin{aligned}
&\mathrm{Sp}\{A(t)B(t')\rho_0\}\\
&\quad= \mathrm{Sp}\Big\{\exp\Big(\frac{i}{\hbar}\mathscr{H}_0 t\Big)A\exp\Big(-\frac{i}{\hbar}\mathscr{H}_0 t\Big)\\
&\quad\times\exp\Big(\frac{i}{\hbar}\mathscr{H}_0 t'\Big)B\exp\Big(-\frac{i}{\hbar}\mathscr{H}_0 t'\Big)\rho_0\Big\}\\
&\quad= \mathrm{Sp}\Big\{A\exp\Big(-\frac{i}{\hbar}\mathscr{H}_0(t-t')\Big)B\exp\Big(\frac{i}{\hbar}\mathscr{H}_0(t-t')\Big)\rho_0\Big\}\\
&= f(t-t')
\end{aligned}$$

2.6. *Explicit Form of Linear Response of a System to External Field*

Expression (2.5.7) for $\chi_{\alpha\beta}^{(1e)}(\mathbf{r},\mathbf{r}';\omega)$ has an invariant form which is independent of the choice of specific representation for the calculation of the trace Sp. To find the explicit formula, it is advisable to use the energy representation which is determined by the eigenfunctions of the nonperturbed Hamiltonian $\mathscr{H}_0$ of the system.

In the energy representation, the elements of the equilibrium (Gibbs) density matrix ρ_0 are

$$(\rho_0)_{nn'} = \Big(\exp\Big(\frac{\mathcal{F}-\mathscr{H}_0}{kT}\Big)\Big)_{nn'} = \exp\Big(\frac{\mathcal{F}-E_n}{kT}\Big)\delta_{nn'}$$

The matrix elements of the operator $j_{0\alpha}$ in the interaction representation follow from (2.2.3):

$$(j_{0\alpha}(\mathbf{r},t))_{nn'} = (j_\alpha(\mathbf{r}))_{nn'}\exp\Big(\frac{i}{\hbar}(E_n-E_{n'})t\Big)$$

where $j_{0\alpha}(\mathbf{r})$ is the Schrödinger current density operator (2.4.3). For the sake of brevity, we will henceforth drop the "0" subscript with the current density operators.

After substituting the explicit dependence of matrix elements on time into (2.5.7), we obtain

$$\chi_{\alpha\beta}^{(1e)}(\mathbf{r},\mathbf{r}';\omega) = -\sum_{s} \frac{e_s^2 n_s(\mathbf{r})}{m_s \omega^2} \delta(\mathbf{r}-\mathbf{r}')\delta_{\alpha\beta}$$

$$+ \frac{1}{\hbar\omega^2} \int_0^\infty d\tau\, e^{i\omega\tau} \sum_{nn'n''} \exp\left(\frac{\mathcal{F}-E_n}{kT}\right) \delta_{nn'} (j_\alpha(\mathbf{r}))_{n''n'} (j_\beta(\mathbf{r}'))_{n'n}$$

$$\times \exp\left(\frac{i}{\hbar}(E_{n''}-E_{n'})t\right) \exp\left(\frac{i}{\hbar}(E_{n'}-E_n)t'\right)$$

$$- (j_\beta(\mathbf{r}'))_{n''n'} (j_\alpha(\mathbf{r}))_{n'n} \exp\left(\frac{i}{\hbar}(E_{n''}-E_{n'})t'\right)$$

$$\times \exp\left(\frac{i}{\hbar}(E_{n'}-E_n)t\right)$$

Since the matrix ρ_0 is diagonal, the summation over n'' vanishes and the time factors are functions of $\tau = t - t'$ only.

The integral of the exponential factors over τ does not converge in the limit $\tau = \infty$. The convergence is provided by introducing into the integrand the factor $\exp(-\delta\tau/\hbar)$, where $\delta \to +0$.

This gives

$$\chi_{\alpha\beta}^{(1e)}(\mathbf{r},\mathbf{r}';\omega) = -\sum_{s} \frac{e_s^2 n_s(\mathbf{r})}{m_s \omega^2} \delta(\mathbf{r}-\mathbf{r}')\delta_{\alpha\beta} + \frac{1}{\omega^2} \sum_{nn'} \exp\left(\frac{\mathcal{F}-E_n}{kT}\right)$$

$$\times \left\{ \frac{(j_\alpha(\mathbf{r}))_{nn'} (j_\beta(\mathbf{r}'))_{n'n}}{E_{n'}-E_n-\hbar\omega-i\delta} + \frac{(j_\beta(\mathbf{r}'))_{nn'} (j_\alpha(\mathbf{r}))_{n'n}}{E_{n'}-E_n+\hbar\omega+i\delta} \right\} \tag{2.6.1}$$

The introduction of the factor $\exp(-\frac{\delta}{\hbar}\tau)$ into the integral over τ generates purely imaginary additional terms $i\delta$ in the denominators in (2.6.1). As we see, these terms yield the rule of circling the poles of $\chi_{\alpha\beta}^{(1e)}(\omega)$ in the integration over ω. Since $\delta > 0$, we conclude that the contour encircling the poles in $\chi^{(1e)}(\omega)$ must be circled from above.

Physically, the introduction of the factor $\exp(-\frac{\delta}{\hbar}\tau)$ can be justified by the following arguments.

We can assume that this factor is produced by the field $\mathcal{E}_\beta(\mathbf{r}', t')$ which vanishes at $t' = -\infty$ and is turned on slowly (adiabatically); this is the process described by $\exp(-\frac{\delta}{\hbar}\tau)$.

Another, more consistent method of justification is to generate perturbation theory expansion using not the equation

$$i\hbar \frac{\partial \rho}{\partial t} = [\mathcal{H}', \rho]$$

for the density matrix but using (2.1.9) to take relaxation into account in the simplest fashion:

$$i\hbar \frac{\partial \rho}{\partial t} = [\mathcal{H}', \rho] - i\delta(\rho - \rho_0)$$

Here the physical meaning carried by $\hbar\delta^{-1}$ is the relaxation time. Additional factors appear in the integrals over time, of the type $\exp\left(-\frac{\delta}{\hbar}(t-t')\right)$, when we take into account the relaxation term in the equation for the density matrix in such expressions as (2.3.4) and (2.3.7) (these are obtained by using perturbation theory). The factor $\exp(-\frac{\delta}{\hbar}\tau)$ thus takes into account the infinitely small relaxation in the system.

By virtue of the factor $\exp(-\frac{\delta}{\hbar}\tau)$, the initial conditions in distant past cease to affect the response of the system at the time t when we integrate from $-\infty$ to t'.

A question about the linewidth then arises in connection with the introduction of the infinitely short relaxation time: is it necessary now to define finite relaxation times in (2.6.1)? When the simplest model systems are considered (e.g., a gas consisting of weakly interacting molecules), the role of this interaction can be taken into account phenomenologically, by introducing relaxation parameters into the equation for the density matrix of one molecule. It is known that the linewidth is then determined by the inverse relaxation time.

Another method of analysis, chosen here, is to consider a macroscopically large system as a whole. It can always be regarded as isolated, so that its exact levels do not decay, or rather, they decay at an arbitrarily slow rate ($\delta_n \to +0$). On the other hand, these are levels that correspond to states of very complex nature. For instance, if we take the above example of rarefied gas, these are not the levels of individual molecules but of the entire macroscopic ensemble of molecules. However, if the interaction between molecules is in some sense weak, the total energy of the system is formed of the sum of energies of individual molecules plus a small contribution of the interaction, which changes depending on the relative positions and motions of molecules, that is, it depends on the configuration of the entire ensemble. Consequently, the exact levels of the system are rearranged into densely spaced groups, i.e., intervals of practically continuous spectrum in the neighborhood of the energies corresponding to the sum of energies of the noninteracting molecules. The width of these intervals is determined by the interaction energy. However, when we describe the system using the density matrix for an individual molecule, this factor manifests itself only in the arising relaxation terms $\mathbf{\Gamma}$ in (2.1.7), in other words, the true distribution of exact energy levels is "simulated" by the broadening of the levels of each molecule.

Of course, the calculation of exact levels E_n is an unfeasible task. Hence, any specific calculation always resorts to simplified models which reduce the problem of a macroscopic ensemble to problems with a small number of particles; the price for this is the appearance of relaxation and broadening of levels defined in this way. In the general theory, however, and when studying general relations, it is always possible and necessary to use the concept of exact and nondamped levels ($\delta \to +0$).

Expression (2.6.1) is an explicit formula for the linear susceptibility of an arbitrary system to an external field. The ordinary susceptibility $\chi^{(1)}$ can be found from the known value of the tensor $\chi^{(1e)}$. The anti-Hermitian component of $\chi^{(1)}$ or $\chi^{(1e)}$ determines, as we have demonstrated in Chapter 1, the losses in the medium. Using (2.6.1), we can write an explicit expression for the anti-Hermitian component of $\chi^{(1e)}$, which we will denote by $\chi^{(e)''}$:

$$\begin{aligned}
\chi^{(e)''}_{\alpha\beta}(\mathbf{r},\mathbf{r}';\omega) &= \frac{1}{2\mathrm{i}}(\chi^{(1e)}_{\alpha\beta}(\mathbf{r},\mathbf{r}';\omega) - \chi^{(1e)*}_{\beta\alpha}(\mathbf{r}',\mathbf{r};\omega)) \\
&= \frac{1}{2\mathrm{i}\omega^2}\sum_{nn'}\exp\left(\frac{\mathcal{F}-E_n}{kT}\right)\left\{\frac{(j_\alpha(\mathbf{r}))_{nn'}(j_\beta(\mathbf{r}'))_{n'n}}{E_{n'}-E_n-\hbar\omega-\mathrm{i}\delta}\right. \\
&\quad - \frac{(j_\beta(\mathbf{r}'))^*_{nn'}(j_\alpha(\mathbf{r}))^*_{n'n}}{E_{n'}-E_n-\hbar\omega+\mathrm{i}\delta} + \frac{(j_\beta(\mathbf{r}'))_{n'n}(j_\alpha(\mathbf{r}))_{n'n}}{E_{n'}-E_n+\hbar\omega+\mathrm{i}\delta} \\
&\quad \left. - \frac{(j_\alpha(\mathbf{r}))^*_{nn'}(j_\beta(\mathbf{r}'))^*_{n'n}}{E_{n'}-E_n+\hbar\omega-\mathrm{i}\delta}\right\}
\end{aligned}$$

Since the operator j_α is Hermitian,

$$(j_\alpha(\mathbf{r}))^*_{nn'} = (j_\alpha(\mathbf{r}))_{n'n}$$

The following equality holds:

$$\lim_{\delta\to 0}\left(\frac{1}{x+\mathrm{i}\delta} - \frac{1}{x-\mathrm{i}\delta}\right) = -2\pi\mathrm{i}\delta(x)$$

For this reason,

$$\begin{aligned}
\chi^{(e)*}_{\alpha\beta}(\mathbf{r},\mathbf{r}';\omega) &= \frac{\pi}{\omega^2}\sum_{nn'}\exp\left(\frac{\mathcal{F}-E_n}{kT}\right) \\
&\quad \times \{(j_\alpha(\mathbf{r}))_{nn'}(j_\beta(\mathbf{r}'))_{n'n}\delta(E_{n'}-E_n-\hbar\omega) \\
&\quad - (j_\beta(\mathbf{r}'))_{nn'}(j_\alpha(\mathbf{r}))_{n'n}\delta(E_{n'}-E_n+\hbar\omega)\}
\end{aligned}$$

Two δ functions have appeared, corresponding to the energy conservation law. The last formula needs to be transformed. We replace the summation subscripts $n \leftrightarrow n'$ in the second term and, remembering that owing to the δ function in this term,

$$E_{n'} = E_n + \hbar\omega$$

we find

$$\begin{aligned}
\chi^{(e)''}_{\alpha\beta}(\mathbf{r},\mathbf{r}';\omega) &= \frac{\pi}{\omega^2}\left[1-\exp\left(-\frac{\hbar\omega}{kT}\right)\right] \\
&\times \sum_{nn'}\exp\left(\frac{\mathcal{F}-E_n}{kT}\right)(j_\alpha(\mathbf{r}))_{nn'}(j_\beta(\mathbf{r}'))_{n'n}\delta(E_{n'}-E_n-\hbar\omega) \quad (2.6.2)
\end{aligned}$$

The meaning of this expression for $\chi^{(e)''}_{\alpha\beta}$ can be understood if we calculate the losses in the medium using perturbation theory and compare them with (2.6.2).

According to standard rules of perturbation theory, the probability for the system to go from state n to state n' in the presence of an external field at a frequency ω is

$$w_{n'n} = \frac{2\pi}{\hbar} |\mathscr{H}'_{n'n}|^2 \delta(E_{n'} - E_n - \hbar\omega) \tag{2.6.3}$$

As the standard perturbation $\mathscr{H}'$, we must use expression (2.4.6) without the term quadratic in the field:

$$\mathscr{H}' = -\frac{1}{c}\int j_\alpha(\mathbf{r},t) A^{(e)}_\alpha(\mathbf{r},t)\,\mathrm{d}^3r + \int \rho(\mathbf{r},t)\varphi^{(e)}(\mathbf{r},t)\,\mathrm{d}^3r \tag{2.6.4}$$

We assume that $A^{(e)}_\alpha$ and $\varphi^{(e)}$ are proportional to $\mathrm{e}^{-\mathrm{i}\omega t}$. This gives

$$\mathscr{H}'_{n'n} = -\frac{1}{2c}\int (j_\alpha(\mathbf{r}))_{n'n} A^{(e)}_{\omega\alpha}(\mathbf{r})\,\mathrm{d}^3r + \frac{1}{2}\int (\rho(\mathbf{r}))_{n'n}\varphi^{(e)}_\omega(\mathbf{r})\,\mathrm{d}^3r$$

Since the operators ρ and $\mathbf{j}$ in the interaction representation satisfy the continuity equation, we have

$$\frac{\mathrm{i}}{\hbar}(E_{n'} - E_n)(\rho(\mathbf{r}))_{n'n} = -\left(\frac{\partial j_\alpha(\mathbf{r})}{\partial x_\alpha}\right)_{n'n}$$

Owing to the presence of a δ function in (2.6.3), we can set

$$E_{n'} - E_n = \hbar\omega$$

Hence, integrating by parts in that term of (2.6.4) which includes $\varphi^{(e)}$, we obtain

$$\mathscr{H}'_{n'n} = \frac{\mathrm{i}}{2\omega}\int (j_\alpha(\mathbf{r}))_{n'n} E^{(e)}_{\omega\alpha}\,\mathrm{d}^3r$$

The losses in the medium are

$$\overline{\dot{Q}} = \hbar\omega w_{n'n} V^{-1} = \frac{\pi}{2\omega V}\int E^{(e)*}_{\omega\alpha}(\mathbf{r})(j_\alpha(\mathbf{r}))_{nn'}(j_\beta(\mathbf{r}'))_{n'n}$$
$$\times E^{(e)}_{\omega\beta}(\mathbf{r}')\,\mathrm{d}^3r\,\mathrm{d}^3r'\,\delta(E_{n'} - E_n - \hbar\omega)$$

If we compare this expression with the relation between the losses and the anti-Hermitian part of $\chi^{(e)}$ derived in Chapter 1, namely,

$$\overline{\dot{Q}} = \frac{\omega}{2V}\int E^{(e)*}_{\omega\alpha}(\mathbf{r})\chi^{(e)''}_{\alpha\beta}(\mathbf{r},\mathbf{r}';\omega)E^{(e)}_{\omega\beta}(\mathbf{r}')\,\mathrm{d}^3r\,\mathrm{d}^3r'$$

we find

$$\chi^{(e)''}_{\alpha\beta}(\mathbf{r},\mathbf{r}';\omega) = \frac{\pi}{\omega^2}(j_\alpha(\mathbf{r}))_{nn'}(j_\beta(\mathbf{r}'))_{n'n}\delta(E_{n'} - E_n - \hbar\omega) \qquad (2.6.5)$$

This is the part of $\chi^{(e)''}_{\alpha\beta}$ which is connected with the transitions of the system from state n to state n'. To find total losses and, correspondingly, the complete expression for $\chi^{(e)''}_{\alpha\beta}$, we need to average (2.6.5) over the initial state with the Gibbs distribution $\exp[(\mathcal{F} - E_n)/kT]$ and take the sum over the final states.

As a next step, we need to take into account not only the transitions $n \to n'$ but also the reverse ones, $n' \to n$, which are connected with the stimulated emission of field quanta. Their probability is

$$w_{nn'} = w_{n'n}$$

Averaging this probability over the initial state n' with the weight $\exp[(\mathcal{F} - E_{n'})/kT]$ and summing up over the final state n leads, by virtue of the equality $E_{n'} - E_n = \hbar\omega$, to the same expression as averaging and summing of $w_{n'n}$ do, except that the factor $e^{-\hbar\omega/kT}$ appears. "Down" transitions from state n' to state n give a negative contribution to losses, so that formula (2.6.5) is faithfully reproduced if all transitions are taken into account. At the same time, the physical meaning of all terms in this formula becomes clear.

2.7. Fluctuation–Dissipation Theorem

We have derived above the general expression (2.5.7) for the linear susceptibility of the system. The nontrivial part of this expression is the commutator of currents taken at different moments of time at different points in space. We know from quantum mechanics that the commutator of the operators of two physical quantities is a measure of the degree to which they influence each other. In our case the current produced by the field at a point $\mathbf{r}'$ at a time t' affects the current at a point $\mathbf{r}$ at a time t.

This is the reason for the relationship between susceptibility $\chi^{(1)}$ and the fluctuations of currents in the system. This relationship is formulated in the fluctuation–dissipation theorem (FDT) which we consider here for electromagnetic fluctuations but which represents a general relation between fluctuations of physical quantities and dissipative properties of a system subject to external factors; in other words, between fluctuations and the generalized susceptibility of the system.

Let us consider equilibrium current density fluctuations in a system. Assume that external fields are absent and the system is left isolated. The mean current and polarization then vanish but the mean square values of these quantities do not, owing to the microscopic fluctuations.

By definition, the correlation function of currents in the system is

$$\Phi_{\alpha\beta}(\mathbf{r},\mathbf{r}';t-t') = \overline{j_\alpha(\mathbf{r},t)j_\beta(\mathbf{r}',t')} \tag{2.7.1}$$

Averaging here can be interpreted either as averaging over the (Gibbs) ensemble or as averaging over time; owing to the ergodicity, the two are identical. If $\mathbf{r}=\mathbf{r}'$ and $\tau = t-t' = 0$, we obtain the mean-square current density.

In the classical case, the order of the quantities $j_\alpha(\mathbf{r},t)$ and $j_\beta(\mathbf{r}',t')$ in (2.7.1) is immaterial. In the quantum case, the current densities are put in correspondence with operators and one has to average the product of operators in the quantum-mechanical sense (with the density matrix). We have already mentioned that the current density operators referring to different spacetime points do not commute; hence, instead of the product, one needs to take the Hermitian symmetrized product operator:

$$\Phi_{\alpha\beta}(\mathbf{r},\mathbf{r}';\tau) = \tfrac{1}{2}\,\mathrm{Sp}\{(j_\alpha(\mathbf{r},t)j_\beta(\mathbf{r}',t') + j_\beta(\mathbf{r}',t')j_\alpha(\mathbf{r},t))\rho_0\} \tag{2.7.2}$$

The averaging is done using the equilibrium time-independent density matrix ρ_0,

$$\rho_0 = \exp\left(\frac{\mathcal{F}-\mathcal{H}_0}{kT}\right)$$

and the Heisenberg current density operators $j_\alpha(\mathbf{r},t)$.

Since the external field is absent and the system's Hamiltonian is the nonperturbed Hamiltonian $\mathcal{H}_0$, the Heisenberg representation is in this case identical to the interaction representation; the time dependence of the matrix elements of operators in the energy representation fixed by the eigenfunctions of $\mathcal{H}_0$ is given by the expression

$$(j_\alpha(\mathbf{r},t))_{nn'} = (j_\alpha(\mathbf{r}))_{nn'}\exp\left(\frac{\mathrm{i}}{\hbar}\,(E_n - E_{n'})t\right)$$

Let us substitute into (2.7.2) explicit expressions for the matrix elements of the operators $j_\alpha(\mathbf{r},t)$, $j_\beta(\mathbf{r}',t')$ and ρ_0 in the energy representation, and find the Fourier transform

$$\Phi_{\alpha\beta}(\mathbf{r},\mathbf{r}';\omega) = \int_{-\infty}^{+\infty} \exp(\mathrm{i}\omega\tau)\Phi_{\alpha\beta}(\mathbf{r},\mathbf{r}';\tau)\,d\tau$$

Subsequent calculations are similar to those of Section 2.6 used to find $\chi^{(1e)}$. The main difference lies in the integration over τ being between infinite limits. We therefore need to use the formula

$$\int_{-\infty}^{+\infty} \exp\left(\frac{\mathrm{i}}{\hbar}(E_{n'} - E_n - \hbar\omega)\tau\right) = 2\pi\hbar\delta(E_{n'} - E_n - \hbar\omega)$$

After calculations, we obtain

$$\Phi_{\alpha\beta}(\mathbf{r},\mathbf{r}';\omega) = \pi\hbar\Big[1+\exp\Big(-\frac{\hbar\omega}{\mathrm{k}T}\Big)\Big]\sum_{nn'}\exp\Big(\frac{\mathcal{F}-E_n}{\mathrm{k}T}\Big)$$
$$\times (j_\alpha(\mathbf{r}))_{nn'}(j_\beta(\mathbf{r}'))_{n'n}\delta(E_{n'}-E_n-\hbar\omega)$$

Comparing this expression with (2.6.5), we find a relation of $\Phi_{\alpha\beta}$ to the anti-Hermitian part of $\chi^{(1e)}$ which determines the dissipation of the energy of the electromagnetic field in the system:

$$\Phi_{\alpha\beta}(\mathbf{r},\mathbf{r}';\omega) = \hbar\omega^2\coth\frac{\hbar\omega}{2\mathrm{k}T}\,\chi^{(e)''}_{\alpha\beta}(\mathbf{r},\mathbf{r}';\omega) \tag{2.7.3}$$

Here

$$\coth x = \frac{\cosh x}{\sinh x} = \frac{\mathrm{e}^x+\mathrm{e}^{-x}}{\mathrm{e}^x-\mathrm{e}^{-x}} = \frac{1+\mathrm{e}^{-2x}}{1-\mathrm{e}^{-2x}}$$

Formula (2.7.3) is known as the fluctuation–dissipation theorem (FDT).

Using (2.7.3), we can also find the correlation function of charge density

$$\Phi(\mathbf{r},\mathbf{r}';\tau) = \tfrac{1}{2}\mathrm{Sp}\{(\rho(\mathbf{r},t')\rho(\mathbf{r}',t)+\rho(\mathbf{r}',t')\rho(\mathbf{r},t))\rho_0\}$$

where $\rho(\mathbf{r},t)$ is the charge density and ρ_0 is the equilibrium charge density matrix.

Its Fourier transform is, according to the Wiener–Khinchin theorem, the spectral density of charge density fluctuations:

$$\Phi(\mathbf{r},\mathbf{r}';\omega) = \int_{-\infty}^{+\infty}\mathrm{e}^{\mathrm{i}\omega\tau}\Phi(\mathbf{r},\mathbf{r}';\tau)\,\mathrm{d}\tau$$

By virtue of the continuity equation

$$\frac{\partial j_{\omega\alpha}}{\partial x_\alpha} - \mathrm{i}\omega\rho_\omega = 0$$

and also because

$$\overline{\rho_\omega(\mathbf{r})\rho^*_{\omega'}(\mathbf{r}')} = \int \mathrm{e}^{\mathrm{i}\omega t}\mathrm{e}^{-\mathrm{i}\omega' t'}\,\overline{\rho(\mathbf{r},t)\rho(\mathbf{r}',t')}\,\mathrm{d}t\,\mathrm{d}t'$$
$$= \int \mathrm{e}^{\mathrm{i}\omega\tau}\mathrm{e}^{\mathrm{i}(\omega-\omega')t'}\,\Phi(\mathbf{r},\mathbf{r}';\tau)\,\mathrm{d}\tau\,\mathrm{d}t'$$
$$= 2\pi\delta(\omega-\omega')\Phi(\mathbf{r},\mathbf{r}';\omega)$$

we have

$$\Phi(\mathbf{r},\mathbf{r}';\omega) = \frac{1}{\omega^2}\left(\frac{\partial j_\alpha(\mathbf{r})}{\partial x_\alpha}\frac{\partial j_\beta(\mathbf{r}')}{\partial x'_\beta}\right)_\omega$$

$$= \frac{1}{\omega^2}\frac{\partial^2}{\partial x_\alpha \partial x'_\beta}\Phi_{\alpha\beta}(\mathbf{r},\mathbf{r}';\omega)$$

$$= \hbar\coth\left(\frac{\hbar\omega}{2kT}\right)\frac{\partial^2}{\partial x_\alpha \partial x'_\beta}\chi^{(e)''}_{\alpha\beta}(\mathbf{r},\mathbf{r}';\omega)$$

Therefore, the spectral density of current density fluctuations is also expressed in terms of the anti-Hermitian part of the linear susceptibility. The correlation of currents and charges in the medium are imposed by two physically different mechanisms:

(a) charge motion and current propagation in the medium, and
(b) correlation of currents and charges via the fields they produce in the medium.

The FDT in formula (2.7.3) relates fluctuations of the true currents in the medium with the response to the external field $\chi^{(1e)}$. Sometimes it is convenient to present the FDT as a relationship between the fluctuations of the so-called extraneous currents and the ordinary susceptibility $\chi^{(1)}$ of the system. The extraneous currents are formally introduced as sources of fluctuations, that is, as such external currents (with respect to the system) which, being substituted into the righthand side of Maxwell's equations (equations (1.3.3)), produce the same fluctuating fields as do the true fluctuations in the system. In other words, the extraneous currents are something like "bare" fluctuations, while true fluctuations are a response of the medium to these currents.

In terms of the extraneous currents $\mathbf{j}^{(\mathrm{ex})}$, these currents produce induced currents $\mathbf{j}^{(\mathrm{ind})}$ in the system, while the sum of currents produces a field which is equivalent to the field created by the true currents $\mathbf{j}$ in the system.

Therefore,

$$\mathbf{j} = \mathbf{j}^{(\mathrm{ex})} + \mathbf{j}^{(\mathrm{ind})} \tag{2.7.4}$$

According to formulas of Section 2.1.6, the extraneous current produces an electric field

$$E^{(\mathrm{ex})}_\omega = \frac{4\pi\omega i}{c^2}\mathcal{D}_0 j^{(\mathrm{ex})}_\omega$$

which induces the current

$$j^{(\mathrm{ind})}_\omega = \sigma^{(e)}E^{(\mathrm{ex})}_\omega = -i\omega\chi^{(e)}E^{(\mathrm{ex})}_\omega = \frac{4\pi\omega^2}{c^2}\chi^{(e)}\mathcal{D}_0 j^{(\mathrm{ex})}_\omega$$

Consequently, condition (2.7.4) implies

$$\begin{aligned} j_\omega &= \left(1 + \frac{4\pi\omega^2}{c^2}\chi^{(e)}\mathcal{D}_0\right) j_\omega^{(ex)} \\ j_\omega^{(ex)} &= \left(1 + \frac{4\pi\omega^2}{c^2}\chi^{(e)}\mathcal{D}_0\right)^{-1} j_\omega \end{aligned} \tag{2.7.5}$$

This relation makes it possible to express the correlation functions of the extraneous currents in terms of the correlation functions of true currents:

$$\begin{aligned} \Phi^{(ex)}(\omega) &= \overline{\left(j^{(ex)}j^{(ex)}\right)_\omega} \\ &= \left(1 + \frac{4\pi\omega^2}{c^2}\chi^{(e)}\mathcal{D}_0\right)^{-1} \overline{(jj)_\omega} \left(1 + \frac{4\pi\omega^2}{c^2}\chi^{(e)}\mathcal{D}_0\right)^{-1+} \end{aligned}$$

Using FDT in the form of (2.7.3), we have

$$\begin{aligned} \Phi^{(ex)}(\omega) &= \hbar\omega^2 \coth\left(\frac{\hbar\omega}{2kT}\right)\left(1 + \frac{4\pi\omega^2}{c^2}\chi^{(e)}\mathcal{D}_0\right)^{-1} \frac{1}{2i}\left(\chi^{(e)} - \chi^{(e)+}\right) \\ &\times \left(1 + \frac{4\pi\omega^2}{c^2}\chi^{(e)}\mathcal{D}_0\right)^{-1+} \end{aligned}$$

$$\begin{aligned} &= \frac{1}{2i}\hbar\omega^2 \coth\left(\frac{\hbar\omega}{2kT}\right)\left[\chi^{(1)}\left(1 + \frac{4\pi\omega^2}{c^2}\chi^{(e)}\mathcal{D}_0\right)^{-1+} \right. \\ &\left. - \left(1 + \frac{4\pi\omega^2}{c^2}\chi^{(e)}\mathcal{D}_0\right)^{-1}\chi^{(1)+}\right] \end{aligned}$$

This formula uses the relation between $\chi^{(1)}$ and $\chi^{(e)}$:

$$\chi^{(1)} = \left(1 + \frac{4\pi\omega^2}{c^2}\chi^{(e)}\mathcal{D}_0\right)^{-1}\chi^{(e)}$$

Let us make use of the identity

$$\begin{aligned} \left(1 + \frac{4\pi\omega^2}{c^2}\chi^{(e)}\mathcal{D}_0\right)^{-1} &= \left(1 + \frac{4\pi\omega^2}{c^2}\chi^{(e)}\mathcal{D}_0\right)^{-1} \\ \times \left(1 + \frac{4\pi\omega^2}{c^2}\chi^{(e)}\mathcal{D}_0 - \frac{4\pi\omega^2}{c^2}\chi^{(e)}\mathcal{D}_0\right) &= 1 - \frac{4\pi\omega^2}{c^2}\chi^{(1)}\mathcal{D}_0 \end{aligned}$$

We obtain

$$\begin{aligned} \Phi^{(ex)}(\omega) &= \frac{1}{2i}\hbar\omega^2 \coth\left(\frac{\hbar\omega}{2kT}\right) \\ &\times \left[\chi^{(1)}\left(1 - \frac{4\pi\omega^2}{c^2}\mathcal{D}_0\chi^{(1)+}\right) - \left(1 - \frac{4\pi\omega^2}{c^2}\chi^{(1)}\mathcal{D}_0\right)\chi^{(1)+}\right] \\ &= \hbar\omega^2 \coth\left(\frac{\hbar\omega}{2kT}\right)\chi^{(1)''} \end{aligned} \tag{2.7.6}$$

In this form, FDT yields the relation between the fluctuations of putative extraneous currents to the conventional linear susceptibility (with its anti-Hermitian part which also determines the dissipative properties of the system).

Another remark on the relative values of the quantities $\chi^{(1)}$ and $\chi^{(1e)}$ is relevant here. Relation (1.6.13) between the tensors $\chi^{(1)}$ and $\chi^{(1e)}$ is simple only in an unbounded spatially uniform medium. If the medium is not uniform or is bounded in space, relation (2.6.13) becomes an integral equation because $\mathcal{D}_0$ in the coordinate representation is an integral operator. The problem gets much more complicated then because the effect of boundaries on the field configuration in the sample is so strong that the nonlocality radius for $\chi^{(1e)}$ becomes of the order of sample size, even if the nonlocality radius for $\chi^{(1)}$ is small (of the order of atomic size a_0). As a result, $\chi^{(1)}$ describes the properties of the medium while $\chi^{(1e)}$ depends essentially on the sample geometry. Consequently, $\chi^{(1)}$ and the correlation of extraneous, not true, currents in the sample give a more fundamental characterization of the medium. In fact, $\chi^{(1)}$ is calculated in the microscopic theory via calculating $\chi^{(1e)}$, using formula (2.5.7) or (2.6.1). Having found $\chi^{(1e)}$ for the infinite medium and from it, $\chi^{(1)}$ from (1.6.13), we can subsequently use the values of $\chi^{(1)}$ obtained in this way to analyze field propagation in any bounded sample of the same medium

2.8. Linear Susceptibility in Dipole Approximation

Expression (2.6.1) is the general formula for $\chi^{(1e)}$ of an arbitrary system in the nonrelativistic approximation. Expression (2.6.1) has poles at

$$\hbar\omega = E_{n'} - E_n$$

when the radiation frequency is in resonance with the transition frequency.

It is slightly unexpected that expression (2.6.1) for $\chi^{(e)}_{\alpha\beta}(\mathbf{r}, \mathbf{r}'; \omega)$ has an ω^2 pole at low frequencies as $\omega \to 0$; in fact, magnetic properties of substances indeed produce this type of singularity in χ, as we have already discussed in Chapter 1. The magnetic susceptibility χ_m contributes to χ a term of the type of $c^2k^2\omega^{-2}\chi_m$. Typically, this contribution to χ is small, and it even vanishes for a longitudinal electric field because the magnetization currents are determined by the vortex (transverse) component of the electric field.

It can be demonstrated that ω^{-2}-type singularity vanishes from (2.6.1) if we project it onto the longitudinal field. The singularity also vanishes if we neglect spatial dispersion, including the magnetic effects, since in our approach the magnetic properties of matter are taken into account only as spatial dispersion effects.

To neglect spatial dispersion, we need to set $\mathbf{k} = 0$ in a spatially uniform medium. This is equivalent to averaging over a physically infinitely small volume V. Small $\mathbf{k}$ corresponds to large $(\mathbf{r} - \mathbf{r}')$ since in a spatially uniform medium we have

$$\chi^{(e)}_{\alpha\beta}(\mathbf{r}, \mathbf{r}'; \omega) = \int \frac{d^3k}{(2\pi)^3}\, \chi^{(e)}_{\alpha\beta}(\mathbf{k}, \omega) \exp(i\mathbf{k}(\mathbf{r} - \mathbf{r}'))$$

The derivation given below is in fact independent of whether the medium is spatially uniform or not.

Let us average the polarization over a physically infinitely small volume V whose linear dimensions l satisfy, by definition, the inequality

$$\lambda \gg l \gg r_0$$

where λ is the wavelength of the field, or a characteristic length on which the field varies, and r_0 is the nonlocality radius of the medium.

The field can be considered uniform within the volume V, so that the averaged polarization is

$$\overline{P}_{\omega\alpha} = \frac{1}{V} \int\limits_V P_{\omega\alpha}(\mathbf{r}) d^3r = \frac{1}{V} \int\limits_V d^3r \int\limits_V d^3r'\, \chi^{(e)}_{\alpha\beta}(\mathbf{r}, \mathbf{r}'; \omega) E^{(e)}_{\omega\beta}$$

Integration in r' can also be carried out over the volume V only since $\chi^{(e)}_{\alpha\beta}(\mathbf{r}, \mathbf{r}'; \omega) \neq 0$ only if $|\mathbf{r} - \mathbf{r}'| \lesssim r_0 \ll l$. For this reason the field $E^{(e)}_{\omega\beta}$ can be factored out of the integral because the field is assumed to be uniform within the range of integration. Having integrated in r and r', we obtain the ordinary susceptibility ignoring the nonlocality of the response, that is, ignoring the spatial dispersion:

$$\overline{\chi}_{\alpha\beta}(\omega) = \frac{1}{V} \int\limits_V \int\limits_V \chi^{(e)}_{\alpha\beta}(\mathbf{r}, \mathbf{r}'; \omega) d^3r\, d^3r'$$

Let us integrate expression (2.6.1) over V. The first term is easily integrated simply because the δ function is present. To integrate the second term, we need to find integrals of the matrix elements of the current density operator

$$\int\limits_V (j_\alpha(\mathbf{r}))_{nn'} d^3r$$

Let us use expression (2.4.3) for the operator $\mathbf{j}_{0\alpha}$ which is denoted here as $\mathbf{j}_\alpha$. This gives

$$\int\limits_V (j_\alpha(\mathbf{r}))_{nn'} d^3r = \sum_{i \in V} \frac{e_i}{m_i} \left(p_{i\alpha} - \frac{e_i}{c} A_\alpha(\mathbf{r}_i)\right)_{nn'} \tag{2.8.1}$$

where summation is carried over the particles inside the volume V.

We can make use now of the relation between the matrix elements of the kinematic momentum operator $\left(p_{i\alpha} - \frac{e_i}{c} A_\alpha(\mathbf{r}_i)\right)$ and the matrix elements of the operator of the coordinate $r_{i\alpha}$:

$$\left(p_{i\alpha} - \frac{e_i}{c} A_\alpha(\mathbf{r}_i)\right)_{nn'} = \frac{\mathrm{i}}{\hbar} m_i (E_n - E_{n'})(r_{i\alpha})_{nn'} \tag{2.8.2}$$

This universal relation is implied by the Hamiltonian of the system without an external field, given by (2.4.5):

$$\begin{aligned}
[\mathscr{H}_0, r_{i\alpha}] &= \frac{1}{2m_i}\left[\left(p_{i\alpha} - \frac{e_i}{c} A_\alpha(\mathbf{r}_i)\right)^2, r_{i\alpha}\right] \\
&= \frac{1}{2m_i}\left(p_{i\alpha} - \frac{e_i}{c} A_\alpha(\mathbf{r}_i)\right)\left[p_{i\alpha} - \frac{e_i}{c} A_\alpha(\mathbf{r}_i), r_{i\alpha}\right] \\
&\quad + \frac{1}{2m_i}\left[p_{i\alpha} - \frac{e_i}{c} A_\alpha(\mathbf{r}_i), r_{i\alpha}\right]\left(p_{i\alpha} - \frac{e_i}{c} A_\alpha(\mathbf{r}_i)\right) \\
&= -\frac{i\hbar}{m_i}\left(p_{i\alpha} - \frac{e_i}{c} A_\alpha(\mathbf{r}_i)\right)
\end{aligned}$$

We can now write for the matrix element $\left(p_{i\alpha} - \frac{e_i}{c} A_\alpha(\mathbf{r}_i)\right)_{nn'}$

$$\begin{aligned}
\left(p_{i\alpha} - \frac{e_i}{c} A_\alpha(\mathbf{r}_i)\right)_{nn'} &= \frac{im_i}{\hbar}([\mathscr{H}_0, r_{i\alpha}])_{nn'} \\
&= \frac{\mathrm{i}}{\hbar} m_i (E_n - E_{n'})(r_{i\alpha})_{nn'}
\end{aligned}$$

Let us substitute (2.8.1) into the integral of (2.6.1) in r and r' over the volume V and express the matrix elements of $\left(p_{i\alpha} - \frac{e_i}{c} A_\alpha(\mathbf{r}_i)\right)_{nn'}$ in terms of $(r_{i\alpha})_{nn'}$ using (2.8.2). After some algebraic manipulations, we obtain

$$\begin{aligned}
\overline{\chi}^{(1)}_{\alpha\beta} = \frac{1}{V}\sum_{nn'} \exp\left(\frac{\mathcal{F} - E_n}{kT}\right)\bigg\{&\frac{(d_\alpha)_{nn'}(d_\beta)_{n'n}}{E_{n'} - E_n - \hbar\omega - i\delta} \\
&+ \frac{(d_\beta)_{nn'}(d_\alpha)_{n'n}}{E_{n'} - E_n + \hbar\omega + i\delta}\bigg\}
\end{aligned} \tag{2.8.3}$$

Here d is the operator of the dipole moment of all particles in the volume V:

$$d_\alpha = \sum_{i \in V} e_i r_{i\alpha}$$

Indeed,

$$\begin{aligned}
\frac{1}{V}\int_V\int_V \chi^{(\mathrm{e})}_{\alpha\beta}(\mathbf{r}, \mathbf{r}'; \omega)\, \mathrm{d}^3r\, \mathrm{d}^3r' &= -\sum_{\mathrm{s}} \frac{\bar{n}_{\mathrm{s}} e_{\mathrm{s}}^2}{m_{\mathrm{s}}\omega^2}\delta_{\alpha\beta} + \frac{1}{V}\sum_{nn'} \exp\left(\frac{\mathcal{F} - E_n}{kT}\right) \\
\times \frac{1}{\hbar^2\omega^2}\sum_{i,i' \in V} e_i e_{i'}&\left\{\frac{(r_{i\alpha})_{nn'}(r_{i'\beta})_{n'n}}{E_{n'} - E_n - \hbar\omega - i\delta} + \frac{(r_{i'\beta})_{nn'}(r_{i\alpha})_{n'n}}{E_{n'} - E_n + \hbar\omega + i\delta}\right\} \\
\times (E_{n'} - E_n)^2 &
\end{aligned} \tag{2.8.4}$$

We will make use of the identity

$$(E_{n'} - E_n)^2 = (E_{n'} - E_n + \hbar\omega)(E_{n'} - E_n - \hbar\omega) + (\hbar\omega)^2$$

Then, since $r_{i\alpha}$ and $r_{i'\beta}$ commute, the following sum appears in (2.8.4):

$$\hbar\omega \sum_{n'} ((r_{i\alpha})_{nn'}(r_{i'\beta})_{n'n} - (r_{i'\beta})_{nn'}(r_{i\alpha})_{n'n}) = \hbar\omega([r_{i\alpha}, r_{i'\beta}])_{nn} = 0$$

Further on, another sum appears:

$$\begin{aligned}
&\sum_{n'} \{(E_{n'} - E_n)(r_{i\alpha})_{n'n}(r_{i'\beta})_{nn'} + (E_{n'} - E_n)(r_{i'\beta})_{nn'}(r_{i\alpha})_{n'n}\} \\
&= \frac{i\hbar}{m_i} \sum_{n'} \left\{ \frac{im_i}{\hbar} (E_{n'} - E_n)(r_{i\alpha})_{nn'}(r_{i'\beta})_{n'n} - (r_{i'\beta})_{nn'} \frac{im_i}{\hbar} \right. \\
&\quad \left. \times (E_{n'} - E_n)(r_{i\alpha})_{n'n} \right\} = \frac{i\hbar}{m_i} \sum_{n'} \{(p_{i\alpha})_{nn'}(r_{i'\beta})_{n'n} \\
&\quad - (r_{i'\beta})_{nn'}(p_{i\alpha})_{n'n}\} = \frac{i\hbar}{m_i}([p_{i\alpha}, r_{i'\beta}])_{nn} = \frac{\hbar^2}{m_i} \delta_{ii'}\delta_{\alpha\beta}
\end{aligned}$$

Therefore,

$$\begin{aligned}
\overline{\chi}^{(1)}_{\alpha\beta}(\omega) &= -\sum_{s} \frac{\bar{n}_s e_s^2}{m_s \omega^2} + \frac{1}{\omega^2 \hbar^2 V} \sum_{ii' \in V} \delta_{ii'}\delta_{\alpha\beta} e_i e_{i'} \\
&\quad + \frac{1}{V} \sum_{nn'} \exp\left(\frac{\mathcal{F} - E_n}{kT}\right) \left\{ \frac{(d_\alpha)_{nn'}(d_\beta)_{n'n}}{E_{n'} - E_n - \hbar\omega - i\delta} \right. \\
&\quad \left. + \frac{(d_\beta)_{nn'}(d_\alpha)_{n'n}}{E_{n'} - E_n + \hbar\omega + i\delta} \right\}
\end{aligned}$$

The first two terms of this formula cancel each other out because

$$\frac{1}{V} \sum_{i \in V} \frac{e_i^2}{m_i} = \sum_{s} \frac{e_s^2 \bar{n}_s}{m_s}$$

We obtain expression (2.8.3) which does not contain a singularity at low frequencies.

Formula (2.8.3) could be obtained immediately if we were solving the problem for a small volume V of the medium and from the start neglected the spatial dispersion, that is, the nonuniformity of the field. In the uniform field, we can use the dipole approximation and assume that the vector potential is independent of the spatial variable:

$$\mathbf{A}^{(e)}(\mathbf{r}, t) \simeq \mathbf{A}^{(e)}(t)$$

The transverse part of the electric field is then

$$\mathcal{E}_{\perp}^{(e)}(t) = -\frac{1}{c}\frac{\partial \mathbf{A}^{(e)}(t)}{\partial t}$$

The dependence on $\mathbf{r}$ in the scalar potential cannot be ignored since the longitudinal component of the field is

$$\mathcal{E}_{\parallel}^{(e)} = -\operatorname{grad}\varphi^{(e)}$$

and

$$\varphi^{(e)} = -\mathcal{E}_{\parallel}^{(e)}\mathbf{r}$$

After a gauge transformation with the function

$$\chi(\mathbf{r}, t) = -\mathbf{A}^{(e)}(t)\bar{\mathbf{r}}$$

we obtain

$$\mathbf{A}^{(e)\prime}(t) = \mathbf{A}^{(e)}(t) + \nabla\chi = 0$$

$$\varphi^{(e)\prime}(\mathbf{r}, t) = \varphi^{(e)}(\mathbf{r}, t) - \frac{1}{c}\frac{\partial\chi}{\partial t} = -\mathcal{E}_{\parallel}^{(e)}\mathbf{r} - \mathcal{E}_{\perp}^{(e)}r = -\mathcal{E}^{(e)}\mathbf{r}$$

This transformation eliminates the vector potential; the field is described by the scalar potential only. The perturbation operator $\mathcal{H}'$ (2.4.6) has a simple form:

$$\mathcal{H}' = -d_\alpha \mathcal{E}_\alpha^{(e)}(t) \tag{2.8.5}$$

This is the so-called dipole approximation. Using this expression for $\mathcal{H}'$, we can immediately obtain the mean value of the dipole moment of the volume V and the mean polarization

$$\overline{\mathcal{P}}_\alpha^{(1)}(t) = \frac{1}{V}\bar{d}_\alpha^{(1)} = \frac{1}{-i\hbar V}\int_{-\infty}^{t} \mathrm{Sp}\{[d_\alpha(t), d_\beta(t')]\rho_0\}\mathcal{E}_\beta^{(e)}(t')\,dt' \tag{2.8.6}$$

Formula (2.8.6) corresponds to expression (2.5.6) which was derived with the spatial dispersion taken into account. Taking the Fourier transform and rewriting the trace via the energy representation immediately gives expression (2.8.3) for susceptibility. Hence, (2.8.3) is the linear susceptibility in the dipole approximation.

If the spatial dispersion is ignored, all results relevant to the interaction between the external field and the matter can be obtained using the perturbation operator (2.8.5). Calculations with operator (2.8.5) are considerably simpler than those with $\mathcal{H}'$ of (2.4.6). In optics, the dipole approximation is usually quite adequate, since $\lambda \gg a_0$, so that hereafter

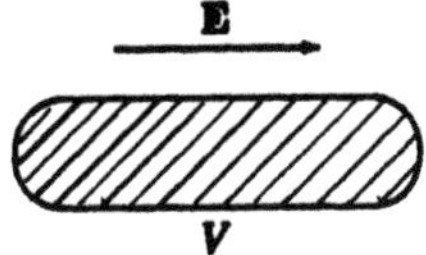

Figure 2.1. Choosing the volume V for the calculation of the mean polarization.

we will use expression (2.8.5) to calculate nonlinear susceptibilities. If one needs to derive formulas taking into account spatial dispersion, it is necessary to use operator $\mathscr{H}'$ of (2.4.6).

The aspect that we have not yet clarified is what response is described by formula (2.8.3), to the external field or to the true one. In general, the above derivation of $\overline{\chi}^{(1)}_{\alpha\beta}(\omega)$ implies that this is the response to the external field; we have already discussed this point. In the dipole approximation, however, this difference becomes unimportant

Indeed, let us single out a volume V in which the field can be regarded as uniform. The external field for the volume V is the field produced in V by particles outside this volume. Let us choose V as shown in Figure 2.1 such that V is elongated in the direction of the field. The equality of tangential components of electric field implies that the fields inside and outside the volume are identical. Hence, the field is the same whether the volume is empty or filled. If the volume V is chosen in this manner, we need not think of the depolarization factor: the external field coincides with the mean field in the volume. Obviously, an arbitrary shape of the volume could be chosen. If the volume is ellipsoidal, the expressions for $\chi^{(1e)}$ and $\chi^{(1)}$ in the dipole approximation differ in depolarization factor only.

The index "e" is therefore omitted in (2.8.3); note that an elongated volume, as shown in Figure 2.1, must be taken for V. In such volumes, $\chi^{(1e)} = \chi^{(1)}$. This cannot be done if spatial dispersion is taken into account since the field is a function of coordinates (it is impossible to choose an elongated volume V oriented along a variable-direction field).

Expressions (2.8.3) and (2.8.6) contain the volume V, even though the quantity $\chi^{(1)}$ is obviously independent of the choice of V. This is easy to see from formula (2.8.6). Let the matter filling V be divided into subsystems, and let

$$d_\alpha(t) = \sum_k d_{k\alpha}(t)$$

where $d_{k\alpha}(t)$ are the dipole moment operators of the kth subsystem. Assume that the interaction between subsystems is negligible; this division is possible for molecules of the medium or, if intermolecular interactions are essential, for macroscopic volumes V_k into which V can be subdivided. In this case,

$$\mathcal{H}_0 = \sum_k \mathcal{H}_{0k}, \quad d_{k\alpha}(t) = \exp\left(\frac{i}{\hbar}\mathcal{H}_{0k}t\right) d_{k\alpha} \exp\left(-\frac{i}{\hbar}\mathcal{H}_{0k}t\right)$$

and $d_{k\alpha}(t)$ commute with $d_{k'\alpha}(t')$ for any t and t' if $k \neq k'$. Therefore,

$$[d_\alpha(t), d_\beta(t')] = \sum_{kk'} [d_{k\alpha}(t), d_{k'\beta}(t')] = \sum_k [d_{k\alpha}(t), d_{k\beta}(t')]$$

and the contribution (2.8.6) to polarization by subsystems is additive. We will ignore the interaction between the molecules of the medium and estimate $\chi^{(1)}$ using (2.8.3). Let volume V contain N atoms (molecules). For each of them, $d_k \sim ea_0$. The energy difference $(E_{n'} - E_n)$ is of the order of the characteristic atomic energy

$$I_0 = me^4/2\hbar^2 = e^2/2a_0$$

An estimate of $\chi^{(1)}$ using (2.8.3) is

$$\chi^{(1)} \sim \frac{N}{V} \frac{(ea_0)^2}{I_0} \sim na_0^3$$

where $n = N/V$ is the number density of atoms (molecules).

In a gas, $n \sim 3 \times 10^{19}\,\mathrm{cm}^{-3}$, so that $\chi^{(1)} \sim 10^{-5}$. In a condensed medium, $na_0^3 \sim 1$, since the distances between atoms are of the order of the atomic size a_0. Indeed, in condensed media (liquids, solids), $\chi^{(1)} \sim 1$.

The evaluation was, in fact, made for the electron contribution to susceptibility. It will be shown, when we consider molecules and crystals, that the ionic contribution is also of this order of magnitude.

The result of this evaluation is not valid for free charges and for the plasma, since these are characterized by continuous spectra.

The estimate was also made for low frequencies, when $\hbar\omega \ll E_{n'} - E_n$. Let us now consider the limiting case of high frequencies when $\hbar\omega \gg I_0$; to be more exact, when $\hbar\omega$ is much greater than the binding energy of electrons in atoms. This last quantity for internal electrons is of the order of $Z^2e^2/a_0 \sim Z^2 I_0$, where Ze is the charge of the nucleus, and may be several orders of magnitude greater than I_0.

In order to take the limit as $\omega \to \infty$ in expression (2.8.3), we need to use the inequalities

$$\hbar\omega \gg |E_{n'} - E_n|$$

which would seemingly be wrong to impose since there are states in the continuous spectrum with arbitrarily high energies E_n.

In fact, an appreciable contribution to the sum over n in (2.8.3) is made only by the states for which $E_n - E_0 \sim \mathrm{k}T$, owing to the factor $\exp\left((\mathcal{F} - E_n)/\mathrm{k}T\right)$; E_0 denotes the energy of the ground (the lowest) state. When the energy difference $|E_{n'} - E_n|$ becomes large, the matrix elements of dipole moment become small; this creates the possibility of retaining only the values of $E_{n'}$ of the order of I_0 or $Z^2 I_0$.

Indeed, let $E_{n'} \gg I_0$. Then the wave functions corresponding to this energy are, in zero approximation, the wave functions of free particles, that is, plane waves

$$\psi_{n'} = \frac{1}{\sqrt{\Omega}} \exp\left(\frac{\mathrm{i}}{\hbar}\mathbf{p}_{n'}\mathbf{r}\right)$$

where $\mathbf{p}_{n'}$ is the momentum of a particle and Ω is the normalization volume. At high values of $E_{n'} = p_{n'}^2/2m$ this wave function rapidly oscillates, with the oscillation period equal to

$$\lambda_{n'} = \frac{2\pi\hbar}{p_{n'}} = \frac{2\pi\hbar}{\sqrt{2mE_{n'}}}$$

The matrix element for one particle,

$$\mathbf{d}_{nn'} = \int \psi_{n'}^{*} e\mathbf{r}\psi_n \mathrm{d}^3 r = \frac{e}{\sqrt{\Omega}} \int \exp\left(-\frac{\mathrm{i}}{\hbar}\mathbf{p}_{n'}\mathbf{r}\right) \mathbf{r}\psi_n \mathrm{d}^3 r$$

becomes low if $\lambda_{n'} \ll a_0/Z$ since ψ_n is characterized by the length scale a_0/Z. If $\lambda_{n'} \ll a_0/Z$, the value of $\mathbf{d}_{n'n}$ falls off not slower than $p_{n'}^{-1}$ as energy $E_{n'}$ increases, while its square decreases steeper (in fact, much steeper) than $p_{n'}^{-2}$. Therefore, the sum over n' is truncated when

$$\frac{p_{n'}}{\hbar} \gg \frac{Z}{a_0}$$

that is, when

$$E_{n'} = \frac{p_{n'}^2}{2m} \gg \frac{\hbar^2 Z^2}{2ma_0^2} = I_0 Z^2$$

Therefore, the estimate $\hbar\omega \gg I_0 Z^2$ leads to the estimate $\hbar\omega \gg |E_{n'} - E_n|$, and the energy differences $(E_{n'} - E_n)$ can be considered small.

If we now neglect $(E_{n'} - E_n)$ in comparison with $\hbar\omega$ in the denominators, the result is zero. We therefore make use of the following linear approximation in the parameter $(E_{n'} - E_n)/\hbar\omega$:

$$\frac{1}{E_{n'} - E_n \pm \hbar\omega} = \pm\frac{1}{\hbar\omega} \frac{1}{1 \pm (E_{n'} - E_n)/\hbar\omega} \simeq \pm\frac{1}{\hbar\omega} - \frac{E_{n'} - E_n}{\hbar^2\omega^2}$$

At high frequencies, we obtain from (2.8.3)

$$\chi^{(1)}_{\alpha\beta}(\omega \to \infty) \simeq \frac{1}{\hbar\omega^2} \sum_{nn'} \exp\left(\frac{\mathfrak{F} - E_n}{kT}\right) \left\{ (d_\alpha)_{nn'} \frac{E_n - E_{n'}}{\hbar} (d_\beta)_{n'n} - (d_\beta)_{nn'} \frac{E_{n'} - E_n}{\hbar} (d_\alpha)_{n'n} \right\}$$

$$= -\frac{\mathrm{i}}{\hbar\omega^2 V} \sum_{nn'} \exp\left(\frac{\mathfrak{F} - E_n}{kT}\right) \times \sum_{i,i' \in V} \frac{e_i e_{i'}}{m_i} ((p_{i\alpha})_{nn'} (r_{i'\beta})_{n'n} - (r_{i'\beta})_{nn'} (p_{i\alpha})_{n'n})$$

A transformation to the matrix elements of momentum operators has been performed in the last equality using (2.8.2). The expression in brackets is the diagonal matrix element of the commutator $[p_{i\alpha}, r_{i'\beta}]$.

Finally,

$$\chi^{(1)}_{\alpha\beta}(\omega \to \infty) \simeq \frac{(-1)}{\omega^2 V} \sum_n \exp\left(\frac{\mathfrak{F} - E_n}{kT}\right) \sum_{i \in V} \frac{e_i^2}{m_i} \delta_{\alpha\beta}$$

$$= -\sum_s \frac{\bar{n}_s e_s^2}{m_s \omega^2} \delta_{\alpha\beta} \tag{2.8.7}$$

We see that only the term depending on frequency as ω^{-2} thus survives at high frequencies. Expression (2.8.7) gives the susceptibility of a set of free charges. This is the same form we find for the susceptibility of the plasma; at high photon energies, all charges behave as free if we neglect the binding energy. The main contribution obviously comes from electrons since the masses of nuclei m_s are large. Expressions (2.8.3) for linear susceptibility satisfy Kramers–Kronig relations; this can be confirmed by straightforward calculation. It can also be shown that expression (2.8.3) possesses the symmetry

$$\chi^{(1)}_{\alpha\beta}(\omega) = \chi^{(1)}_{\beta\alpha}(\omega)$$

which is a particular case of the symmetry of kinetic coefficients (Onsager's principle).

Indeed, if no external magnetic field is present, the wave functions of the system can be chosen as real. This follows from the reality of the Hamiltonian in zero magnetic field. The Schrödinger equation implies that

$$-\mathrm{i}\hbar \frac{\partial \psi^*}{\partial t} = \mathscr{H}_0^* \psi^*$$

and, since $\mathscr{H}_0^* = \mathscr{H}_0$, ψ^* satisfies the same equation as ψ but with time reversed, $t \to -t$. This immediately implies that if ψ_n is an eigenfunction of $\mathscr{H}_0$, then ψ_n^* is also an eigenfunction which corresponds to the same value of energy.

If the level E_n is not degenerate, then ψ_n and ψ_n^* can differ only in a phase factor; multiplying ψ by a phase factor, we can have $\psi = \psi^*$. If, however, E_n is a degenerate level, then we arrive at a real basis, provided we introduce, instead of ψ_n and ψ_n^*, new real functions which correspond to the same value of energy,

$$\psi_n' = \tfrac{1}{2}(\psi_n + \psi_N^*) \qquad \psi_n'' = \tfrac{1}{2\mathrm{i}}(\psi_n - \psi_n^*)$$

and normalize them.

Expression (2.8.3) is invariant with respect to a transformation of the basis; this follows from the invariance of the trace and of the expression (2.8.6). As a result, we can consider (2.8.3) in a real basis.

Since $\mathbf{d}$ is a Hermitian operator,

$$(d_\alpha)_{nn'} = (d_\alpha)^*_{n'n} = (d_\alpha)_{n'n}$$

In view of this, equation (2.8.3) yields

$$\chi^{(1)}_{\alpha\beta}(\omega) = \chi^{(1)}_{\beta\alpha}(\omega)$$

In the general formula (2.6.1) which was obtained taking into account spatial dispersion, similar arguments lead to the symmetry

$$\chi^{(\mathrm{e})}_{\alpha\beta}(\mathbf{r}, \mathbf{r}'; \omega) = \chi^{(\mathrm{e})}_{\beta\alpha}(\mathbf{r}', \mathbf{r}; \omega)$$

or, in a spatially uniform medium where $\chi^{(\mathrm{e})}_{\alpha\beta}$ is a function of $(\mathbf{r} - \mathbf{r}')$,

$$\chi^{(\mathrm{e})}_{\alpha\beta}(\mathbf{k}, \omega) = \chi^{(\mathrm{e})}_{\beta\alpha}(-\mathbf{k}, \omega)$$

In the last case, the matrix elements of current density are not real but they enter the expression for $\chi^{(\mathrm{e})}$ in pairs, so that the imaginary component vanishes.

Since the matrix elements $(d_\alpha)_{nn'}$ are real, we can now rewrite (2.8.3) as

$$\chi^{(1)}_{\alpha\beta}(\omega) = \frac{1}{V}\sum_{nn'} \exp\left(\frac{\mathfrak{F} - E_n}{\mathrm{k}T}\right) \frac{2(E_{n'} - E_n)(d_\alpha)_{nn'}(d_\beta)_{n'n}}{(E_{n'} - E_n)^2 - (\hbar\omega)^2} \tag{2.8.8}$$

Before the advent of quantum mechanics, one would use Lorentz's oscillatory model and write

$$\chi = \sum_k \frac{f_k e^2}{m(\omega_k^2 - \omega^2)} \tag{2.8.9}$$

where f_k was known as the oscillator strength. This term still survives in spectroscopy but expression (2.8.8), having the same form as (2.8.9), allows a representation of oscillator strengths in terms of the matrix elements of transitions. If now we take the limit $\omega \to \infty$ in (2.8.8), we arrive at (2.8.7). The transformations preceding (2.8.7) were indeed meant to produce the familiar "sum rule" for oscillator strengths: the sum of oscillator strengths equals the number of electrons.

2.9. Quadratic Susceptibility. Symmetry Properties

Let us derive now explicit expressions for nonlinear susceptibilities in the dipole approximation, that is, for spatial dispersion neglected. Sufficiently compact and clear expressions are obtained in the dipole approximation, which are readily generalized if spatial dispersion must be taken into account.

Expression (2.8.5) can be used in the dipole approximation. This operator is linear in the external field so that we have no problems using the general perturbations theory, outlined in Section 2.3, for the density matrix and for the mean values of an arbitrary physical quantity.

We will choose the polarization

$$\mathcal{P}_\alpha(t) = \frac{1}{V} d_\alpha(t)$$

as the quantity for which nonlinear contributions are calculated. Formula (2.3.7) gives

$$\overline{\mathcal{P}_\alpha^{(2)}} = \frac{1}{V}\overline{d_\alpha^{(2)}}(t) = \frac{1}{(\mathrm{i}\hbar)^2 V}\int\limits_{-\infty}^{t} \mathrm{d}t'$$

$$\times \int\limits_{-\infty}^{t'} \mathrm{d}t'' \mathrm{Sp}\{[[d_\alpha(t), d_\beta(t')], d_\gamma(t'')]\rho_0\}\mathcal{E}_\beta(t')\mathcal{E}_\gamma(t'') \qquad (2.9.1)$$

In (2.9.1), β and γ are summation indices and t' and t'' are integration variables. Therefore, we can transpose the dummy variables, $\beta \leftrightarrow \gamma$, $t' \leftrightarrow t''$, and rewrite (2.9.1) in a symmetrized form.

Let us turn in (2.9.1) to the Fourier transform representation. Since ρ_0 is the stationary density matrix, the trace in the integrand is a function of time differences

$$\tau' = t - t' \qquad \tau'' = t' - t''$$

Consequently,

$$P^{(2)}_{\omega\alpha} = \int\limits_{-\infty}^{+\infty} e^{i\omega t}\overline{\mathcal{P}^{(2)}_{\alpha}}(t)dt = \frac{1}{(i\hbar)^2 V}\int\limits_{-\infty}^{+\infty} dt\, e^{i\omega t}\int\limits_0^{\infty} d\tau' \int\limits_0^{\infty} d\tau''$$
$$\times \operatorname{Sp}\{[[d_\alpha(t), d_\beta(t')], d_\gamma(t'')]\rho_0\}\int \frac{d\omega' d\omega''}{(2\pi)^2} E_{\omega'\beta}E_{\omega''\gamma}$$
$$\times e^{-i\omega(t-\tau')}\, e^{-i\omega''(t-\tau'-\tau'')}$$

Integrating in t and symmetrizing in the variables $\beta\omega \leftrightarrow \gamma\omega''$, we obtain

$$P^{(2)}_{\omega\alpha} = \frac{1}{2(i\hbar)^2 V}\int \frac{d\omega' d\omega''}{(2\pi)^2}\,\delta(\omega-\omega'-\omega'')\int\limits_0^{\infty} d\tau' \int\limits_0^{\infty} d\tau'' e^{i(\omega'+\omega'')\tau'}$$
$$\times e^{i\omega''\tau''}\operatorname{Sp}\{[[d_\alpha(t), d_\beta(t-\tau')], d_\gamma(t-\tau'-\tau'')]\rho_0\}E_{\omega'\beta}E_{\omega''\gamma}$$
$$+ (\beta\omega' \leftrightarrow \gamma\omega'')$$

On the other hand, by virtue of the definition given in Chapter 1,

$$P^{(2)}_{\omega\alpha} = \int \frac{d\omega' d\omega''}{2\pi}\,\delta(\omega-\omega'-\omega'')\chi^{(2)}_{\alpha\beta\gamma}(\omega=\omega'+\omega'')E_{\omega'\beta}E_{\omega''\gamma}$$

so that

$$\chi^{(2)}_{\alpha\beta\gamma}(\omega=\omega'+\omega'') = \frac{1}{2(i\hbar)^2 V}\int\limits_0^{\infty} d\tau' \int\limits_0^{\infty} d\tau'' e^{i(\omega'+\omega'')\tau'+i\omega''\tau''}$$
$$\times \operatorname{Sp}\{[[d_\alpha(t), d_\beta(t-\tau')], d_\gamma(t-\tau'-\tau'')]\rho_0\}$$
$$+ (\beta\omega' \leftrightarrow \gamma\omega'') \tag{2.9.2}$$

Strictly speaking, expression (2.9.2) describes the response to the external field $\chi^{(2e)}$. However, as we saw in the linear case, it differs from the susceptibility $\chi^{(2)}$ in the dipole approximation only in factors that depend on the geometry of the volume V, that is, on depolarization factors for all three fields. As in the case of linear susceptibility, we will calculate the trace in the energy representation, using the known time dependence of the matrix elements of the operators d_α in the interaction representation, and the Gibbs density matrix. The double commutator in (2.9.2) produces four terms. We will write in full only one, which corresponds to that order of operators in which they are arranged in the commutator:

$$\chi^{(2)}_{\alpha\beta\gamma}(\omega = \omega' + \omega'') = \frac{1}{2(\mathrm{i}\hbar)^2 V} \int_0^\infty \mathrm{d}\tau' \int_0^\infty \mathrm{d}\tau'' \exp\left(\mathrm{i}(\omega' + \omega'')\tau' + \mathrm{i}\omega''\tau''\right)$$

$$\times \sum_{nn'n''} \exp\left(\frac{\mathcal{F} - E_n}{\mathrm{k}T}\right) (d_\alpha)_{nn'} (d_\beta)_{n'n''} (d_\gamma)_{n''n} \exp\left(\frac{\mathrm{i}}{\hbar}(E_n - E_{n'})t\right)$$

$$\times \exp\left(\frac{\mathrm{i}}{\hbar}(E_{n'} - E_{n''})(t - \tau')\right) \exp\left(\frac{\mathrm{i}}{\hbar}(E_{n''} - E_n)(t - \tau' - \tau'')\right)$$

$$+ 7 \ \text{terms}$$

The time factor containing t drops out since the trace is cyclic; the integrals in τ' and τ'' are again calculated as in the linear case, by introducing the factors $\exp(-\frac{\delta}{\hbar}\tau)$.

Finally,

$$\chi^{(2)}_{\alpha\beta\gamma}(\omega = \omega' + \omega'') = \frac{1}{2V} \sum_{nn'n''} \exp\left(\frac{\mathcal{F} - E_n}{\mathrm{k}T}\right)$$

$$\times \frac{(d_\alpha)_{nn'} (d_\beta)_{n'n''} (d_\gamma)_{n''n}}{(E_{n'} - E_n - \hbar(\omega' + \omega'') - \mathrm{i}\delta)(E_{n''} - E_n - \hbar\omega'' - \mathrm{i}\delta)}$$

$$+ 7 \ \text{terms} \tag{2.9.3}$$

The seven terms not written in the righthand side of (2.9.3) differ in the order of operators d_α, d_β, d_γ or are obtained from the first four terms by the replacement $\beta\omega' \leftrightarrow \gamma\omega''$. They play an essential role in deriving the symmetry properties of the expression obtained.

We can immediately estimate the quantity $\chi^{(2)}$. As in the case of linear susceptibility, (2.9.3) is in fact independent of the volume V and $\chi^{(2)} \sim (N/V)(ea_0)^3/I_0^2$.

The characteristic quantity is the ratio

$$\chi^{(2)}/\chi^{(1)} \sim ea_0/I_0 \sim a_0^2/e \sim 1/\mathcal{E}_{\mathrm{at}}$$

We have used here the fact that

$$E_{n'} - E_n \sim I_0 \sim e^2/a_0$$

and that $\mathcal{E}_{\mathrm{at}} = e/a_0^2$ is the characteristic atomic field.

It is convenient to compare not the susceptibilities $\chi^{(2)}$ and $\chi^{(1)}$ but the polarizations

$$\mathcal{P}^{(2)}/\mathcal{P}^{(1)} = \chi^{(2)}\mathcal{E}^2/\chi^{(1)}\mathcal{E} \sim \mathcal{E}/\mathcal{E}_{\mathrm{at}}$$

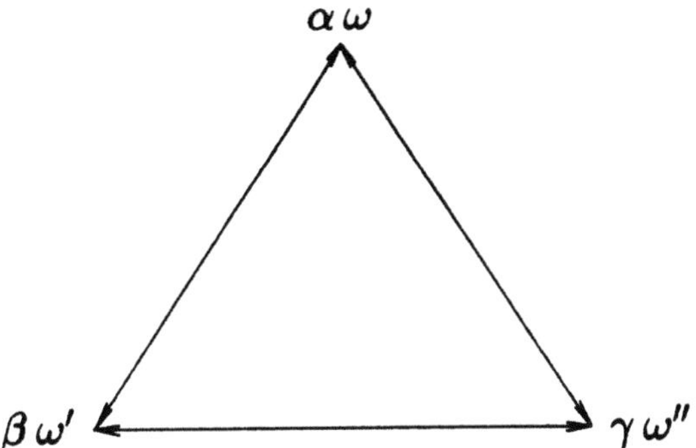

Figure 2.2. Permutation of frequencies and polarization indices in $\chi^{(2)}_{\alpha\beta\gamma}(\omega = \omega' + \omega'')$.

This estimate is not valid for metals and plasmas, owing to the continuous spectrum in both cases. It is also wrong in those cases in which the first excited level is anomalously close to the ground level, as we find in semiconductors. The estimate is quite good for the electron part of $\chi^{(2)}$ in dielectrics.

We have already mentioned in Chapter 1 that $\chi^{(2)}$ has the symmetry

$$\chi^{(2)}_{\alpha\beta\gamma}(\omega = \omega' + \omega'') = \chi^{(2)}_{\beta\alpha\gamma}(-\omega' = -\omega + \omega'') = \chi^{(2)}_{\gamma\beta\alpha}(-\omega'' = \omega' - \omega)$$

Polarization indices are swapped together with frequencies. The expression $\omega = \omega' + \omega''$ in $\chi^{(2)}$ can be treated as an algebraic equality, transferring terms from one side to another and simultaneously transposing the indices α, β, and γ (see Figure 2.2).

If we write explicitly all terms of (2.9.3), it satisfies the commutation relations given earlier. In fact, it is not easy to verify this statement, so cumbersome (2.9.3) is. To prove the symmetry, we will directly use the general expression (2.9.2) with nonexpanded double commutators. It is explicitly symmetric with respect to the transposition $\beta\omega' \leftrightarrow \gamma\omega''$ but symmetry with respect to the transpositions of $\alpha\omega$ with $\beta\omega'$ and $\gamma\omega''$ is not so obvious. Let us transform (2.9.2) to an explicitly symmetric form by introducing the Fourier transforms of the operators $d_\alpha(t)$:

$$d_{\Omega_\alpha} = \int_{-\infty}^{+\infty} d_\alpha(t) \mathrm{e}^{\mathrm{i}\Omega t} \mathrm{d}t$$

Conversely,

$$d_\alpha(t) = \int_{-\infty}^{+\infty} d_{\omega_\alpha} \mathrm{e}^{-\mathrm{i}\omega t} \frac{\mathrm{d}\Omega}{2\pi}$$

These last expressions must be interpreted in the sense that the Fourier transform is calculated for each matrix element of the operator, and the obtained quantity is taken as the matrix element of the operator in the lefthand side.

Therefore,

$$
\begin{aligned}
(d_{\Omega_\alpha})_{nn'} &= \int_{-\infty}^{+\infty} (d_\alpha(t))_{nn'} \exp(\mathrm{i}\Omega t)\mathrm{d}t \\
&= \int_{-\infty}^{+\infty} (d_\alpha)_{nn'} \exp\left(\frac{\mathrm{i}}{\hbar}(E_n - E_{n'} + \hbar\Omega)t\right) \mathrm{d}t \\
&= 2\pi\hbar (d_\alpha)_{nn'} \delta(E_n - E_{n'} + \hbar\Omega)
\end{aligned}
\tag{2.9.4}
$$

Let us write all operators d_α, d_β, d_γ in (2.9.2) in terms of their Fourier transforms

$$
\begin{aligned}
\chi^{(2)}_{\alpha\beta\gamma}(\omega = \omega' + \omega'') &= \frac{1}{2(\mathrm{i}\hbar)^2 V} \int_0^\infty \mathrm{d}\tau' \int_0^\infty \mathrm{d}\tau'' \int \frac{\mathrm{d}\Omega\, \mathrm{d}\Omega'\, \mathrm{d}\Omega''}{(2\pi)^3} \\
&\times \mathrm{Sp}\{[[d_{\Omega_\alpha}, d_{\Omega'\beta}], d_{\Omega''\gamma}]\rho_0\} \exp[-\mathrm{i}\Omega t - \mathrm{i}\Omega'(t - \tau') - \mathrm{i}\Omega''(t - \tau' - \tau'')] \\
&\times \exp[\mathrm{i}(\omega' + \omega'')\tau' + \mathrm{i}\omega''\tau''] + (\beta\omega' \leftrightarrow \gamma\omega'')
\end{aligned}
\tag{2.9.5}
$$

Owing to the δ functions present in (2.9.4), we have

$$
\begin{aligned}
\hbar\Omega &= E_{n'} - E_n \\
\hbar\Omega' &= E_{n''} - E_{n'} \\
\hbar\Omega'' &= E_n - E_{n''}
\end{aligned}
$$

whence

$$
\hbar(\Omega + \Omega' + \Omega'') = 0
$$

This last equality is again a corollary of the cyclic nature of the trace, so that owing to this equality t splits off (2.9.5) and integration over τ' and τ'' becomes possible:

$$
\begin{aligned}
\chi^{(2)}_{\alpha\beta\gamma}(\omega = \omega' + \omega'') &= \frac{1}{2\hbar^2 V} \int \frac{\mathrm{d}\Omega\, \mathrm{d}\Omega'\, \mathrm{d}\Omega''}{(2\pi)^3} \\
&\times \mathrm{Sp}\left\{\rho_0 \frac{[[d_{\Omega_\alpha}, d_{\Omega'\beta}], d_{\Omega''\gamma}]}{(\Omega' + \Omega'' + \omega' + \omega'')(\Omega'' + \omega'')}\right. \\
&\left. + \rho_0 \frac{[[d_{\Omega_\alpha}, d_{\Omega''\gamma}], d_{\Omega'\beta}]}{(\Omega' + \Omega'' + \omega' + \omega'')(\Omega' + \omega')}\right\}
\end{aligned}
\tag{2.9.6}
$$

The Fourier transform for operators was required in order to integrate over τ' an τ'' without expanding the commutators.

Equation (2.9.6) can be written in an explicitly symmetric form, by using several identities. Let A, B, and C be arbitrary operators; the following identities hold, easily proved by straightforward expansion of commutators:

$$[A, B] + [B, A] = 0$$

For the cyclic permutation of three operators we have

$$[[A, B], C] + [[B, C], A] + [[C, A], B] = 0$$

In addition, we make use of identities of the type

$$\frac{1}{(\Omega' + \omega')(\Omega'' + \omega'')} = \frac{1}{(\Omega' + \Omega'' + \omega' + \omega'')(\Omega' + \omega')} + \frac{1}{(\Omega' + \Omega'' + \omega' + \omega'')(\Omega'' + \omega'')}$$
$$= -\frac{1}{(\Omega + \tilde{\omega})(\Omega' + \omega')} - \frac{1}{(\Omega + \tilde{\omega})(\Omega'' + \omega'')}$$

To ensure obvious symmetry, we have introduced the notation $\tilde{\omega} = -\omega$. The condition $\omega = \omega' + \omega''$ is then written in a form which is explicitly symmetric in frequency,

$$\tilde{\omega} + \omega' + \omega'' = 0$$

in complete analogy to

$$\Omega + \Omega' + \Omega'' = 0$$

Using the identities, we can recast (2.9.6) to an explicitly symmetric form:

$$\chi^{(2)}_{\alpha\beta\gamma}(\tilde{\omega} + \omega' + \omega'' = 0) = \frac{1}{\hbar^2 V\, 3!} \sum_P P_{\Omega\tilde{\omega}\alpha,\Omega'\omega'\beta,\Omega''\omega''\gamma}$$
$$\times \int \frac{d\Omega\, d\Omega'\, d\Omega''}{(2\pi)^3} \mathrm{Sp}\left\{ \frac{\rho_0[[d_{\Omega_\alpha}, d_{\Omega'\beta}], d_{\Omega''\gamma}]}{(\Omega' + \omega')(\Omega'' + \omega'')} \right\} \tag{2.9.7}$$

Here $P_{\Omega\tilde{\omega}\alpha,\Omega'\omega'\beta,\Omega''\omega''\gamma}$ denotes a permutation of the triple of indices $\Omega\tilde{\omega}\alpha$, $\Omega'\omega'\beta$, and $\Omega''\omega''\gamma$, and the summation is carried out over all 3! permutations.

The permutations given above are valid if the denominators contain no imaginary terms of the type of $i\delta$. These additional terms can be dropped if neither of the frequencies ω, ω', ω'' coincide with any of the transition frequencies $E_{n'} - E_n$. In other words, the conclusion is valid if all three frequencies ω, ω', ω'' fall within the transparency band of the system. The phenomenological derivation of the symmetry of quadratic susceptibility in Chapter 1 was also based on the assumption of zero true absorption at all the frequencies ω, ω', ω''.

2.10. Cubic Susceptibility. Degeneration Cases. Dynamic Kerr Effect

In third order, perturbation theory gives, in analogy to (2.9.1),

$$\overline{\mathcal{P}_{\alpha}^{(3)}(t)} = \frac{1}{V}\overline{d_{\alpha}^{(3)}(t)} = \frac{(-1)^3}{(\mathrm{i}\hbar)^3 V}\int\limits_{-\infty}^{t}\mathrm{d}t'\int\limits_{-\infty}^{t'}\mathrm{d}t''\int\limits_{-\infty}^{t''}\mathrm{d}t'''$$
$$\times\,\mathrm{Sp}\{[[[d_\alpha(t), d_\beta(t')], d_\gamma(t'')], d_\delta(t''')]\rho_0\}\mathcal{E}_\beta(t')\mathcal{E}_\gamma(t'')\mathcal{E}_\delta(t''') \tag{2.10.1}$$

We again can rewrite (2.10.1) in a symmetrized form, using the symmetrization operator

$$S = \frac{1}{3!}\sum P_{t'\beta,t''\gamma,t'''\delta}$$

where P are the permutations of pairs of indices.

Skipping over intermediate algebras (quite similar to those in the calculation of quadratic susceptibility), we obtain

$$\chi^{(3)}_{\alpha\beta\gamma\delta}(\omega = \omega' + \omega'' + \omega''') = \frac{1}{V}\sum_{nn'n''n'''}\exp\left(\frac{\mathcal{F} - E_n}{\mathrm{k}T}\right)$$
$$\times\left\{\frac{(d_\alpha)_{nn'}(d_\beta)_{n'n''}}{(E_{n'} - E_n - \hbar(\omega' + \omega'' + \omega''') - \mathrm{i}\delta)(E_{n''} - E_n - \hbar(\omega'' + \omega''') - \mathrm{i}\delta)}\right.$$
$$\left.\times\,\frac{(d_\gamma)_{n''n'''}(d_\delta)_{n'''n}}{(E_{n'''} - E_n - \hbar\omega''' - \mathrm{i}\delta)} + 7\ \ \text{terms}\right\} \tag{2.10.2}$$

Eight terms were produced by expanding the triple commutator. This expression must now be symmetrized in frequencies ω', ω'', and ω''' using the operator

$$\widetilde{\mathrm{S}} = \frac{1}{3!}\sum_{P} P_{\omega'\beta,\omega''\gamma,\omega'''\delta}$$

so that the number of terms is actually $8 \times 6 = 48$.

Numerical estimates must be made now. By analogy to the linear and quadratic susceptibilities,

$$\chi^{(3)} \simeq 1/\mathcal{E}^2_{\mathrm{at}} \simeq 10^{-14}\,\mathrm{CGS}$$

since $\mathcal{E}_{\mathrm{at}} \sim 3 \times 10^9\,\mathrm{V/cm} = 10^7\,\mathrm{CGS}$.

The spread of $\chi^{(3)}$ values in various media may reach several orders of magnitude. The value of $\chi^{(3)}$ may be considerably higher than 10^{-14} CGS if there are low-lying excited levels (this is the case of semiconductors). $\chi^{(3)}$ reaches high values under resonance conditions, when the resonance

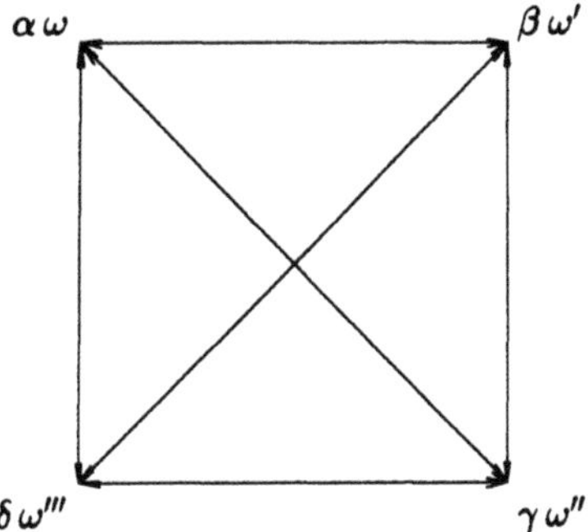

Figure 2.3. Permutation of frequencies and polarization indices in $\chi^{(3)}_{\alpha\beta\gamma\delta}(\omega = \omega' + \omega'' + \omega''')$.

denominator $E_{n'} - E_n - \hbar\omega - i\delta$ is of the order of $\delta = \hbar T_2^{-1} \ll I_0$ (T_2 is the relaxation time). The estimate $\chi^{(3)} \sim \mathcal{E}_{\rm at}^{-2}$ applies to nonresonance conditions, when $(E_{n'} - E_n - \hbar\omega) \sim I_0$.

By analogy to the case of quadratic susceptibility, we can derive symmetry relations for the cubic susceptibility when no absorption occurs at either of the frequencies ω', ω'', and ω'''. The expression for $\chi^{(3)}$ remains unchanged under a transposition of the pairs $\alpha\omega$, $\beta\omega'$, $\gamma\omega''$, $\delta\omega'''$ (Figure 2.3). The frequencies ω', ω'', ω''' can be taken with a "plus" and a "minus" sign, so that $\omega' + \omega'' + \omega'''$ are the sum, the difference, and all possible combination frequencies.

The quadratic susceptibility makes it possible to describe the following physical phenomena:

(a) Second harmonic generation, and also the generation of the sum and difference frequencies:

$$\omega = \omega' \pm \omega''$$

(b) Generation of static polarization and direct current

$$0 = \omega - \omega$$

and also conjugate processes involving direct current

$$\omega = \omega + 0$$

This last case is the well-known Pockels effect, that is, linear variation of the refractive index of the crystal in the electric field.

The cubic susceptibility determines a much greater number of phenomena. In addition to the general case of third harmonic generation, already mentioned above, and generation of various combination frequencies, a number of degenerate cases are possible. These degenerate cases correspond to equalities of the type

$$\omega' = -\omega''$$

In fact, this means that there is a field at a frequency ω'. Indeed, with the field being real, there automatically exists a Fourier component at a frequency $-\omega'$. What is meant is a process described by the susceptibility

$$\chi^{(3)}(\omega = -\omega' + \omega' + \omega)$$

The susceptibilities of orders higher than three are rarely required; as a rule, they provide corrections to lower-order nonlinearities. The lowest-order nonlinearity in media without the center of inversion is the quadratic susceptibility; in media with the center of inversion this is the cubic susceptibility. If necessary, it is possible to write general expressions of the type of (2.10.1) or (2.10.2) for higher-order susceptibilities.

Degenerate cases may, in principle, occur for even-order susceptibilities. For instance, a process described by the susceptibility

$$\chi^{(4)}(\omega = \omega' + \omega'' + \omega''' + \omega'''')$$

where $\omega' + \omega'' + \omega''' = 0$ and $\omega'''' = \omega$ is a possible case. This is, however, an exceptional scenario, requiring a synchronization of radiation sources with the frequencies

$$\omega', \omega'', \omega'''$$

In the case of cubic susceptibility, however, and also of odd-higher-order susceptibilities, degenerate cases occur automatically, owing to the field being real.

In these degenerate cases, $\chi^{(3)}$, $\chi^{(5)}$, etc. give corrections to the susceptibility at a frequency ω, which depend on the field at ω'.

Indeed,

$$\begin{aligned}
P_{\omega\alpha} &= \chi^{(1)}_{\alpha\beta}(\omega)E_{\omega\beta} + \sum_{\omega'} \chi^{(3)}_{\alpha\gamma\delta\beta}(\omega = -\omega' + \omega' + \omega) \\
&\quad \times E^*_{\omega'\gamma}E_{\omega'\delta}E_{\omega\beta} \\
&\quad + \sum_{\omega'\omega''} \chi^{(5)}_{\alpha\gamma\delta\mu\varepsilon\beta}(\omega = -\omega' + \omega' - \omega'' + \omega'' + \omega) \\
&\quad \times E^*_{\omega'\gamma}E_{\omega'\delta}E^*_{\omega''\mu}E_{\omega''\varepsilon}E_{\omega\beta} + \dots \\
&= \{\chi^{(1)}_{\alpha\beta}(\omega) + \sum_{\omega'} \chi^{(3)}_{\alpha\gamma\delta\beta}(\omega = -\omega' + \omega' + \omega)E^*_{\omega'\gamma}E_{\omega'\delta} + \dots\}E_{\omega\beta} \\
&= \chi_{\alpha\beta}(\omega, \{E\})E_{\omega\beta} \qquad (2.10.3)
\end{aligned}$$

This equation takes into account that $E^*_{\omega'\gamma} = E_{-\omega'\gamma}$ and operates with the effective susceptibility $\chi_{\alpha\beta}(\omega, \{E\})$, which depends on fields at frequencies ω', ω'' etc. The dependence of $\chi_{\alpha\beta}$ on field strengths is an analogue of the Kerr effect: the variation of the susceptibility and refractive index in electric field. In the narrow sense, the Kerr effect is the dependence of dielectric permittivity on the static field:

$$\varepsilon = \varepsilon_0 + \alpha \mathcal{E}^2$$

What we consider here is a variable field and its effect on susceptibility; this phenomenon is usually referred to as the dynamic Kerr effect. To describe this effect, it is sufficient to take into account $\chi^{(3)}$, that is, the quadratic dependence of $\chi_{\alpha\beta}$ on the field; actually, higher-order corrections have to be taken into account sometimes. If there are several frequencies, then the expression for $\chi_{\alpha\beta}(\omega, \{E\})$ implies summation over ω', as in (2.10.3).

We have already mentioned in Chapter 1 that the losses in the medium are characterized by the imaginary part of $\chi^{(1)}$ (in the general case, by its anti-Hermitian part):

$$\overline{\dot{Q}_\omega} = \frac{\omega}{2} E^*_\alpha \chi^{(1)''}_{\alpha\beta} E_\beta$$

where $\chi^{(1)''}$ is the anti-Hermitian part of $\chi^{(1)}$. The same is true for the effective susceptibility $\chi_{\alpha\beta}(\omega, \{E\})$. What is the meaning of nonlinear field-dependent corrections to the anti-Hermitian part $\chi''_{\alpha\beta}(\omega, \{E\})$? We will show that the contribution to $\chi''_{\alpha\beta}(\omega, \{E\})$ related to $\chi^{(3)}$ describes nonlinear (two-photon) absorption.

Let both frequencies ω and ω' for any n and n' fall within the transparency band of the medium, that is,

$$E_{n'} - E_n - \hbar\omega \neq 0 \qquad E_{n'} - E_n - \hbar\omega' \neq 0$$

This condition can be satisfied only by ignoring the continuous spectrum. It is, however, unimportant that these inequalities are violated for strongly excited states since their population is very low. It is only important that they be satisfied for the n and n' corresponding to the ground state and the states lying quite close to it.

Only the factor $E_{n''} - E_n - \hbar(\omega'' + \omega''')$ in (2.10.2) can vanish at $\delta = 0$. Therefore, taking into account that in the degenerate case $\omega''' = \omega$, we find

$$\begin{aligned} \chi''_{\alpha\beta}(\omega, \{E\}) = \frac{\pi}{V} \sum_{n''} \exp\left(\frac{\mathcal{F} - E_n}{kT}\right) \\ \times \left(\sum_{n'} \frac{(\mathbf{dE}_{\omega'})_{n''n'}(d_\alpha)_{n'n}}{E_{n'} - E_n - \hbar\omega}\right)^* \sum_{n'''} \frac{(\mathbf{dE}_{\omega'})_{n''n'''}(d_\beta)_{n'''n}}{E_{n'''} - E_n - \hbar\omega} \\ + 7 \text{ terms} \end{aligned} \qquad (2.10.4)$$

In deriving (2.10.4), we made use of the equality

$$\operatorname{Im}\frac{1}{E_{n''}-E_n-\hbar(\omega+\omega')-i\delta}=\pi\delta(E_{n''}-E_n-\hbar(\omega+\omega'))$$

The two sums in (2.10.4) over n' and n'' are obtained one from the other by the substitution $\alpha \leftrightarrow \beta$.

The meaning of formula (2.10.4) becomes perfectly lucid if we find, using perturbation theory, the probability of the two-photon transition $n \to n''$ in the presence of the fields $\mathbf{E}_{\omega'}$ and $\mathbf{E}_{\omega}$. According to perturbation theory, we obtain in second order (by analogy to (2.6.5))

$$w^{(2)}_{n''n}=\frac{2\pi}{\hbar}|\mathscr{H}^{(2)}_{n''n}(\omega,\omega')|^2\delta(E_{n''}-E_n-\hbar\omega'-\hbar\omega) \tag{2.10.5}$$

where

$$\mathscr{H}^{(2)}_{n''n}(\omega,\omega')=\sum_{n'}\left(\frac{\mathscr{H}'_{n''n'}(\omega)\mathscr{H}'_{n'n}(\omega')}{E_{n'}-E_n-\hbar\omega'}+\frac{\mathscr{H}'_{n''n'}(\omega')\mathscr{H}'_{n'n}(\omega)}{E_{n'}-E_n-\hbar\omega}\right)$$

is the composite matrix element for the transition $n \to n''$ via intermediate states n' and $\mathscr{H}'_{n''n}(\omega,\omega') = -(\mathbf{d}\mathbf{E}_{\omega'})_{n'n}$. A comparison of (2.10.4) and (2.10.5) shows that $\chi''_{\alpha\beta}(\omega,\{E\})$ indeed corresponds to the two-photon absorption $\overline{\dot{Q}}_{\omega} = \hbar\omega w^{(2)}_{n''n}$ averaged over the initial state n with the Gibbs density matrix and summed up over the final states n''. Furthermore, $\chi''_{\alpha\beta}$ also takes into account the induced two-photon emission, which enters with the weight $\exp[-\hbar(\omega+\omega')/\varkappa T]$, since the δ function in (2.10.4) and (2.10.5) gives

$$E_{n''}-E_n=\hbar(\omega+\omega')$$

The contribution to $\chi''(\omega,\{E\})$ is made not only by the two-photon absorption but also by the (two-photon) induced Raman scattering. Expressions in the last case are the same but $\omega > 0$ and $\omega' < 0$, so the formulas include $\delta(E_{n''} - E_n - \hbar\omega + \hbar|\omega'|)$. The same is true for the factor $[1 - \exp(-\hbar(\omega + \omega')/kT)]$, which becomes negative at $|\omega'| > \omega$. The medium becomes nonabsorbing and amplifying for radiation at a frequency ω. This is the well-known effect of amplification of the Stokes wave in the induced Raman scattering of light.

2.11. Dynamic Stark Effect and Balance Equations

Expression (2.10.2) for the cubic susceptibility in the degenerate case has second-order poles. Indeed,

$$\chi^{(3)}_{\alpha\beta\gamma}(\omega=-\omega'+\omega'+\omega)=\frac{1}{V}\sum_{nn'n''n'''}\exp\left(\frac{\mathcal{F}-E_n}{kT}\right)(d_\alpha)_{nn'}(d_\beta)_{n'n''}$$
$$\times(d_\gamma)_{n''n'''}(d_\delta)_{n'''n}((E_{n'}-E_n-\hbar\omega)(E_{n''}-E_n$$
$$-\hbar(\omega'+\omega))(E_{n'''}-E_n-\hbar\omega))^{-1}+7\ \text{terms} \tag{2.11.1}$$

If we now take the terms with $n' = n'''$, they have a pole $(E_{n'} - E_n - \hbar\omega)^{-2}$. Similarly, $\chi^{(5)}$ has third-order poles, etc. In fact, these high-order poles are a corollary of the series expansion of expressions with first-order poles; in reality, they correspond to the shifts of the levels of the system in the field $\mathcal{E}$.

The origin of such singularities is seen from the expansion (a geometric progression)

$$\frac{1}{x-a} = \frac{1}{x} + \frac{a}{x^2} + \frac{a^2}{x^3} + \dots$$

The expression in the lefthand side has a first-order pole at a point a, and that in the righthand side has terms corresponding to poles of all orders at a point $x = 0$.

In our case, the situation is of the same type. Let us show that if we collect all "dangerous" terms (terms having poles of higher than first order) in $\chi_{\alpha\beta}(\omega, \{E\})$ which originate from $\chi^{(3)}$, $\chi^{(5)}$, and so forth, they will form an expression with a first-order pole. Collecting terms with $n' = n'''$ in (2.11.1), and also similar terms in $\chi^{(5)}$, etc. we obtain

$$\begin{aligned}\chi_{\alpha\beta}(\omega, \{E\}) &\simeq \frac{1}{V} \sum_{nn'} \exp\left(\frac{\mathcal{F} - E_n}{kT}\right) \frac{(d_\alpha)_{nn'}(d_\beta)_{n'n}}{E_{n'} - E_n - \hbar\omega - i\delta} \\ &\times \left[1 + \left(-\frac{\Delta E_{n'}}{E_{n'} - E_n - \hbar\omega - i\delta}\right) + \left(-\frac{\Delta E_{n'}}{E_{n'} - E_n - \hbar\omega - i\delta}\right)^2 + \dots\right] \\ &= \frac{1}{V} \sum_{nn'} \exp\left(\frac{\mathcal{F} - E_n}{kT}\right) \frac{(d_\alpha)_{nn'}(d_\beta)_{n'n}}{(E_{n'} + \Delta E_{n'}) - E_n - \hbar\omega - i\delta} \end{aligned} \tag{2.11.2}$$

Here the first term in brackets originates from $\chi^{(1)}$, the second from $\chi^{(3)}$, the third from $\chi^{(5)}$, etc. By $\Delta E_{n'}$ we denoted the quantity

$$\Delta E_{n'} = -\sum_{n''} \frac{|(\mathbf{dE}_{\omega'})_{n''n'}|^2}{E_{n''} - E_n - \hbar(\omega + \omega')} \simeq -\sum_{n''} \frac{|(\mathbf{dE}_{\omega'})_{n''n'}|^2}{E_{n''} - E_{n'} - \hbar\omega'}$$

This last approximate equality is based on the fact that (2.11.2) holds near the resonance,

$$E_{n'} \simeq E_n + \hbar\omega$$

since it is in this region that the pole terms selected above give the decisive contribution to $\chi''_{\alpha\beta}(\omega, \{E\})$.

Summing up now over all frequencies ω', we obtain

$$\Delta E_{n'} = -\sum_{n''\omega'} \frac{|(\mathbf{dE}_{\omega'})_{n''n'}|^2}{E_{n''} - E_{n'} - \hbar\omega' - i\delta} \tag{2.11.3}$$

Actually, we need to take into account in (2.11.2) all terms with transposed order of operators d_α and with transposed indices n, n', and n''. Hence, shifts of the levels E_n are also appreciable. Formula (2.11.3) is nothing else but the expression for the shift of the level in the second order of perturbation theory in response to the field $\mathbf{E}_{\omega'}$. This is the Stark effect, that is, the shift of levels not in the static but in high-frequency (optical) field. The typically used term is the "dynamic Stark effect."

Under the conditions of the double resonance, when

$$E_{n''} - E_{n'} \sim \hbar\omega'$$

it is necessary to take the imaginary additional terms into account in the denominators of (2.11.3); $\Delta E_{n'}$ has an imaginary part. The meaning of the imaginary part in the energy of a level is obvious: it determines the lifetime of the quasistationary level. Indeed, since the wave function is time-dependent,

$$\psi \sim \exp(-\frac{\mathrm{i}}{\hbar} E_{n'}t)$$

and hence,

$$|\psi|^2 \simeq \exp(\frac{2}{\hbar} \mathrm{Im}\Delta E_{n'}t)$$

so the system decays with the probability per unit time

$$w = -2\mathrm{Im}\Delta E_{n'}/\hbar$$

By virtue of (2.11.3), the decay probability is

$$w = \frac{2\pi}{\hbar} \sum_{n''\omega'} |(\mathbf{dE}_{\omega'})_{n''n'}|^2 \, \delta(E_{n''} - E_{n'} - \hbar\omega) \tag{2.11.4}$$

since

$$(E_{n''} - E_{n'} - \hbar\omega' - \mathrm{i}\delta)^{-1} = P\,(E_{n''} - E_{n'} - \hbar\omega')^{-1} + \mathrm{i}\pi\delta(E_{n''} - E_{n'} - \hbar\omega')$$

where P is the symbol of the principal value, that is, integrals with the function $(E_{n''} - E_{n'} - \hbar\omega')^{-1}$ must be regarded as integrals in the sense of the principal value.

Expression (2.11.4) coincides with the probability of transition of the system from a state n' to a state n'', which is calculated by perturbation theory and summed up over all n''. It is not zero if the resonance conditions $E_{n''} - E_{n'} = \hbar\omega'$ are satisfied. This situation can be realized when $E_{n''}$ falls within the continuous spectrum, corresponding, for instance, to an ionized atom. Therefore, an additional broadening of the level $E_{n'}$ (2.11.4) is often referred to as ionization-induced broadening, even though this broadening is not necessarily caused by ionization.

We have thus clarified the role of those terms in the expression for $\chi^{(3)}$ which lead to high-order poles. They can be ignored but then one has to take into account Stark shifts of levels.

Even more peculiar terms can be seen in expression (2.10.2) for the cubic susceptibility. If we set in (2.10.2) $\omega'' = -\omega'''$ and $n = n''$, the second factor in the denominator either vanishes or becomes infinitesimal at $\delta \neq 0$. This means that an expansion using perturbation theory is not correct if $Ed \sim \delta \sim \hbar/\tau$, where τ is the relaxation time. Earlier we assumed that perturbation theory expansions hold up to fields $\mathcal{E} \sim \mathcal{E}_{\rm at}$ or $\mathcal{E}d \sim I_0$; in this particular case, however, they fail at much lower fields of the order of δd^{-1}.

Let us try to clarify the physical meaning of these anomalous terms.

If we separately consider those terms of (2.10.2) which produce these complications in the third order of perturbation theory, and then analyze their origin, we find that when the density matrix is expanded in powers of field (2.3.2), then secular terms (i.e., terms growing with time) appear already in the second order. These are diagonal matrix elements of the form $\rho_{nn}^{(2)}(t)$, whose meaning is the probability of finding the system at the nth level (i.e., the occupancy of the state with index n).

By virtue of (2.3.4),

$$\rho_{nn}^{(2)}(t) = \frac{1}{(\mathrm{i}\hbar)^2} \int\limits_{-\infty}^{t} \mathrm{d}t' \int\limits_{-\infty}^{t'} \mathrm{d}t'' [\mathcal{H}'(t'), [\mathcal{H}'(t''), \rho_0]]_{nn} \tag{2.11.5}$$

If we use the equations with relaxation for the density matrix, then integrals in t' and t'' display factors $\mathrm{e}^{-(t-t')/\tau}$ and $\mathrm{e}^{-(t'-t'')/\tau}$ which will lead to finite values of $\rho_{nn}^{(2)}$ at any t; however, these values will be higher at higher τ, that is, at small $\delta = \hbar/\tau$.

If the Gibbs density matrix

$$(\rho_0)_{nn} = \exp\left(\frac{\mathcal{F} - E_n}{\mathrm{k}T}\right)$$

and the matrix elements $\mathcal{H}'(t)$ in the interaction representation are substituted into (2.11.5), we obtain

$$\rho_{nn}^{(2)}(t) = -\frac{2\pi}{\hbar} \sum_{n'} |(\mathbf{dE}_{\omega''})_{n'n}|^2 \delta(E_n - E_{n'} - \hbar\omega'')$$
$$\times \left[\exp\left(\frac{\mathcal{F} - E_n}{\mathrm{k}T}\right) - \exp\left(\frac{\mathcal{F} - E_{n'}}{\mathrm{k}T}\right)\right] \int\limits_{-\infty}^{t} \mathrm{d}t' \tag{2.11.6}$$

We have used here the dipole approximation for $\mathcal{H}'$ and assumed the field to be harmonic, at a frequency ω''. Expression (2.11.6) is derived from

(2.11.5) after integration in t'', which produced the δ function. Integration over t' produces the very infinity that we have come across in the expression for $\chi^{(3)}$. If δ is finite, integration in t' gives a finite but very large value, of the order of $\hbar\delta^{-1}$.

Expression (2.11.6) can be rewritten in the form

$$\rho_{nn}^{(2)}(t) = -\sum_{n'}(w_{n'n}\rho_{nn}^{(0)} - w_{nn'}\rho_{n'n'}^{(0)}) \int_{-\infty}^{t} \mathrm{d}t' \tag{2.11.7}$$

where $w_{n'n} = w_{nn'}$ is the probability per unit time of transition of the system from a state n to a state n' in response to the applied field of frequency ω'', or of a transition from n' to n.

The meaning of the infinity which we found in (2.11.6) has now become clear. The occupancies of the levels $\boldsymbol{\rho}_{nn}$ over an infinitely long period can change quite substantially even in a very weak field.

If the populations at a time t_0 are $\boldsymbol{\rho}_{nn}(t_0)$, then at a time $t = t_0 + \Delta t$ they change to

$$\rho_{nn}(t) = \boldsymbol{\rho}_{nn}(t_0) - \sum_{n'}(w_{n'n}\boldsymbol{\rho}_{nn}(t_0) - w_{nn'}\boldsymbol{\rho}_{n'n'}(t_0))\Delta t$$

This means that ρ_{nn} satisfy the differential equations (balance equations)

$$\frac{\mathrm{d}\rho_{nn}(t)}{\mathrm{d}t} = -\sum_{n'} w_{n'n}(\boldsymbol{\rho}_{nn}(t) - \boldsymbol{\rho}_{n'n'}(t)) \tag{2.11.8}$$

The balance equations (2.11.8) make it possible to calculate, at any moment of time, diagonal elements of the density matrix, of which we can construct the density matrix $\boldsymbol{\rho}$ in the energy representation, which differs from the Gibbs matrix $\boldsymbol{\rho}_0$. If we use this diagonal density matrix instead of $\boldsymbol{\rho}_0$ in expressions of the type (2.9.2), we thereby take into account the "heating" of the system in the electromagnetic field and the effect of this "heating" on the susceptibilities χ. The matrix elements $\boldsymbol{\rho}_{nn}$ are slow functions of time, so that the corresponding susceptibilities are also slow functions of time. All other matrix elements involved in the calculation of χ vary with characteristic atomic frequencies of the order of $I_0\hbar^{-1}$; in the calculation of integrals over time, the diagonal density matrix $\boldsymbol{\rho}$ can be regarded as stationary.

We conclude that the terms of $\chi^{(3)}$ connected with the impossibility of using perturbation theory in the form in which it was formulated in Section 2.3, reflect the changes in the probabilities for the system to be in a specific state n; this factor must be taken into account separately, after which we can make use of the resulting expressions for the susceptibilities, without their "dangerous" terms but with (if required) corrected density matrix $\boldsymbol{\rho}_0$. The density matrix $\boldsymbol{\rho}$ may greatly differ from the equilibrium value $\boldsymbol{\rho}_0$, even for the effective temperature, if the resonance conditions $E_{n'} - E_n \simeq \hbar\omega''$ are satisfied.

Some of the qualitatively new effects which then arise will be conveniently illustrated using the so-called two-level system as an example.

2.12. Two-Level System Approximation

When calculating susceptibility or, in the general case, the response of the medium at resonance (including nonstationary response), it is expedient to use the two-level system approximation. The two-level system model is applicable when the frequency of the field is close to that of one of the transitions in the system.

This model allows a number of exact and nearly exact solutions and is therefore widely used in quantum electronics. The two-level model and models with a small number of levels are at the basis of the theory of lasers and masers. These topics are treated in courses of quantum electronics; here we will only give the most basic information and derive the most important formulas for susceptibilities under resonance conditions.

Let the resonance be met for only one pair of levels of an atom, or molecule, or a system. Then the populations of just these two levels differ substantially from the equilibrium values, since these two values are coupled via the electromagnetic field. In this case, equations (2.1.10) for the density matrix must be written for the matrix elements ρ_{11}, ρ_{22}, $\rho_{12} = \rho_{21}^*$. The other matrix elements $\rho_{nn'}$ can be set equal to zero, or they can be taken into account using perturbation theory, as we have already done above in the general theory. The matrix elements ρ_{11} or ρ_{22} are related by the normalization condition $\mathrm{Sp}\{\rho\} = \rho_{11}+\rho_{22} = 1$, or, to be precise, $\rho_{11}+\rho_{22} = \rho_{11}^{(0)}+\rho_{22}^{(0)}$. We can therefore write the equations for the nondiagonal matrix element ρ_{12} and for the difference between the level populations, $p = \rho_{22} - \rho_{11}$. As we have already mentioned in connection with equation (2.1.10), the relaxation constants generally have different values for a nondiagonal matrix element (relaxation time T_2) and for diagonal matrix elements, or for the difference between the populations (relaxation time T_1). Equations for the density matrix can be written in the interaction representation, in accordance with (2.1.10), as

$$
\begin{aligned}
i\hbar \frac{d\rho_{12}}{dt} &= [\mathscr{H}', \rho]_{12} - \frac{i\hbar}{T_2}\rho_{12} = -(\rho_{22} - \rho_{11})\boldsymbol{\mu} \\
&\times (\mathbf{E}\exp(-i(\omega_0 + \omega)t) + \mathbf{E}^*\exp(i(\omega - \omega_0)t)) - \frac{i\hbar}{T_2}\rho_{12} \\
i\hbar \frac{dp}{dt} &= i\hbar \frac{d\rho_{22}}{dt} - i\hbar \frac{d\rho_{11}}{dt} \\
&= -2\rho_{12}\boldsymbol{\mu}^*(\mathbf{E}\exp(-i(\omega - \omega_0)t) + \mathbf{E}^*\exp(i(\omega + \omega_0)t)) \\
&+ 2\rho_{21}\boldsymbol{\mu}(\mathbf{E}^*\exp(i(\omega + \omega_0)t) + \mathbf{E}\exp(-i(\omega - \omega_0)t)) - \frac{i\hbar}{T_1}(p - p_0)
\end{aligned}
\tag{2.12.1}
$$

These equations make use of the perturbation operator $\mathscr{H}'$ in the dipole approximation

$$
\mathscr{H}' = -\mathbf{d}\mathcal{E} = -\mathbf{d}(\mathbf{E}e^{-i\omega t} + \mathbf{E}^* e^{i\omega t})
$$

where ω is the mean frequency, $\mathbf{E}$ is the slowly varying complex field amplitude, and ω_0 denotes the transition frequency; also,

$$
\hbar\omega_0 = E_2 - E_1
$$

and $\boldsymbol{\mu} = \mathbf{d}_{12} = \mathbf{d}_{12}^*$ is the matrix element of the dipole moment operator. It is assumed that $\mathbf{d}_{11} = \mathbf{d}_{22} = 0$, that is, there is no permanent dipole moment. The difference between populations relaxes to the value p_0, determined by the conditions under which the system is excited. If the system is at the lowest level when the field is zero, then $p_0 = -1$; in the case of complete inversion, $p_0 = 1$. In the general case, p_0 is the stationary difference between the populations in zero field.

It is assumed that the detuning $\Delta = \omega - \omega_0$ is small. This means that the righthand sides of (2.12.1) have rapidly oscillating terms which vary at a frequency $\omega + \omega_0 \simeq 2\omega_0$. They make a small contribution at $\Delta \ll \omega_0$ and can be ignored. This is the resonance approximation, or the rotating-field approximation (also known as the approximation of rotating-polarization waves). This terminology dates back to the spin resonance theory, in which rapidly oscillating terms correspond to the interaction of spin with circularly polarized field of the sense reversed with respect to the sense of circular polarization of the resonant field. A linearly polarized field is presented as a sum of circularly polarized fields (left and righthanded), but only one circularly polarized component efficiently interacts with spin: the other component gives rapidly oscillating terms of the equations and can be neglected.

In the rotating-polarization-waves approximation, equations (2.12.1) can be rewritten as

$$\begin{aligned}\frac{\mathrm{d}r}{\mathrm{d}t} &= \frac{\mathrm{i}}{\hbar}p\,(\boldsymbol{\mu}\mathbf{E}^*) - \left(\frac{1}{T_2} + \mathrm{i}\Delta\right)r \\ \frac{\mathrm{d}p}{\mathrm{d}t} &= -\frac{4}{\hbar}\,\mathrm{Im}(\boldsymbol{\mu}^*\mathbf{E}r) - \frac{p - p_0}{T_1}\end{aligned} \tag{2.12.2}$$

In equations (2.12.2), r denotes a modified nondiagonal matrix element of the density matrix:

$$r = \rho_{12}\mathrm{e}^{-\mathrm{i}\Delta t}$$

The mean dipole moment of the system (or molecule, or atom) is

$$\overline{\mathbf{d}} = \mathrm{Sp}\{\mathbf{d}\rho\} = \mathbf{d}_{12}\rho_{21}\mathrm{e}^{-\mathrm{i}\omega_0 t} + \mathbf{d}_{21}\rho_{12}\mathrm{e}^{\mathrm{i}\omega_0 t} = \boldsymbol{\mu}r^*\mathrm{e}^{-\mathrm{i}\omega t} + \boldsymbol{\mu}^* r\mathrm{e}^{\mathrm{i}\omega t}$$

The susceptibility $\chi_{\alpha\beta}$ of a medium consisting of such molecules and containing n molecules in unit volume is found from the equation

$$\chi_{\alpha\beta}(\omega)E_\beta = n\mu_\alpha r^* \tag{2.12.3}$$

We will first consider the behavior of a two-level system over times less than the relaxation times T_1 and T_2. We can now set $T_1^{-1} = T_2^{-1} = 0$ in equations (2.12.2). The first integral of the resulting equations is

$$|2r|^2 + p^2 = \mathrm{const}$$

because the equations imply

$$\frac{2\,\mathrm{d}}{\mathrm{d}t\,(rr^*) + p\,\frac{\mathrm{d}p}{\mathrm{d}t} = 0}$$

Referring to this first integral, one usually says that the length of the Bloch vector is conserved. The Bloch vector is the vector in the three-dimensional space, with coordinates $\mathrm{Re}(2r)$, $\mathrm{Im}(2r)$, p.

The initial conditions $r(0) = 0$, $p(0) = -1$, which signify that the system is in the lowest state at $t = 0$, yield at $\Delta = 0$ and $\mathbf{E} = \mathrm{const}$ the solution

$$p(t) = -\cos\frac{2|\boldsymbol{\mu}^*\mathbf{E}|}{\hbar}t = -\cos\Omega t$$

where $\Omega = 2|\boldsymbol{\mu}^*\mathbf{E}|/\hbar$ is the so-called Rabi frequency. The population of the upper level then is

$$\rho_{22} = \frac{1}{2}(1 + p) = \sin^2\frac{\Omega t}{2}$$

If $\Delta \neq 0$, we obtain

$$\rho_{22} = \frac{\Omega^2}{\Omega^2 + \Delta^2} \sin^2 \frac{\sqrt{\Omega^2 + \Delta^2}}{2} t$$

This means that the population of the upper level oscillates at a frequency $(\Omega^2 + \Delta^2)^{1/2}$, and its maximum value is $\Omega^2/(\Omega^2 + \Delta^2)$. If $\Delta = 0$, the system periodically occupies the upper level.

If we take relaxation into account, the oscillations $p(t)$ and ρ_{22} damp out, and $p(t)$ tends to a constant value which depends on the value of the field and the relaxation times T_1 and T_2. Relaxation times may vary in a wide range; typically, $T_2 \ll T_1$. Their values depend on the specific relaxation mechanism but the upper limit is dictated by the spontaneous decay:

$$(T_{1\max})^{-1} = 2(T_{2\max})^{-1} = 4|\boldsymbol{\mu}|^2\omega_0^2/3\hbar c^3$$

Assuming $\dot{r} = \dot{p} = 0$ in (2.12.2), we obtain algebraic equations for stationary values of p and r. Their solution gives

$$p = \frac{p_0}{1+F} \qquad r = \frac{\mathrm{i}(\boldsymbol{\mu}\mathbf{E}^*)p}{\hbar(T_2^{-1} + \mathrm{i}\Delta)}$$
$$E = \frac{4|(\boldsymbol{\mu}^*\mathbf{E})|^2 T_1}{\hbar^2 T_2(T_2^{-2} + \Delta^2)} = \frac{\Omega^2 T_1 T_2}{T_2^2(T_2^{-2} + \Delta^2)} \tag{2.12.4}$$

According to the solutions of (2.12.4), $p \to 0$ in the strong field, in other words, populations of the levels equal out. This is the familiar saturation effect, which reduces both the effective susceptibility of the system and the absorption.

By virtue of (2.12.3),

$$\chi_{\alpha\beta}(\omega, \{E\})E_\beta = n\mu_\alpha r^* = -\frac{\mathrm{i}pn\mu_\alpha\mu_\beta^*}{\hbar(T_2^{-1} - \mathrm{i}\Delta)} E_\beta$$

whence

$$\chi_{\alpha\beta}(\omega, \{E\}) = -\frac{\mathrm{i}pn\mu_\alpha\mu_\beta^*}{\hbar(T_2^{-1} - \mathrm{i}\Delta)}$$

Since we can choose real μ_α, the result is $\chi_{\alpha\beta} = \chi_{\beta\alpha}$. Here p is a function of the field $\mathbf{E}$, as follows from (2.12.4). At the exact resonance, i.e., at $\Delta = 0$, χ is purely imaginary.

In a weak field ($\mathbf{E} \to 0$) at $\Delta = 0$, we have

$$\chi_{\alpha\beta}(\omega_0) = \chi_{0\alpha\beta} = -\mathrm{i}n\mu_\alpha\mu_\beta p_0 T_2 \hbar^{-1} = \mathrm{i}\chi''_{0\alpha\beta}$$

The imaginary part of χ also determines the losses at $\mathbf{E} \neq 0$, $\Delta \neq 0$:

$$\begin{aligned}\chi_{\alpha\beta}(\omega) &= \chi_{0\alpha\beta}\frac{T_2^{-1} + \mathrm{i}\Delta}{T_2(\Delta^2 + T_2^{-2}(1 + \Omega^2 T_1 T_2))} \\ \mathrm{Im}\chi_{\alpha\beta}(\omega) &= \chi''_{0\alpha\beta}\frac{1}{T_2^2(\Delta^2 + T_2^{-2}(1 + \Omega^2 T_1 T_2))}\end{aligned} \tag{2.12.5}$$

As we see from equation (2.12.5), the shape of the absorption line is still Lorentzian but the linewidth increases with increasing field intensity, since the Rabi frequency Ω is proportional to the field strength. In resonance, absorption is reduced by a factor of $(1 + \Omega^2 T_1 T_2)^{-1}$.

An expression of the type (2.12.5) with the saturation effect ignored can be derived from the general formula for $\chi^{(1)}$ in the dipole approximation (2.8.3) if we retain in it only the resonance terms and also set $\delta = \hbar T_2^{-1}$.

The two-level approximation can be generalized to the case of several resonance fields and several resonance transitions. This case requires the use of a three-level system approximation, or rather that of a system with a finite (not too large) number of levels. One particular case here is that of a system with degenerate levels, which is typical for atoms and molecules: these are freely oriented systems for which we always have degeneracy of levels in the quantum number of the projection of momentum onto an arbitrary axis. In this approximation, equations for the density matrix reduce to a finite (small) number of equations for the matrix elements of the density matrix.

The two-level approximation is used not only for allowed but also for forbidden transitions when one studies the resonant two-photon absorption or induced Raman scattering of light. Let us briefly consider the ordinary method of solution, using as example the Raman light scattering.

Let a system have a pair of levels: $n' = 2$ (upper level) and $n = 1$ (lower level). As before, ω_0 is the transition frequency; the transition, however, is forbidden, that is, $\mathbf{d}_{12} = 0$. The system is placed in a biharmonic field

$$\mathcal{E}(t) = \mathbf{E}_\omega \mathrm{e}^{-\mathrm{i}\omega t} + \mathbf{E}^*_\omega \mathrm{e}^{\mathrm{i}\omega t} + \mathbf{E}_{\omega'} \mathrm{e}^{-\mathrm{i}\omega' t} + \mathbf{E}^*_{\omega'} \mathrm{e}^{\mathrm{i}\omega' t}$$

where ω' is the pumping frequency and ω is the frequency of the Stokes wave. These frequencies are related as

$$\omega' - \omega = \omega_0 + \Delta$$

where $\Delta \ll \omega_0$ is a small detuning.

The equations for the density matrix make it possible to express the nondiagonal elements of the density matrix of the form $\rho_{1k} = \rho^*_{k1}$ and $\rho_{2k} = \rho^*_{k2}$ ($k \neq 1, 2$) in terms of ρ_{11}, ρ_{22}, $\rho_{12} = \rho^*_{21}$, assuming $|\rho_{1k}|$, $|\rho_{2k}| \ll \rho_{11}$, ρ_{22} and $|\rho_{12}|$. Substituting the expressions for the matrix elements ρ_{1k} and ρ_{2k} into the equations for ρ_{11}, ρ_{22}, $\rho_{12} = \rho_{21}^*$, we obtain a closed set of equations of the type of (2.12.1). Having solved it, we find the dipole moment of the system and its susceptibility.

The equation for the matrix element ρ_{1k} in the interaction representation is

$$\begin{aligned} i\hbar \frac{d\rho_{1k}}{dt} &= -\rho_{12}\mathscr{H}'_{2k} - \rho_{11}\mathscr{H}'_{1k} \\ &= (\rho_{12} d_{2k} \exp(i\omega_{2k}t) + \rho_{11} d_{1k} \exp(i\omega_{1k}t))\boldsymbol{\mathcal{E}}(t) \end{aligned}$$

where only terms linear in the field have been retained in the righthand part. Here and in what follows ω_{kl} denotes transition frequencies:

$$\hbar\omega_{kl} = E_k - E_l$$

We integrate the equation for ρ_{1k} in time, assuming ρ_{12} and ρ_{11} to be constant since they vary slowly, and we also ignore the value assumed by the function at the lower limit of integration, taking into account thereby the small relaxation in the system. As a result, the matrix element ρ_{1k} (and similarly ρ_{2k}) is expressed through ρ_{12}, ρ_{11}, ρ_{22}. Substituting these expressions into the equations for ρ_{12}, ρ_{21}, ρ_{11}, and ρ_{22}, we arrive at the equations for $p = \rho_{22} - \rho_{11}$ and $r = \rho_{12}\,e^{i\Delta t}$:

$$\begin{aligned} i\hbar \frac{dr^*}{dt} &= (E^*_{\omega\alpha}(\varkappa^{(1)}_{\alpha\beta}(\omega) - \varkappa^{(2)}_{\alpha\beta}(\omega))E_{\omega\beta} + E^*_{\omega'\alpha}(\varkappa^{(1)}_{\alpha\beta}(\omega') \\ &\quad - \varkappa^{(2)}_{\alpha\beta}(\omega'))E_{\omega'\beta})r^* + E^*_{\omega\alpha}\varkappa_{\alpha\beta}(\omega')E_{\omega'\beta}p - i\hbar(T_2^{-1} - i\Delta)r^* \qquad (2.12.6) \\ i\hbar \frac{dp}{dt} &= -4i\,\mathrm{Im}(E^*_{\omega\alpha}\varkappa_{\alpha\beta}(\omega')E_{\omega'\beta}r) - i\hbar(p - p_0)T_1^{-1} \qquad (2.12.7) \end{aligned}$$

We have used above the notation

$$\varkappa^{(1)}_{\alpha\beta}(\omega) = \frac{1}{\hbar}\sum_k \left(\frac{(d_\alpha)_{1k}(d_\beta)_{k1}}{\omega_{k1} - \omega} + \frac{(d_\beta)_{1k}(d_\alpha)_{k1}}{\omega_{k1} + \omega}\right) \qquad (2.12.8)$$

$$\varkappa_{\alpha\beta}(\omega') = \frac{1}{\hbar}\sum_k \left(\frac{(d_\alpha)_{2k}(d_\beta)_{k1}}{\omega_{k1} - \omega'} + \frac{(d_\beta)_{2k}(d_\alpha)_{k1}}{\omega_{k2} + \omega'}\right) \qquad (2.12.9)$$

The expressions for $\varkappa^{(2)}_{\alpha\beta}(\omega)$, $\varkappa^{(1)}_{\alpha\beta}(\omega')$, and $\varkappa^{(2)}_{\alpha\beta}(\omega')$ are similar to equality (2.12.8) for $\varkappa^{(1)}_{\alpha\beta}(\omega)$ after the replacement $1 \to 2$, $\omega \to \omega'$.

If we compare the expression for $\varkappa^{(1)}_{\alpha\beta}(\omega)$ with (2.8.3), it becomes clear that $\varkappa^{(1)}_{\alpha\beta}(\omega)E_{\omega\beta}$ is the dipole moment induced in the system (which is in its ground state $n = 1$) by the field $\mathbf{E}_\omega$ at a frequency ω. Hence, $\varkappa^{(1)}_{\alpha\beta}(\omega)$ is the polarizability of the system (e.g., a molecule) in the ground state $n = 1$ at the frequency ω.

According to (2.12.6), the fields $\mathbf{E}_\omega$ and $\mathbf{E}_{\omega'}$ produce the Stark level shifts ε_1 and ε_2:

$$\varepsilon_1 = -E^*_{\omega\alpha}\varkappa^{(1)}_{\alpha\beta}(\omega)E_{\omega\beta} - E^*_{\omega'\alpha}\varkappa^{(1)}_{\alpha\beta}(\omega')E_{\omega'\beta}$$
$$\varepsilon_2 = -E^*_{\omega\alpha}\varkappa^{(2)}_{\alpha\beta}(\omega)E_{\omega\beta} - E^*_{\omega'\alpha}\varkappa^{(2)}_{\alpha\beta}(\omega')E_{\omega'\beta}$$

which change the transition frequency by $\hbar^{-1}\varepsilon = \hbar^{-1}(\varepsilon_2 - \varepsilon_1)$.

Formula (2.12.9) is the well-known Raman scattering tensor, or the Raman polarizability tensor of the system. Using the expressions for the matrix elements ρ_{1k} and ρ_{2k}, we can find the polarization at the frequencies ω, ω', $2\omega' - \omega$, and $2\omega - \omega'$, that is, at the Stokes frequency, the pumping frequency, the anti-Stokes frequency, and the second Stokes frequency. Indeed,

$$\mathcal{P} = n\mathrm{Sp}\{\rho\mathbf{d}\} = n\sum_k(\rho_{1k}\mathbf{d}_{k1}\exp(\mathrm{i}\omega_{k1}t) + \rho_{2k}\mathbf{d}_{k2}\exp(\mathrm{i}\omega_{k2}t)$$
$$+ \rho_{k1}\mathbf{d}_{1k}\exp(\mathrm{i}\omega_{1k}t) + \rho_{k2}\mathbf{d}_{2k}\exp(\mathrm{i}\omega_{2k}t))$$

where n is the number of molecules per unit volume.

For example, the polarization at the Stokes frequency ω under stationary conditions, when $\dot{p} = \dot{r} = \dot{r}^* = 0$, is

$$P_{\omega\alpha} = \varkappa_{\alpha\beta}(\omega')E_{\omega'\beta}r = \frac{\mathrm{i}pn}{\hbar(T_2^{-1} + \mathrm{i}(\Delta - \hbar^{-1}\varepsilon))}\varkappa_{\alpha\beta}(\omega')$$
$$\times E_{\omega'\beta}\,(E^*_{\omega\delta}\varkappa_{\delta\gamma}(\omega')E_{\omega'\gamma})^*$$

Hence, the Raman susceptibility is

$$\chi^{(3)}_{\alpha\beta\gamma\delta}(\omega = \omega' - \omega' + \omega)$$
$$= \frac{\mathrm{i}pn}{\hbar(T_2^{-1} + \mathrm{i}(\Delta - \hbar^{-1}\varepsilon))}\varkappa_{\alpha\beta}(\omega')\varkappa^*_{\delta\gamma}(\omega') \qquad (2.12.10)$$

and is expressed in terms of the Raman scattering tensor $\chi_{\alpha\beta}(\omega')$.

In general, the stationary difference between populations, p, is not equal to the equilibrium value p_0 and is found from equation (2.12.7). By analogy to the case of one-photon resonance, the saturation effect may be observed. In relatively weak fields, when $p \simeq p_0$ and we can ignore Stark shifts ε_1 and ε_2, expression (2.12.10) coincides with the expression (2.10.2) for $\chi^{(3)}$ if it retains only the resonance terms, that is, the terms whose denominators contain the factor $\omega_0 + \omega - \omega' = -\Delta$.

By virtue of (2.12.10), $\chi''_{\alpha\beta}(\omega, \{E\}) < 0$, that is, amplification takes place for $p < 0$, when there is no population inversion of levels, for example, when the system is in thermal equilibrium. This is a well-known effect of amplification of the Stokes wave in induced Raman scattering of light.

The intensity and the cross section of the spontaneous Raman scattering of light are also expressed in terms of the tensor $\chi_{\alpha\beta}(\omega')$. The intensity of the spontaneous scattering can be given in terms of the Raman susceptibility $\chi^{(3)}_{\alpha\beta\gamma\delta}(\omega = \omega' - \omega' + \omega)$ and the pumping intensity, or in terms of $\chi''_{\alpha\beta}(\omega, \{E\})$, which is a generalization of the fluctuation–dissipation theorem to scattering.

3

Plasma in Electromagnetic Fields

3.1. Interaction between Free Charges and Electromagnetic Fields

We have mentioned several times in the preceding chapters that plasma, or partially or completely ionized gas, is a special case, so the interaction of the electromagnetic field with plasma must be treated separately. Even the estimates of the linear and nonlinear susceptibilities given in Chapter 2 are only valid for the contribution of bound electrons.

A number of specific features of the interaction between the electron gas and the field manifest themselves even in the simplest problem of the interaction between free charges and the electromagnetic field.

In contrast to all other cases of interaction of the field with atoms, molecules or condensed media, the classical theory is adequate for treating free charges and the plasma in an electromagnetic field. The conditions of applicability of the classical theory, given below, are satisfied in most cases that are of interest for radiophysics and laser physics.

The classical equation of motion of an electron or some other charged particle in electromagnetic field has the form

$$\frac{d\mathbf{p}}{dt} = e\boldsymbol{\mathcal{E}}(\mathbf{r}, t) + \frac{e}{c}[\mathbf{v}\boldsymbol{\mathcal{H}}(\mathbf{r}, t)]$$

where $\boldsymbol{\mathcal{E}}$ and $\boldsymbol{\mathcal{H}}$ are the electric and magnetic fields and $\mathbf{p}$ is the electron momentum.

Neglecting relativistic terms of the order of v^2/c^2, we obtain

$$\mathbf{p} = \frac{m\mathbf{v}}{\sqrt{1 - v^2/c^2}} \simeq m\mathbf{v}$$

$$m\ddot{\mathbf{r}} = e\boldsymbol{\mathcal{E}}(\mathbf{r}, t) + \frac{e}{c}[\mathbf{v}, \boldsymbol{\mathcal{H}}(\mathbf{r}, t)] \tag{3.1.1}$$

The Lorentz force is of the order of v/c and is taken into account in equation (3.1.1).

Let us consider the behavior of a free charge in the field of a standing wave. Let

$$\mathcal{E} = \mathbf{E}(\mathbf{r}) \cos \omega t \qquad \mathcal{H} = \mathbf{H}(\mathbf{r}) \sin \omega t$$

By virtue of Maxwell's equations,

$$\operatorname{curl} \mathbf{E}(\mathbf{r}) = -\frac{\omega}{c} \mathbf{H}(\mathbf{r}) \tag{3.1.2}$$

The Lorentz force in the lowest order in v/c can be neglected and the dependence of the electric field on the coordinate $\mathbf{r}$ of the electron can be ignored. Hence,

$$m\ddot{\mathbf{r}} = e\mathbf{E} \cos \omega t$$

so that

$$\mathbf{v} = \dot{\mathbf{r}} = \frac{e\mathbf{E}}{m\omega} \sin \omega t + \mathbf{v}_0$$

where $\mathbf{v}_0$ is determined by the initial velocity of the electron. Correspondingly,

$$\mathbf{r} = \mathbf{r}_0 + \mathbf{v}_0 t - \frac{e\mathbf{E}}{m\omega^2} \cos \omega t = \mathbf{r}_0 + \delta\mathbf{r}$$

The ratio of the displacement $\delta\mathbf{r}$ over one period of the field to the wavelength is of the order of v/c since $\delta r \sim v\lambda/c$. Therefore, the displacement $\delta\mathbf{r}$ over one period of the field is always small in nonrelativistic problems as compared with the wavelength; furthermore, the field acting on an electron differs only slightly from a harmonic field of frequency ω.

An electron oscillates in the field, and its mean kinetic energy of oscillatory motion is

$$\epsilon_k = \frac{m v_k^2}{2} = \frac{e^2 E^2}{4m\omega^2}$$

We will now take into account in equation (3.1.1) only the terms of first order in v/c. To do this, we expand the electric field in powers of $\delta\mathbf{r}$ and retain only the linear term. After averaging it over the field period, we obtain

$$m\overline{\ddot{\mathbf{r}}} = -\frac{e^2}{2mw^2}\Big((\mathbf{E}(\mathbf{r}_0)\nabla)\mathbf{E}(\mathbf{r}_0) - \frac{w}{c}[\mathbf{E}(\mathbf{r}_0)\mathbf{H}(\mathbf{r}_0)]\Big)$$

Using Maxwell's equation (3.1.2) and the identity

$$[\mathbf{E} \operatorname{curl} \mathbf{E}] = \frac{1}{2}\nabla E^2 - (\mathbf{E}\nabla)\mathbf{E}$$

we find

$$m\overline{\ddot{\mathbf{r}}} = -\frac{e^2}{4m\omega^2} \nabla E^2(\mathbf{r})$$

This equation shows that on average, electrons feel a force which pushes them out of the region of strong field and makes them form bunches around the nodes of the electric field.

Let us consider now the motion of electrons in the field of a traveling wave:

$$\mathcal{E}(\mathbf{r},t) = E\cos(\omega t - kz)\mathbf{e}_x$$
$$\mathcal{H}(\mathbf{r},t) = E\cos(\omega t - kz)\mathbf{e}_y$$

We have taken into account here that the amplitudes of the electric and magnetic fields in a traveling plane wave equal each other (in the Gaussian system of units), and $\mathbf{e}_x$ and $\mathbf{e}_y$ are unit polarization vectors of the field along the axes x and y.

The equations of motion of an electron (3.1.1) in the coordinate representation are

$$m\ddot{x} = eE\left(1 - \frac{v_z}{c}\right)\cos(\omega t - kz)$$
$$m\ddot{y} = 0$$
$$m\ddot{z} = \frac{ev_x}{c}E\cos(\omega t - kz)$$

In the zero approximation, the acceleration of an electron is directed along x. Electron oscillations are polarized along the x axis, and the electron emits radiation as described by classical electrodynamics. The intensity of this radiation is

$$\dot{\varepsilon} = \frac{2|\ddot{\mathbf{d}}_\omega|^2}{3c^3} = \frac{1}{3c^3}\left|\frac{e^2\mathbf{E}}{m}\right|^2$$

where $\mathbf{d}_\omega$ is the Fourier component of the classical dipole moment of the electron.

The scattering of light by a free electron is the Compton effect if we take into account the changes in the frequency of the scattered light. In the classical analysis, when this frequency shift is ignored, we deal with the so-called Thomson scattering whose cross section is

$$\sigma_\omega = \frac{\dot{\varepsilon}}{I} = \frac{8\pi}{3}\left(\frac{e^2}{mc^2}\right)^2 = \frac{8\pi}{3}r_0^2 = 6.65 \times 10^{-25}\ \mathrm{cm}^2 \qquad (3.1.3)$$

Here we have taken into account that the energy flux in the traveling wave is

$$I = \frac{c}{8\pi}E^2$$

and $r_0 = e^2/mc^2$ is the classical electron radius.

In the next order of v/c, we introduce the Lorentz force in the equations of motion along the z axis:

$$m\ddot{z} = \frac{e^2E^2}{m\omega c}\cos\omega t\sin\omega t = \frac{e^2E^2}{2m\omega c}\sin 2\omega t$$

This equation shows that in the first approximation in v/c an electron oscillates along z at twice the field frequency, 2ω. On the whole, an electron describes a Lissajous' figure of eight in the x–z plane. This figure is strongly elongated in the direction of x, since the ratio of oscillation amplitudes along the z and x axes is of the order of v/c.

Higher harmonics appear in higher-order approximations in v/c but these must be calculated on the basis of the relativistic equation of motion since they are of orders v^2/c^2 and higher. It is of interest that nonlinear effects vanish for the circular polarization of the field: in this case electron's velocity is invariably parallel to the direction of the magnetic field, and the Lorentz force is always zero. The scattering cross section for the second harmonic can be found using the equation of motion of the electron moving along the z axis:

$$\sigma_{2\omega} = \sigma_\omega \frac{\varepsilon_k}{mc^2} \ll \sigma_\omega$$

For a free electron interacting with the electromagnetic field, the classical theory is valid as long as the inequality

$$\hbar\omega \ll mc^2$$

holds. If this inequality is satisfied, the scattering of electromagnetic waves by free electrons is described by the Thomson formula (3.1.3) both in the quantum and the classical theories, and the quantum corrections are of the order of $\hbar\omega/mc^2$. The ratio $\hbar\omega/mc^2$ is small down to the gamma radiation range, when the wavelength becomes of the order of the so-called Compton wavelength $\hbar/mc$ (about 10^{-10} cm).

A free electron can scatter electromagnetic field (in quantum terms, it scatters photons) but cannot absorb or emit light quanta. Physically, this constraint is implied by the fact that energy and momentum conservation laws

$$\varepsilon = \varepsilon' + \hbar\omega \qquad \mathbf{p} = \mathbf{p}' + \hbar\mathbf{k}$$

(where $\hbar k$ and $\hbar\omega$ are the momentum and energy of the photon, $\mathbf{p}$ and ε are the momentum and energy of the electron before the emission of a photon, and $\mathbf{p}'$ and ε' are those after the photon emission) cannot be satisfied as long as

$$\omega/k > v$$

In vacuum, however, the phase velocity $\omega/k = c$ of the wave is always greater than the velocity v of the electron. An electron is allowed to emit and absorb light only when it moves in matter at a velocity exceeding the phase velocity of light (the Vavilov–Čerenkov effect).

The situation is very different when electrons collide with other particles. In this case electrons can transfer a part of their momentum to a colliding particle so that a process accompanied by emission or absorption of a photon can satisfy the conservation laws. Hence, a collision of electrons results in absorption or emission of light (the well-known bremsstrahlung emission). If the field is nonzero, the emission is induced, not spontaneous; furthermore, an inverse process also occurs: absorption whose probability can be related, as in the ordinary case, with that of the direct process (the bremsstrahlung and induced bremsstrahlung emission). The net result of all these processes is the absorption of the electromagnetic energy by electrons.

Electrons in plasma can collide both with other electrons and with ions and neutral atoms. However, electron–electron collisions make almost no contribution to absorption since they do not change the total momentum of electrons. Moreover, the total current $\mathbf{j}$ in the system also remains unchanged, since it differs from the total momentum only in the universal factor e/m. However, the work done by the field on the system, averaged over one period, $\overline{\mathcal{E}\mathrm{j}}$, is zero which means no absorption (or emission). In contrast to this, collisions of electrons with ions and neutral atoms (or molecules) change the total current owing to the difference in the masses of the colliding particles, despite the conservation of the total momentum; they are the main cause of absorption, as we will show below.

In addition to the electromagnetic wave, an electron moving in a gas also feels the force exerted by the particles of gas. This is the interaction of three bodies: a photon, an electron and an atom; this interaction is responsible for the absorption of the electromagnetic radiation.

A free electron in the field of the wave is accelerated and decelerated at different moments of time and average energy transfer is zero. Energy is absorbed, on average, only at the expense of collisions in which momentum changes jumpwise; this absorption means that energy of electrons increases. Let us illustrate this by a simple model of elastic collisions with backscattering.

We assume that collisions occur over a time $\tau \ll 2\pi/\omega$. Let the masses M of the scattering particles be large as compared with the electron mass m. Collisions are then elastic and the electron energy is conserved.

The momentum of a free electron in a high-frequency field of frequency ω varies harmonically:

$$\dot{\mathbf{p}} = e\mathbf{E}\cos\omega t$$

$$\mathbf{p} = \mathbf{p}_0 + \frac{e\mathbf{E}}{\omega}\sin\omega t$$

The average electron energy thus is

$$\bar{\varepsilon} = \frac{\overline{p^2}}{2m} = \frac{1}{2m}\overline{\left(p_0^2 + \frac{2e}{\omega}(\mathbf{p}_0\mathbf{E})\sin\omega t + \frac{e^2}{\omega^2}E^2\sin^2\omega t\right)}$$

$$= \frac{p_0^2}{2m} + \frac{e^2E^2}{4m\omega^2} = \varepsilon_0 + \varepsilon_k$$

Let a collision occur at a time t_0 and result in reversing the momentum of the electron:

$$\mathbf{p}(t_0 + 0) = -\mathbf{p}(t_0 - 0)$$

At $t > t_0$, we have

$$\mathbf{p}(t) = -\mathbf{p}_0 - \frac{2e\mathbf{E}}{\omega}\sin\omega t_0 + \frac{e\mathbf{E}}{\omega}\sin\omega t$$

The average electron energy after collision is

$$\bar{\varepsilon}' = \frac{1}{2m}\left(\mathbf{p}_0 + \frac{2e\mathbf{E}}{\omega}\sin\omega t_0\right)^2 + \varepsilon_k$$

and the change in energy is

$$\Delta\varepsilon = \bar{\varepsilon}' - \bar{\varepsilon} = \frac{2e}{m\omega}(\mathbf{p}_0\mathbf{E})\sin\omega t_0 + \frac{2e^2E^2}{m\omega^2}\sin^2\omega t_0$$

This change in energy depends on the phase of the field at the time t_0 and can be positive or negative. However, the average energy transfer averaged over phases ωt_0 is

$$\overline{\Delta\varepsilon} = \frac{e^2E^2}{m\omega^2} = 4\varepsilon_k > 0$$

If all collisions resulted in backscattering, then

$$\frac{d\bar{\varepsilon}}{dt} = \frac{e^2E^2}{m\omega^2}\nu(p) \tag{3.1.4}$$

where $\nu(p)$ is the collision frequency which in general depends on the electron's momentum (velocity).

Formula (3.1.4) is obtained by applying a qualitative analysis to a very simplified model. It is thus obvious that the assumption $\nu \ll \omega$ is chosen in order to have the effect of each individual collision averaged over one period of the field. Nevertheless, it gives a result of correct order of magnitude when all collisions, not only those with backscattering and $\nu \ll \omega$, are considered. This conclusion follows from the expression for field energy dissipation in the plasma, derived using the general formula of Chapter 1,

$$\overline{\dot{Q}} = \frac{\omega}{8\pi}\varepsilon''_{\alpha\beta}E^*_\alpha E_\beta$$

and from the expression for the complex dielectric permittivity of the plasma which we will derive in the section that follows.

3.2. Elementary Theory of Dielectric Permittivity of Plasma

Formally, collisions with electrons can be taken into account by adding to the equation of motion of an electron in the field of the electromagnetic wave the terms describing friction:

$$\ddot{\mathbf{r}} + \nu\dot{\mathbf{r}} = e\mathcal{E}(t)/m \tag{3.2.1}$$

Here ν is the frequency of collisions which determines the rate of momentum transfer in collisions. In general, it is a function of velocity; however, equation (3.2.1) uses the mean (effective) collision frequency.

The frequency of collisions can be given in terms of the collision cross section σ_t and scatterer concentration n':

$$\nu(v) = n' v \sigma_t(v) \tag{3.2.2}$$

The cross section σ_t is the so-called transport cross section which is not identical to the total cross section σ. If $\sigma(v, \theta)$ is the differential cross section of scattering to an angle θ, then

$$\sigma(v) = 2\pi \int_0^\pi \sigma(v, \theta) \sin\theta \, d\theta$$

$$\sigma_t(v) = 2\pi \int_0^\pi \sigma(v, \theta)(1 - \cos\theta) \sin\theta \, d\theta$$

The factor $(1 - \cos\theta)$ takes into account the fact that in an elastic collision, a change in momentum is maximum at large scattering angles and tends to zero at small angles.

For slow electrons whose energy is less than the atomic scale $I_0 \sim 13.6\,\text{eV}$, the cross section of collisions with neutral atoms (or molecules) σ_t is of the order of $a_0^2 \sim 10^{-16}\,\text{cm}^2$. For fast electrons, this cross section drops as energy increases unless inelastic collisions begin to play a significant role, resulting in excitations of atoms.

The cross section of collisions with ions is much higher for slow ions. In accordance with the well-known Rutherford formula for the differential cross section of scattering of electrons by the Coulomb potential Ze/r,

$$\sigma(v, \theta) = \left(\frac{Ze^2}{2mv}\right)^2 \Big/ \sin^4\frac{\theta}{2}$$

the total cross section σ and the transport cross section σ_t diverge because $\sigma(v, \theta)$ increases at small scattering angles θ. If we take into account the

Debye screening of ion's charge Ze in a plasma with electron density n, the electron–ion interaction energy becomes

$$U(r) = \frac{Ze^2}{r} \exp(-r/r_0)$$

where k is the Boltzmann constant, T is temperature and

$$r_0 = \sqrt{\frac{\mathrm{k}T}{4\pi ne^2}} \tag{3.2.3}$$

is the so-called Debye–Hückel screening radius. The total and transport cross sections taking account of screening are finite, of the order of $r_0^2 \gg a_0^2$ for very slow electrons whose momentum satisfies the inequality $p \ll \hbar/r_0$.

In principle, cross sections $\sigma(v)$ and $\sigma_t(v)$ can be calculated in quantum mechanics. For neutral molecules and atoms which may dominate scattering in weakly ionized plasma, the cross section depends weakly on velocity and angle, and can be set equal to

$$\sigma(v,\theta) = a^2/4 \qquad \sigma_t(v) = \sigma(v) = \pi a^2$$

This means that a molecule behaves as a hard sphere of radius a. By virtue of formula (3.2.2), therefore, the collision frequency is proportional to velocity. If this is taken into account, the dielectric permittivity of the plasma and especially its imaginary part (which determines absorption) depend on frequency in a very complicated manner. Using equation (3.2.1) with the effective collision frequency ν is, however, a good approximation.

For a harmonic field

$$\mathcal{E}(t) = \mathbf{E}\exp(-\mathrm{i}\omega t) + \mathbf{E}^* \exp(\mathrm{i}\omega t)$$

we obtain

$$\mathbf{r}(t) = \mathbf{R}\exp(-\mathrm{i}\omega t) + \mathbf{R}^* \exp(\mathrm{i}\omega t)$$
$$\mathbf{R} = -e\mathbf{E}/m\omega(\omega + \mathrm{i}\nu)$$

Since $e\mathbf{R}$ is the amplitude of the dipole moment per electron, we find the susceptibility and dielectric permittivity of the plasma,

$$\chi(\omega)\mathbf{E} = ne\mathbf{R}$$
$$\chi(\omega) = -ne^2/m\omega(\omega + \mathrm{i}\nu)$$
$$\varepsilon(\omega) = 1 + 4\pi\chi = 1 - 4\pi ne^2/m\omega(\omega + \mathrm{i}\nu)$$

where n is the electron concentration. The real and imaginary parts of $\varepsilon(\omega)$ are

$$\begin{aligned} \varepsilon'(\omega) &= 1 - 4\pi ne^2/m(\omega^2 + \nu^2) \\ \varepsilon''(\omega) &= 4\pi ne^2\nu/m\omega(\omega^2 + \nu^2) \end{aligned} \tag{3.2.4}$$

respectively. The imaginary part of ε determines the losses and is related to the conductivity $\sigma(\omega)$ by the relation

$$\varepsilon''(\omega) = \frac{4\pi}{\omega}\sigma(\omega)$$

If $\nu \ll \omega$, we have

$$\varepsilon(\omega) \simeq 1 - \frac{4\pi ne^2}{m\omega^2} = 1 - \frac{\omega_0^2}{\omega^2} \tag{3.2.5}$$

where $\omega_0 = (4\pi ne^2/m)^{1/2}$ is the so-called plasma frequency.

When taking into account ions of mass M and charge Ze, we obtain

$$\varepsilon(\omega) = 1 - \frac{4\pi ne^2}{m\omega^2} - \frac{4\pi n'(Ze)^2}{M\omega^2}$$

where n' is the ion concentration. The ion contribution is small compared with the electron contribution since $m \ll M$, so we will hereafter ignore it.

The classical treatment of the behavior of electrons in the field, taking collisions into account, is only valid if the energy of radiation quanta is much smaller than the kinetic energy of the electron, that is, if

$$\hbar\omega \ll \mathrm{k}T$$

This becomes clear from the fact that the collision frequency ν depends on velocity, that is, on the energy of particles. If, however, $\hbar\omega \gtrsim \mathrm{k}T$, the particle's energy before and after a quantum was absorbed differ greatly, so that it is not clear at all which frequency should be used in (3.2.1) and (3.2.4). A more complete analysis shows, however, that formulas of the type of (3.2.4) remain valid in many cases even when $\hbar\omega \gtrsim \mathrm{k}T$, but ν must then be treated as a function of ω. We have already indicated that the collision frequency must be calculated by quantum methods but the absorption of electromagnetic waves can be calculated classically as well.

Let us briefly consider electromagnetic waves in a plasma with dielectric permittivity (3.2.5). Maxwell's equations for monochromatic plane waves of the form

$$\mathcal{E}(\mathbf{r}, t) = \mathbf{E}\exp(-\mathrm{i}\omega t + \mathrm{i}\mathbf{kr})$$

are the vector-algebraic equations

$$[\mathbf{kH}] = -\frac{\omega\varepsilon(\omega)}{c}\mathbf{E} \quad [\mathbf{kE}] = -\frac{\omega}{c}\mathbf{H} \quad \varepsilon(\omega)(\mathbf{kE}) = 0 \quad (\mathbf{kH}) = 0 \tag{3.2.6}$$

The last two equations are implied by the first two; if $\varepsilon \neq 0$, these equations imply that the fields $\mathbf{E}$ and $\mathbf{H}$ are transverse.

From the first two equations of (3.2.6) we obtain

$$[\mathbf{k}[\mathbf{kE}]] + \frac{\omega^2}{c^2}\,\varepsilon(\omega)\mathbf{E} = 0$$

that is, owing to the transversality of **E**, we have

$$k^2 = \frac{\omega^2}{c^2}\,\varepsilon(\omega) = \frac{\omega^2}{c^2} - \frac{\omega_0^2}{c^2}$$

This is the law of dispersion of transverse waves in the plasma. With this law known, it is easy to find the phase and group velocities of the wave:

$$v_{\mathrm{ph}} = \frac{\omega}{k} = c\sqrt{1 + \omega_0^2/c^2k^2} > c$$

$$v_{\mathrm{gr}} = \frac{\mathrm{d}\omega}{\mathrm{d}k} = c(1 + \omega_0^2/c^2k^2)^{-1/2} = \frac{c^2}{v_{\mathrm{ph}}} < c$$

If $\omega < \omega_0$, equation (3.2.5) gives $\varepsilon < 0$. Hence, $k^2 < 0$, so k is a purely imaginary quantity. The meaning of this becomes obvious from the expression for the field when $\mathbf{k} = \mathrm{i}\boldsymbol{\kappa}$:

$$\mathcal{E}(\mathbf{r}, t) = \mathbf{E}\exp(-\mathrm{i}\omega t)\exp(\pm\boldsymbol{\kappa}\mathbf{r})$$

If $\omega < \omega_0$, transverse waves do not propagate through plasma. Waves are reflected from the boundaries of the plasma or from the region where the electron concentration is so high that $\omega_0 > \omega$.

The wave damping length in the plasma is of the order of

$$l = \frac{1}{\kappa} = c(\omega_0^2 - \omega^2)^{-1/2} \sim c\omega_0^{-1} = \sqrt{\frac{mc^2}{4\pi ne^2}} \sim 10^6/\sqrt{n}\ \mathrm{cm}$$

In metals, $n \sim 10^{22}\,\mathrm{cm}^{-3}$ and $l \sim 10^{-5}\,\mathrm{cm}$, that is, of the order of the light wavelength. The condition $\omega < \omega_0$ is satisfied in metals up to the energy of ultraviolet radiation. The characteristic sheen of metals is explained by this reflection of the optical radiation.

If $\varepsilon(\omega) = 0$, Maxwell's equations (3.2.6) may have solutions of still another type. The second or fourth equation implies that in this case $\mathbf{H} \perp \mathbf{k}$, and the first equation implies that $\mathbf{H} = 0$. The second equation then says that if $\mathbf{H} = 0$, we have $\mathbf{E}_\perp = 0$, that is, **E** only has a component along the wave vector **k**. In the approximation we are considering now, a longitudinal wave in the plasma can have an arbitrary value of the wave vector **k** but by virtue of the equation (3.2.5), the condition $\varepsilon(\omega) = 0$ implies that the frequency equals the plasma frequency ω_0.

The origin of plasma oscillations at the frequency ω_0 can be reasoned out quite simply. Let all electrons shift by a distance x; this produces a surface charge at the boundary of the medium at a density enx. This charge is negative at one boundary and positive at the other because the medium was neutral, on average, before electrons were displaced. These charges produce an electric field $\mathcal{E} = -4\pi nex$; hence, the charges feel the force

$$F = e\mathcal{E} = -4\pi ne^2x$$

This force, tending to return the charges to their equilibrium positions, is proportional to the displacement x, so that the form of the equations of motion of the displaced layer of electrons is that of the equations of motion of a harmonic oscillator; its solution is the oscillations at the frequency ω_0.

Longitudinal waves are thus the oscillations of charge density and of the longitudinal electric field connected with it. We will show below that if the spatial dispersion is taken into account and we use the equation $\varepsilon(\mathbf{k},\omega) = 0$, the frequency of longitudinal waves becomes dependent on $\mathbf{k}$ and is not equal any more to ω_0.

Equation (3.2.1) describes the mean velocity of directed motion of electrons. For the harmonic field

$$\mathcal{E}(t) = \mathbf{E}\cos\omega t$$

equation (3.1.5) gives the velocity

$$\dot{\mathbf{r}} = \frac{e\mathbf{E}\nu}{m(\omega^2+\nu^2)}\cos\omega t + \frac{e\mathbf{E}\omega}{m(\omega^2+\nu^2)}\sin\omega t$$

and the work done by the field per unit time is

$$\frac{\mathrm{d}\varepsilon}{\mathrm{d}t} = \dot{\mathbf{r}}e\mathbf{E}\cos\omega t$$

The mean work of the field per unit time is

$$\overline{\frac{\mathrm{d}\varepsilon}{\mathrm{d}t}} = \frac{e^2E^2\nu}{2m(\omega^2+\nu^2)} \simeq \frac{e^2E^2\nu}{2m\omega^2} = 2\varepsilon_k\nu \tag{3.2.7}$$

It is proportional to the effective collision frequency and to the oscillatory energy ε_k of an electron (we assume that $\nu \ll \omega$). The work (3.2.7) increases the mean kinetic energy of the electron, that is, it heats up the electron component of the plasma.

In collisions with molecules and ions, the kinetic energy is partly transferred to heavy particles. Since their mass M is much higher than m, elastic collisions transfer very little energy to particles that are nearly at rest. Indeed, the momentum $\Delta\mathbf{p}$ transfer and the energy transfer $\Delta\varepsilon$ in one collision are related by the formula

$$\Delta\varepsilon = \frac{(\Delta\mathbf{p})^2}{2M} = \frac{p^2}{2M}[(1-\cos\theta)^2 + \sin^2\theta] = \frac{p^2}{M}(1-\cos\theta) \ll \frac{p^2}{2m} = \varepsilon$$

where θ is the scattering angle and p is the initial electron momentum. The rate of the collisional energy transfer to molecules is

$$2\pi n' v p^2 M^{-1} \int (1 - \cos\theta)\sigma(v,\theta) \sin\theta \, d\theta$$

since the number of collisions per unit time which scatter to an angle θ is

$$2\pi n' v \sigma(v,\theta) \sin\theta \, d\theta$$

As we had before, $\sigma(v,\theta)$ is the differential cross section and n' is the density of scatterers.

Since

$$\nu = n' \sigma_t v = 2\pi n' v \int (1 - \cos\theta)\sigma(v,\theta) \sin\theta \, d\theta$$

the energy transfer to molecules or ions per unit time is

$$\frac{mv^2}{2} \frac{2m}{M} \sigma_t n' v = \frac{2m}{M} \nu\varepsilon$$

Now we can describe how the electron energy grows in time using the equation

$$\frac{d\varepsilon}{dt} = \frac{e^2 E^2}{2m\omega^2} \nu(v) - \frac{2m}{M} \varepsilon\nu(v) \tag{3.2.8}$$

The classical description is valid when $\hbar\omega < \varepsilon$.

Equation (3.2.8) now gives the maximum energy that an electron can pick up in the field of the electromagnetic wave:

$$\varepsilon_{\max} = \frac{e^2 E^2}{2m\omega^2} \frac{M}{2m}$$

If $\varepsilon_{\max}$ is higher than the ionization potential of molecules, shock ionization of molecules and high-frequency breakdown occur in the gas since the electrons liberated in the process of ionization are accelerated by the field to a potential above the ionization potential and an avalanche is produced.

Obviously, $\varepsilon_{\max}$ is the average energy to which electrons are heated; nevertheless, $\varepsilon_{\max}$ is an estimate of this energy. The kinetic equation gives a distribution in energy of the type of $\exp(-\varepsilon/\varepsilon_{\max})$, which falls off steeply at $\varepsilon > \varepsilon_{\max}$.

3.3. Kinetic Equation for Plasmas

The kinetic equation of the type of the well-known Boltzmann equation for gases is widely used for the description of plasmas. Just as it was for gases, the condition of applicability of the kinetic equation is that the plasma be sufficiently rarefied. The plasma must not deviate too much from the ideal gas characteristics. This is a stronger condition than a similar one for a gas of neutral molecules, because of the long-range nature of the Coulomb force. It can also be reformulated as a constraint that the kinetic energy of particles be higher than the mean electrostatic interaction energy,

$$\mathrm{k}T \gg e^2/\bar{r} \sim e^2 n^{1/3}$$

where $\bar{r} \sim n^{-1/3}$ is the mean distance between electrons and n is their density. In view of (3.2.3), this condition can be written as

$$\bar{r} = n^{-1/3} \ll r_0 \equiv \sqrt{\mathrm{k}T/4\pi n e^2}$$

This last inequality signifies that there must be a large number of electrons within the sphere of a radius equal to the so-called Debye radius r_0.

The possibility of applying the classical theory to the plasma is also connected with the requirement that a plasma can be regarded as a nondegenerate gas of electrons, that is, the plasma temperature must be higher than the temperature of degeneracy T_0 defined as

$$\mathrm{k}T_0 = \frac{\hbar^2}{m\bar{r}^2} = \frac{\hbar^2}{m} n^{2/3}$$

All these conditions are satisfied in a broad class of problems.

The kinetic equation is an equation for the distribution function $f(\mathbf{v}, \mathbf{r}, t)$. By definition, $f\, \mathrm{d}^3 r\, \mathrm{d}^3 v$ is the mean number of electrons in the phase volume $\mathrm{d}^3 r\, \mathrm{d}^3 v$. If no external factors act on the system, the distribution function f is independent of $\mathbf{r}$ and t, while the dependence on $\mathbf{v}$ is the familiar Maxwellian distribution

$$f(\mathbf{v}, \mathbf{r}, t) \equiv f_0(v) = n\left(\frac{m}{2\pi \mathrm{k}T}\right)^{3/2} \exp\left(-\frac{mv^2}{2\mathrm{k}T}\right) \tag{3.3.1}$$

Using f, we can calculate all averaged characteristics of the plasma, such as the charge density $\rho(\mathbf{r}, t)$ and the current density $\mathbf{j}(\mathbf{r}, t)$:

$$\rho(\mathbf{r}, t) = e\int f(\mathbf{v}, \mathbf{r}, t)\, \mathrm{d}^3 v$$

$$\mathbf{j}(\mathbf{r}, t) = e\int \mathbf{v} f(\mathbf{v}, \mathbf{r}, t)\, \mathrm{d}^3 v$$

The kinetic equation for the distribution function f can be written as the law of conservation of the number of particles, which is the continuity equation in the six-dimensional phase space,

$$\frac{\partial f}{\partial t} + \nabla_r f\mathbf{v} + \nabla_v f\dot{\mathbf{v}} = \mathrm{St} f \tag{3.3.2}$$

where the subscripts "r" and "v" indicate derivatives with respect to coordinates and velocities, respectively, and $\mathrm{St} f$ is the so-called collision integral, which takes into account the collisions of electrons among themselves and with other particles. As a result of collisions, an electron transfers jumpwise to a point in the phase space with a different value of velocity $\mathbf{v}$.

For the plasma,

$$\dot{\mathbf{v}} = \frac{1}{m}\mathbf{F} = \frac{1}{m}\left(e\mathcal{E} + \frac{e}{c}[\mathbf{v}\mathcal{H}]\right) \tag{3.3.3}$$

$\mathcal{E}$ and $\mathcal{H}$ must be understood as the mean values of the electric and magnetic fields at a point where the electron is. Since the plasma is rarefied, we can neglect the correlation of the positions of electrons; in this case, $\mathcal{E}$ and $\mathcal{H}$ are the mean self-consistent fields at a given point in space. In a plasma, therefore, the local (acting) field is identical to the mean field.

The force $\mathbf{F}$, defined by (3.3.3) satisfies the condition

$$\nabla_v \mathbf{F} = 0$$

and it is also obvious that $\nabla_r \mathbf{v} = 0$; hence, equation (3.3.2) can be written in the form

$$\frac{\partial f}{\partial t} + \mathbf{v}\nabla_r f + \frac{\mathbf{F}}{m}\nabla_v f = \mathrm{St} f \tag{3.3.4}$$

In the general case, the collision integral is

$$\mathrm{St} f = \int \left(f(\mathbf{v}', \mathbf{r}, t) w_{vv'} - f(\mathbf{v}, \mathbf{r}, t) w_{v'v}\right) \mathrm{d}^3 v'$$

where $w_{vv'}$ is the probability for an electron with velocity $\mathbf{v}'$ to change it to $\mathbf{v}$ as a result of a collision, and $w_{v'v}$ is the probability for an electron with velocity $\mathbf{v}$ to change it to $\mathbf{v}'$. The first term in the collision integral describes the inflow and the second, the outflow of electrons to and from the phase space volume considered.

In a weak field, the distribution function does not deviate considerably from the equilibrium function,

$$f(\mathbf{v}, \mathbf{r}, t) = f_0(v) + \varphi(\mathbf{v}, \mathbf{r}, t)$$

where $|\varphi| \ll f_0$.

It will be clear soon that if the dependence of the field on coordinates is ignored, that is, if effects of spatial dispersion are not considered, then for a weak field we have

$$\varphi(\mathbf{v}, t) = \frac{\mathbf{v}}{v}\,\mathbf{a}(v, t) \tag{3.3.5}$$

where the vector **a** points along the field. In this case the collision integral becomes especially simple. Indeed, if only elastic collisions are taken into account, then $v' = v$ and $w_{v'v} = w_{vv'} = n'\sigma(v, \theta)v$ where θ is the angle between **v** and **v**$'$, $\sigma(v, \theta)$ is the differential cross section of elastic scattering to an angle θ, and n' is the concentration of scatterers, then the collision integral is

$$\mathrm{St} f = n' \int \sigma(v, \theta)(av' - av) \sin\theta \, \mathrm{d}\theta \, \mathrm{d}\varphi$$

We have taken into account that f_0 alone causes the collision integral to vanish since collisions as such do not result in a deviation from equilibrium.

The angles θ and ϕ are the coordinates of the vector **v**$'$ in the spherical system of coordinates with the z axis along the vector **v**, chosen in such a way that the vector **a** lies in the plane x–z.

Obviously,

$$\mathbf{a}\mathbf{v}' = a_x v'_x + a_y v'_y + a_z v'_z = av(\sin\theta \, \cos\varphi \, \sin\theta_1 + \cos\theta \cos\theta_1)$$

where θ_1 is the angle between a and v. Hence,

$$\begin{aligned} \mathrm{St} f &= n' \int \sigma(v, \theta)(av \sin\theta \, \cos\varphi \, \sin\theta_1 + (\mathbf{av}) \cos\theta - \mathbf{av}) \\ &\qquad \times \sin\theta \, \mathrm{d}\theta \mathrm{d}\varphi = 2\pi n' \mathbf{av} \int \sigma(\theta, \varphi)(\cos\theta - 1) \sin\theta \, \mathrm{d}\theta \\ &= -\nu(v)\varphi(\mathbf{v}, t) \end{aligned}$$

The kinetic equation now becomes

$$\frac{\partial \varphi}{\partial t} + \frac{e}{m}\boldsymbol{\mathcal{E}}\nabla_v f_0 + \nu(v)\varphi = 0 \tag{3.3.6}$$

The term $\mathbf{F}\nabla_v \varphi / m$ is omitted because it is of order two with respect to the field. The term containing the magnetic field vanishes since the vector

$$\nabla_v f_0 = -\frac{m}{\mathrm{k}T} f_0 \mathbf{v}$$

is directed along the vector **v** and $[\mathbf{v}\boldsymbol{\mathcal{H}}] = 0$.

The Fourier transform of equation (3.3.6) yields

$$(-\mathrm{i}\omega + \nu(v))\varphi_\omega(\mathbf{v}) + \frac{e}{m}\mathbf{E}_\omega \nabla_v f_0 = 0$$

whence

$$\varphi_\omega(\mathbf{v}) = -\mathrm{i}e\mathbf{E}_\omega \nabla_v f_0 / m(\omega + \mathrm{i}\nu(v))$$

The Fourier components of current density and polarization can be found using the distribution function:

$$-\mathrm{i}\omega \mathbf{P}_\omega = \mathbf{j}_\omega = e \int \mathbf{v}\varphi_\omega(\mathbf{v})\,\mathrm{d}^3v = \frac{\mathrm{i}e^2}{\mathrm{k}T} \int \frac{\mathbf{v}(\mathbf{E}_\omega \mathbf{v}) f_0}{\omega + \mathrm{i}\nu}\,\mathrm{d}^3v \tag{3.3.7}$$

If the collision frequency ν is independent of the velocity v, we can use the identity which holds for the equilibrium distribution function f_0 and an arbitrary vector $\mathbf{A}$:

$$\int \mathbf{v}(\mathbf{A}\mathbf{v}) f_0\,\mathrm{d}^3v = n(\mathrm{k}T/m)\mathbf{A} \tag{3.3.8}$$

This identity is proved by direct calculation of the integral in the lefthand side of (3.3.8). The Cartesian coordinates are

$$\int v_\alpha v_\beta f_0\,\mathrm{d}^3v = n(\mathrm{k}T/m)\delta_{\alpha\beta}$$

If $\alpha \neq \beta$, the integral vanishes owing to the spherical symmetry of the function f_0. If $v_\alpha \to -v_a$, the value of the integral must remain unchanged; on the other hand, its sign is reversed and the integral therefore vanishes. If $\alpha = \beta$ and we use (3.3.1), the integral reduces to a single integral

$$\int v_\alpha^2 f_0\,\mathrm{d}^3v = n\Big(\frac{m}{2\pi \mathrm{k}T}\Big)^{1/2} \int v_\alpha^2 \exp\Big(-\frac{mv_\alpha^2}{2\mathrm{k}T}\Big)\,\mathrm{d}v_\alpha = n(\mathrm{k}T/m)$$

Using (3.3.8), we obtain an expression for $\varepsilon(\omega)$,

$$\varepsilon(\omega) = 1 + 4\pi\chi(\omega) = 1 + 4\pi\,\frac{P_\omega}{E_\omega} = 1 - \frac{4\pi ne^2}{m\omega(\omega + \mathrm{i}\nu)} \tag{3.3.9}$$

which is identical to the result of the elementary theory, (3.2.4), based on equation (3.2.1).

When the dependence of the collision frequency ν on the absolute value of velocity v is taken into account, integration over the solid angle $\mathrm{d}\Omega$ in the velocity space can be performed in (3.3.7). Obviously,

$$\frac{1}{4\pi} \int v_\alpha v_\beta\,\mathrm{d}\Omega = \delta_{\alpha\beta}\overline{v_\alpha^2} = \tfrac{1}{3}\,\delta_{\alpha\beta}\overline{(v_x^2 + v_y^2 + v_z^2)} = \tfrac{1}{3}v^2\delta_{\alpha\beta}$$

Hence,

$$\mathbf{P}_\omega = -\frac{4\pi e^2 \mathbf{E}_\omega}{3\mathrm{k}T\omega} \int\limits_0^\infty \frac{v^4 f_0}{\omega + \mathrm{i}\nu(v)}\,\mathrm{d}v$$

that is,

$$\epsilon(\omega) = 1 - \frac{(4\pi e)^2}{3kT\omega} \int_0^\infty \frac{v^4 f_0(v)}{\omega + i\nu(v)}\, dv = 1 - \frac{32\sqrt{\pi}\, e^2 n}{3m\omega} \int_0^\infty \frac{u^4 \exp(-u^2)}{\omega + i\nu(v)}\, du \tag{3.3.10}$$

where we have introduced a dimensionless variable

$$u = \sqrt{\frac{m}{2kT}}\, v$$

If $\nu = \text{const}$, expression (3.3.10) obviously transforms to (3.3.9) and, correspondingly, ε' and ε'' transform into the expressions (3.2.4). The effective collision frequency ν_{eff} is sometimes introduced, such that if the effective collision frequency ν_{eff} were substituted into the expressions in (3.2.4) for ν, it would give a frequency dependence in agreement with (3.3.10). However, ν_{eff} then depends on ω in a complex manner and, furthermore, different effective collision frequencies have to be introduced for ε' and ε''. As a result, the introduction of ν_{eff} is justified only in the limiting case of $\omega \gg \nu_{\text{eff}}$, when

$$\varepsilon' = 1 - \frac{4\pi n e^2}{m\omega^2} \qquad \varepsilon'' = \frac{4\pi n e^2 \nu_{\text{eff}}}{m\omega^2}$$

where

$$\nu_{\text{eff}} = \frac{8}{3\sqrt{\pi}} \int_0^\infty \nu(u) u^4 \exp(-u^2)\, du$$

We have already indicated that for $kT \ll I_0$ one can assume that the scattering cross section σ_t for molecules and atoms is independent of velocity. For this hard-sphere model we have

$$\nu(v) = n'\sigma_t v = \pi a^2 n' v$$

and the effective collisions frequency is

$$\nu_{\text{eff}} = \frac{4\pi}{3}\, a^2 n' \bar{v}$$

where $\bar{v}$ is the mean electron velocity

$$\bar{v} = \sqrt{\frac{8kT}{\pi m}}$$

In what follows, we only consider the limiting case of low effective collision frequency, that is, we assume $\nu \to 0$. In this case, the collisions integral can be chosen in its simplest form,

$$\text{St} f = -\nu\varphi = -\nu(f - f_0)$$

Here ν is some effective collision frequency which tends to zero. In fact, we consider a collisionless plasma and need to take into account the infinitesimal ν only to interpret correctly the specific features of susceptibilities of the plasma that arise at $\nu = 0$. We have already encountered a similar situation in Chapter 2 when calculating the linear response of the medium in the general case.

3.4. Dielectric Permittivity of Plasmas, with Spatial Dispersion Taken into Account

The kinetic equation for the case of inhomogeneous weak field is, as is clear from the remark at the end of Section 3.3,

$$\frac{\partial \varphi}{\partial t} + \mathbf{v}\nabla_r \varphi + \frac{e}{m}\mathcal{E}\nabla_v f_0 + \nu\varphi = 0$$

where $\varphi(\mathbf{v}, \mathbf{r}, t) = f(\mathbf{v}, \mathbf{r}, t) - f_0(v)$.

Let us find the Fourier transform of this kinetic equation with respect to t and r and, as usual, introduce the Fourier transforms of the functions φ, $\mathbf{j}$, and $\boldsymbol{\mathcal{P}}$. This will give us

$$-\mathrm{i}\,\omega\varphi_{\mathbf{k}\omega}(\mathbf{v}) + \mathrm{i}(\mathbf{kv})\varphi_{\mathbf{k}\omega}(\mathbf{v}) + \nu\varphi_{\mathbf{k}\omega}(\mathbf{v}) = -\frac{e}{m}\,\mathbf{E}_{\mathbf{k}\omega}\nabla_v f_0$$

whence

$$\varphi_{k\omega} = -\frac{\mathrm{i}e\mathbf{E}_{\mathbf{k}\omega}\nabla_v f_0}{m(\omega - \mathbf{kv} + \mathrm{i}\nu)}$$

Using this equation, we can find $\mathbf{j}_{\mathbf{k}\omega}$ and $\mathbf{P}_{\mathbf{k}\omega}$:

$$\mathbf{j}_{\mathbf{k}\omega} = e\int \mathbf{v}\varphi_{\mathbf{k}\omega}\,\mathrm{d}^3v = \frac{\mathrm{i}e^2}{\mathrm{k}T}\int \mathbf{v}\,\frac{(\mathbf{E}_{\mathbf{k}\omega}\mathbf{v})f_0}{\omega - \mathbf{kv} + \mathrm{i}\nu}\,\mathrm{d}^3v$$

We will first calculate the longitudinal components of the current density and polarization. If the x axis points along the vector $\mathbf{k}$, then

$$P_{\mathbf{k}\omega x} = \frac{\mathrm{i}}{\omega}\,j_{\mathbf{k}\omega x} = -\frac{e^2 n}{\omega \mathrm{k}T}\left(\frac{m}{2\pi \mathrm{k}T}\right)^{3/2}\int \frac{v_x^2 E_{\mathbf{k}\omega x}\exp\left(-\frac{mv^2}{2\mathrm{k}T}\right)}{\omega - kv_x + \mathrm{i}\nu}\,\mathrm{d}^3v \quad (3.4.1)$$

The terms with $E_y v_y$ and $E_z v_z$ vanish; this can be established by arguments similar to those used in proving (3.3.8).

The integrand in (3.4.1) depend only on v_x, so that integration in v_y and v_z is performed immediately. We obtain

$$P_{\mathbf{k}\omega x} = -\frac{e^2 n E_{\mathbf{k}\omega x}}{\omega \mathrm{k}T}\left(\frac{m}{2\pi \mathrm{k}T}\right)^{1/2}\int_{-\infty}^{+\infty} \frac{v_x^2\exp\left(-mv_x^2/2\mathrm{k}T\right)}{\omega - kv_x + \mathrm{i}\nu}\,\mathrm{d}v_x$$

$$= -\frac{e^2 n E_{\mathbf{k}\omega x}}{\omega \mathrm{k}Tk}\int_{-\infty}^{+\infty} \frac{v_x(kv_x - \omega - \mathrm{i}\nu + \omega + \mathrm{i}\nu)}{\omega - kv_x + \mathrm{i}\nu}\,\exp\left(-mv_x^2/2\mathrm{k}T\right)\,\mathrm{d}v_x$$

Making use of

$$\int_{-\infty}^{+\infty} v_x \exp\left(-mv_x^2/2kT\right) dv_x = 0$$

we find for $\nu \to 0$ that

$$P_{\mathbf{k}\omega x} = -\frac{e^2 n E_{\mathbf{k}\omega x}}{\omega kT} \int_{-\infty}^{+\infty} \frac{v_x \exp\left(-\frac{mv_x^2}{2kT}\right)}{\omega - kv_x + i\nu} dv_x$$

$$= \frac{e^2 n E_{\mathbf{k}\omega x}}{kTk^2} \int_{-\infty}^{+\infty} \frac{\omega - kv_x + i\nu - \omega - i\nu}{\omega - kv_x + i\nu} \exp\left(-\frac{mv_x^2}{2kT}\right) dv_x$$

Now we define a dimensionless integration variable

$$\xi = \sqrt{\frac{m}{2kT}}\, v_x$$

that is, the ratio of v_x to the thermal velocity, and introduce a dimensionless parameter

$$\mu = \frac{\omega}{k}\sqrt{\frac{m}{2kT}}$$

that is, the ratio of the phase velocity ω/k to the thermal velocity; we obtain

$$P_{\mathbf{k}\omega x} = \frac{E_{\mathbf{k}\omega x}}{4\pi r_0^2 k^2}(1 + F(\mu)) \tag{3.4.2}$$

Here $r_0^{-2} = 4\pi ne^2/kT$ (see (3.2.3)) and

$$F(\mu) = \frac{\mu}{\sqrt{\pi}} \int_{-\infty}^{+\infty} \frac{\exp(-\xi^2)}{\xi - \mu - i0} d\xi$$

In an isotropic medium, and thus in a particular case of the plasma, $\varepsilon_{\alpha\beta}(\mathbf{k}, \omega)$ has the form

$$\varepsilon_{\alpha\beta}(\mathbf{k}, \omega) = \varepsilon_{\perp}(k, \omega)\left(\delta_{\alpha\beta} - \frac{k_\alpha k_\beta}{k^2}\right) + \varepsilon_{\parallel}(k, \omega)\frac{k_\alpha k_\beta}{k^2}$$

where $\varepsilon_{\parallel}$ and $\varepsilon_{\perp}$ are the longitudinal and the transverse dielectric permittivities, which differ if the spatial dispersion is taken into account. In fact, this is a decomposition of an arbitrary invariant **k**-dependent rank-two tensor into a longitudinal and a transverse part, analogous to (1.6.7).

As follows from (3.4.2),

$$\varepsilon_{\parallel}(k,\omega) = 1 + \frac{4\pi P_{\mathbf{k}\omega x}}{E_{\mathbf{k}\omega x}} = 1 + \frac{1}{r_0^2 k^2}\left(1 + F(\mu)\right)$$

By analogy to (3.4.1) and (3.4.2), we can calculate the transverse component of the current density and polarization and the transverse dielectric permittivity of the plasma. It is expressed via the same function $F(\mu)$,

$$\varepsilon_{\perp}(k,\omega) = 1 + \frac{\omega_0^2}{\omega^2} F(\mu)$$

where, as before, $\omega_0^2 = 4\pi ne^2/m$ is the squared plasma (Langmuir) frequency for electrons.

The function $F(\mu)$ is well known and has been carefully tabulated. By using the familiar formula

$$\frac{1}{x - \mathrm{i}0} = P\,\frac{1}{x} + \mathrm{i}\pi\delta(x)$$

where P is the symbol of the principal value of an integral with a singular function, we can immediately separate the imaginary part of the function $F(\mu)$ and, hence, of the dielectric permittivity:

$$\mathrm{Im}F(\mu) = \mu\sqrt{\pi}\,\exp(-\mu^2)$$

The physical meaning of the imaginary part of $\varepsilon(k,\omega)$, which describes absorption, is quite clear. The imaginary part of $F(\mu)$ originates from the pole of the function $(\xi - \mu)^{-1}$ in the integrand of the definition of $F(\mu)$. The condition $\xi = \mu$ is equivalent to the equality

$$v_x = \omega/k$$

that is, the imaginary part of dielectric permittivity is caused by electrons whose longitudinal velocity component (along $\mathbf{k}$) equals the phase velocity ω/k of the wave. These electrons are accelerated by the field and absorb its energy; the damping associated with these electrons is known as the *Landau damping*.

The parameter μ determines the relative role of the effects of spatial dispersion. Spatial dispersion is connected with the nonlocal nature of the response; the nonlocality radius in a plasma is the least of two lengths: $l = v/\nu$ (the free path length) and v/ω (the distance to which an electron is displaced during one period of the field). In view of the assumption $\nu \ll \omega$, the nonlocality radius is determined by the length v/ω, where v is of the order of the thermal velocity; hence, the condition of smallness of spatial dispersion effects, $kv/\omega \ll 1$, is equivalent to the inequality $\mu \gg 1$. This is the limiting case we are now interested in.

If $\mu \gg 1$, the Landau damping is exponentially small, which is a corollary of the fact that in a Maxwellian distribution, the fraction of electrons whose velocity is equal to the phase velocity ω/k of the wave is exponentially small. Moreover, it is nonzero only for longitudinal waves and for the longitudinal dielectric permittivity. The electron velocity can equal the phase velocity only for this wave. We have already seen in Section 3.2 that the phase velocity of the wave is always greater than the speed of light. As a result, there is no Landau damping for transverse waves; formally, $\mathrm{Im}F(\mu) \neq 0$ in this case as well, but only because the Maxwellian distribution is allowed to be used if velocities are less than the velocity of light.

Since the imaginary part of $\varepsilon(k,\omega)$ is small when $\mu \gg 1$, we will hereafter neglect $\varepsilon''(k,\omega)$ and treat the expressions for $\varepsilon(k,\omega)$ as real ones, in which the function $F(\mu)$ has been replaced by the principal value of the integral,

$$F'(\mu) = \frac{\mu}{\sqrt{\pi}} P \int_{-\infty}^{+\infty} \frac{\exp(-\xi^2)}{\xi - \mu} \, \mathrm{d}\xi$$

If $\mu \gg 1$, we can use the expansion

$$\frac{\mu}{\xi - \mu} = -\frac{1}{1 - \xi/\mu} = -\left(1 + \frac{\xi}{\mu} + \frac{\xi^2}{\mu^2} + \ldots\right)$$

and obtain an asymptotic formula

$$F'(\mu) = -1 - \frac{1}{2\mu^2} - \frac{3}{4\mu^4} = \ldots$$

An approximate expression for $\varepsilon_{\parallel}(k,\omega)$ now becomes

$$\varepsilon_{\parallel}(k,\omega) = 1 - \frac{\omega_0^2}{\omega^2}\left(1 + \frac{3kT}{m\omega^2} k^2\right) \tag{3.4.3}$$

This formula holds if

$$\frac{2kT}{m} \frac{k^2}{\omega} = \mu^{-2} \ll 1$$

so that the term proportional to k^2 is small and

$$\varepsilon_{\parallel}(k,\omega) \simeq 1 - \omega_0^2/\omega^2$$

which is identical to the result of the elementary theory.

The law of dispersion of longitudinal waves is obtained from the condition

$$\varepsilon_{\parallel}(k,\omega) = 0$$

As before, we have $\omega = \omega_0$ in the zero approximation, while in the first approximation we take nonlocality into account and substitute ω_0^2 for ω^2 in that term of (3.4.3) which contains k^2. The condition $\varepsilon_{\parallel} = 0$ then yields the law of dispersion

$$\omega^2 = \omega_0^2 \left(1 + \frac{3\mathrm{k}T}{m\omega_0^2} k^2\right) = \omega_0^2(1 + 3r_0^2k^2)$$

It can be compared with the law of dispersion for transverse waves written as

$$\omega^2 = \omega_0^2 \left(1 + \frac{c^2k^2}{\omega_0^2}\right)$$

The dependence of ω^2 on k^2 is of the same form as for transverse waves except that for longitudinal waves, the coefficient with k^2 is much smaller. This occurs because the ratio of the coefficients with k^2 in the dispersion law is

$$3\mathrm{k}T/mc^2 \ll 1$$

the thermal velocity being much smaller than c,

$$\sqrt{\frac{2\mathrm{k}T}{m}} \ll c$$

In the limit $\mu \gg 1$, and if we use the asymptotic formula for $F(\mu)$, the transverse dielectric permittivity of the plasma needs the same correction for spatial dispersion as does (3.4.3), but without the factor 3:

$$\varepsilon_{\perp}(k,\omega) = 1 - \frac{\omega_0^2}{\omega}\left(1 + \frac{\mathrm{k}T}{m\omega^2} k^2\right)$$

Let us consider right away the dielectric permittivity of the plasma at low frequencies. The limit $\omega \to 0$ corresponds to $\mu \to 0$. Consequently, we need an asymptotic formula for $F(\mu)$ as $\mu \to 0$:

$$F(\mu) \simeq -2\mu^2 + \mathrm{i}\sqrt{\pi}\,\mu$$

Now we obtain

$$\varepsilon_{\parallel}(k,0) = 1 + 1/r_0^2k^2$$

Using this expression, we will find the field of a pointlike small probe charge e_1 in the plasma. The electrostatics equation implies

$$\nabla \mathcal{D} = 4\pi e_1 \delta^3(\mathbf{r})$$

which gives, in Fourier transforms,

$$i\mathbf{k}\mathbf{D}_{\mathbf{k}} = 4\pi e_1$$

Since

$$\mathbf{D}_{\mathbf{k}} = \varepsilon_{\parallel}(k,0)\mathbf{E}_{\mathbf{k}} = -i\mathbf{k}\varepsilon_{\parallel}(k,0)\varphi_{\mathbf{k}}$$

the Fourier-transform components of the scalar potential φ are

$$\varphi_1 = \frac{4\pi e_1}{k^2 + r_0^{-2}}$$

whence the scalar potential of the field is

$$\varphi(\mathbf{r}) = \frac{e_1}{|\mathbf{r}|}\exp(-|\mathbf{r}|/r_0)$$

which agrees with the Debye theory of plasma screening of electric charges.

3.5. *Nonlinear Susceptibilities of the Plasma*

If spatial dispersion is neglected, the quadratic susceptibility of the isotropic medium is zero, as it is in any medium with central symmetry. The same is true for nonlinear susceptibilities of a higher even order. As we will see later, odd-order plasma susceptibilities also vanish if spatial dispersion is ignored.

If, however, the magnetic field and, generally, any nonlocality of response in the plasma are taken into account, there appears polarization that is quadratic in field. The role of magnetic field and nonlocality effects has already been clarified in Section 3.1 when we treated the motion of free charges in electromagnetic field. It was shown that the generation of harmonics by free charges can be calculated only if we take into account the magnetic field and its action on the charge.

In the plasma we have $\chi^{(2)}_{\alpha\beta\gamma}(\omega = \omega' + \omega'') = 0$ but if spatial dispersion is taken into account, then we obtain

$$P^{(2)}_{\alpha} = \chi^{(2)}_{\alpha\beta\gamma\delta}E_{\beta}E_{\gamma}k_{\delta} \neq 0$$

Covering terms linear in wave vector means taking into account the magnetic field, which is an effect of order v/c. The terms of polarization quadratic in $\mathbf{k}$ take into account effects of order v^2/c^2, etc. In the nonrelativistic approach, when terms of orders v^2/c^2 and higher are neglected, it would be beyond the accuracy of the nonrelativistic approximation to calculate polarization terms quadratic in $\mathbf{k}$. In view of this, we will calculate nonlinear polarization up to terms linear in $\mathbf{k}$.

We begin with the kinetic equation

$$\frac{\partial f}{\partial t} + \mathbf{v}\nabla_r f + \frac{\mathbf{F}}{m}\nabla_v f + \nu(f - f_0) = 0 \qquad (3.5.1)$$

Assuming that $f = f_0 + \varphi(\mathbf{v}, \mathbf{r}, t)$, we calculate the Fourier transform of φ:

$$\varphi(\mathbf{v}, \mathbf{r}, t) = \int \varphi_{\mathbf{k}\omega}(\mathbf{v}) \exp(\mathrm{i}\mathbf{kr} - \mathrm{i}\omega t) \frac{\mathrm{d}^3k\,\mathrm{d}\omega}{(2\pi)^4}$$

The force $\mathbf{F}$ must also be Fourier-transformed, so as to have

$$\mathbf{F}_{\mathbf{k}\omega} = e\mathbf{E}_{\mathbf{k}\omega} + \frac{e}{c}[\mathbf{v}\mathbf{H}_{\mathbf{k}\omega}]$$

Maxwell's equation

$$\operatorname{curl}\boldsymbol{\mathcal{E}} = -\frac{1}{c}\frac{\partial \boldsymbol{\mathcal{H}}}{\partial t}$$

rewritten in terms of Fourier transforms becomes

$$[\mathbf{k}\mathbf{E}_{\mathbf{k}\omega}] = \frac{\omega}{c}\mathbf{H}_{\mathbf{k}\omega}$$

Therefore, $\mathbf{F}_{\mathbf{k}\omega}$ can be expressed in terms of the electric field:

$$\mathbf{F}_{\mathbf{k}\omega} = e\mathbf{E}_{\mathbf{k}\omega} + \frac{e}{\omega}[\mathbf{v}[\mathbf{k}\mathbf{E}_{\mathbf{k}\omega}]] = e\mathbf{E}_{\mathbf{k}\omega} + \frac{e}{\omega}\mathbf{k}(\mathbf{v}\mathbf{E}_{\mathbf{k}\omega}) - \frac{e}{\omega}\mathbf{E}_{\mathbf{k}\omega}(\mathbf{kv}) \qquad (3.5.2)$$

Note that expression (3.5.2) is linear in $\mathbf{k}$.

Multiplying (3.5.1) by $\exp(-\mathrm{i}\mathbf{kr} + \mathrm{i}\omega t)$ and integrating, we obtain an equation for $\varphi_{\mathbf{k}\omega}(\mathbf{v})$:

$$(-\mathrm{i}\omega + \mathrm{i}(\mathbf{kv}) + \nu)\varphi_{\mathbf{k}\omega} = -\frac{1}{m}\mathbf{F}_{\mathbf{k}\omega}\nabla_v f_0 - \frac{1}{m}\int \mathbf{F}_{\mathbf{k}''\omega''}\nabla_v\varphi_{\mathbf{k}'\omega'}$$
$$\times\, \delta(\omega - \omega' - \omega'')\delta^3(\mathbf{k} - \mathbf{k}' - \mathbf{k}'')\frac{\mathrm{d}\omega'\,\mathrm{d}\omega''\,\mathrm{d}^3k'\,\mathrm{d}^3k''}{(2\pi)^4}$$

We will seek the solution of this equation by expanding the distribution function $\varphi_{\mathbf{k}\omega}(\mathbf{v})$ in powers of field,

$$\varphi_{\mathbf{k}\omega}(\mathbf{v}) = \varphi^{(1)}_{\mathbf{k}}(\mathbf{v}) + \varphi^{(2)}_{\mathbf{k}\omega}(\mathbf{v}) + \ldots$$

where

$$\varphi^{(n)}_{\mathbf{k}\omega} \sim E^n \sim F^n$$

Collecting together terms of equal order in $\mathbf{F}$, we obtain

$$\varphi^{(1)}_{\mathbf{k}\omega}(\mathbf{v}) = \left(-\frac{\mathrm{i}}{m}\right)\frac{1}{\omega - \mathbf{kv} + \mathrm{i}\nu}\mathbf{F}_{\mathbf{k}\omega}\nabla_v f_0$$

This expression was used in Section 3.4 to find the linear response of the plasma.

Now we have

$$\varphi^{(2)}_{\mathbf{k}\omega}(\mathbf{v}) = \left(-\frac{\mathrm{i}}{m}\right)^2 \frac{1}{\omega - \mathbf{kv} + \mathrm{i}\nu} \int \mathbf{F}_{\mathbf{k}''\omega} \nabla_v \varphi^{(1)}_{\mathbf{k}'\omega'}(\mathbf{v}) \delta(\omega - \omega' - \omega'')$$
$$\times\, \delta^3(\mathbf{k} - \mathbf{k}' - \mathbf{k}'') \frac{\mathrm{d}\omega'\, \mathrm{d}\omega''\, \mathrm{d}^3k'\, \mathrm{d}^3k''}{(2\pi)^4} = \left(-\frac{\mathrm{i}}{m}\right)^2 (\omega - \mathbf{kv} + \mathrm{i}\nu)^{-1}$$
$$\times \int \delta(\omega - \omega' - \omega'')\, \delta^3(\mathbf{k} - \mathbf{k}' - \mathbf{k}'')\, \mathbf{F}_{\mathbf{k}''\omega''} \nabla_v (\omega' - \mathbf{k}'\mathbf{v} + \mathrm{i}\nu)^{-1}$$
$$\times\, \mathbf{F}_{\mathbf{k}'\omega'} \nabla_v f_0 \frac{\mathrm{d}\omega'\, \mathrm{d}\omega''\, \mathrm{d}^3k'\, \mathrm{d}^3k''}{(2\pi)^4} \tag{3.5.3}$$

Likewise,

$$\varphi^{(3)}_{\mathbf{k}\omega}(\mathbf{v}) = \left(-\frac{\mathrm{i}}{m}\right)^3 \frac{1}{\omega - \mathbf{kv} + \mathrm{i}\nu} \int \delta(\omega - \omega' - \omega'')$$
$$\times\, \delta^3(\mathbf{k} - \mathbf{k}' - \mathbf{k}'' - \mathbf{k}''')\, \mathbf{F}_{\mathbf{k}'''\omega'''} \nabla_v [(\omega' + \omega'') - (\mathbf{k}' + \mathbf{k}'')\mathbf{v} + \mathrm{i}\nu]^{-1}$$
$$\times\, \mathbf{F}_{\mathbf{k}''\omega''} \nabla_v \frac{1}{\omega' - \mathbf{k}'\mathbf{v} + \mathrm{i}\nu} \mathbf{F}_{\mathbf{k}'\omega'} \nabla_v f_0$$
$$\times\, \frac{\mathrm{d}\omega'\, \mathrm{d}\omega''\, \mathrm{d}\omega'''\, \mathrm{d}^3k'\, \mathrm{d}^3k''\, \mathrm{d}^3k'''}{(2\pi)^8} \tag{3.5.4}$$

The structure of the expressions giving $\varphi^{(n)}_{\mathbf{k}\omega}$ for $n > 3$ is also clear but we do not need them here.

Nonlinear contributions to current density and polarization can be found using (3.5.3) and (3.5.4).

For instance,

$$(-\mathrm{i}\omega)\mathbf{P}^{(2)}_{\mathbf{k}\omega} = \mathbf{j}^{(2)}_{\mathbf{k}\omega} = e \int \mathbf{v} \varphi^{(2)}_{\mathbf{k}\omega}(\mathbf{v})\, \mathrm{d}^3 v \tag{3.5.5}$$

Let us try to neglect spatial dispersion; this means setting $\mathbf{k} = 0$ in (3.5.3). Owing to (3.5.2), $\mathbf{k} = 0$ implies

$$\mathbf{F}_{\mathbf{k}\omega} = e\mathbf{E}_{\mathbf{k}\omega}$$

We will either neglect relaxation, that is, assume $\nu = 0$, or assume that ν is independent of velocity. In this case the factors $(\omega' + \mathrm{i}\nu)^{-1}$ in (3.5.3) are independent of velocity and can be factored out of the differential operator ∇_v.

We thus have

$$(-\mathrm{i}\omega)\mathbf{P}_\omega^{(2)} = \frac{e^2(-\mathrm{i})^2}{m^2(\omega+\mathrm{i}\nu)} \int \frac{\mathrm{d}\omega'\,\mathrm{d}\omega''}{2\pi}\,\delta(\omega-\omega'-\omega'') \times \frac{1}{\omega'+\mathrm{i}\nu} \int \mathrm{d}^3\mathbf{v}\,(\mathbf{E}_{\omega''}\nabla_v)(\mathbf{E}_{\omega'}\nabla_v)\,f_0$$

Integration in velocities can be done by parts, transferring the differential operators Δ_v to the function v. A linear function twice differentiated gives zero. The expression for $\mathbf{P}_\omega^{(2)}$ thus vanishes. The same result is obtained for $\mathbf{P}_\omega^{(3)}$, etc. where triple differentiation of a linear function also gives zero.

Therefore, if we set $\mathbf{k} = 0$ and ignore spatial dispersion, the quadratic and higher-order nonlinear susceptibilities vanish. Note that this reasoning is not valid if the collision frequency ν is assumed to depend on velocity. Another special case is that of solid state plasma. Free charge carriers in semiconductors can be described by a kinetic equation of the same type as the gas plasma. The main difference is that the distribution function must be treated as a function of $f(\mathbf{p}, \mathbf{r}, t)$ (i.e., as a function of momenta and coordinates). In this situation the velocity is known to be a nonlinear function of momentum,

$$\mathbf{v} = \nabla_p E(\mathbf{p})$$

where $E(\mathbf{p})$ is the law of dispersion for electrons in the solid.

All expressions for the current density and polarization are transferable to the case of semiconductors, with ∇_p replacing $m^{-1}\nabla_v$, but integration is done everywhere over momenta, not velocities. The velocity $\mathbf{v}$ is now a nonlinear function of momentum and multiple differentiation of this function over p does not make the expression for $\mathbf{P}_\omega$ vanish. Therefore, in semiconductors the contribution to $\mathbf{P}_\omega^{(2)}$ due to free charge carriers (electrons and holes) is generally nonzero.

Let us now calculate $\mathbf{P}_\omega^{(2)}$ taking into account spatial dispersion; as we have justified above, we only calculate the contribution linear in $\mathbf{k}$. The parameter which determines the smallness of the effects of spatial dispersion is kv_T/ω, where $v_\mathrm{T} = \sqrt{2kT/m}$ is the thermal velocity of electrons. This parameter is equal to the ratio of the thermal velocity to the wave's phase velocity and is much less than unity. Since $v \sim v_\mathrm{T}$, we have $kv/\omega \ll 1$, and factors of the type of $(\omega - \mathbf{kv})^{-1}$ in (3.5.3) can be expanded in powers of $\mathbf{kv}/\omega$:

$$\frac{1}{\omega-\mathbf{kv}} = \frac{1}{\omega}\frac{1}{1-\mathbf{kv}/\omega} \simeq \frac{1}{\omega}\left(1+\frac{\mathbf{kv}}{\omega}\right)$$

If this and a similar expression for $(\omega' - \mathbf{k}'\mathbf{v})^{-1}$ are substituted into (3.5.3), as well as the linear in $\mathbf{k}$ expressions for $\mathbf{F}_{\mathbf{k}'\omega'}$ and $\mathbf{F}_{\mathbf{k}''\omega''}$ of (3.5.2) and if, after the multiplication has been performed, we retain only terms linear in $\mathbf{k}$, then integration over velocities in (3.5.5) gives an expression for $\mathbf{P}^{(2)}_{\mathbf{k}\omega}$:

$$\mathbf{P}^{(2)}_{\mathbf{k}\omega} = -\frac{\mathrm{i}e^3 n}{m^2\omega^2} \int \delta(\omega - \omega' - \omega'')\delta^3(\mathbf{k} - \mathbf{k}' - \mathbf{k}'') \left[\frac{(\mathbf{E}_{\mathbf{k}'\omega'}\mathbf{E}_{\mathbf{k}''\omega''})}{\omega'\omega''}\,\mathbf{k}'' + \left(\frac{\mathbf{E}_{\mathbf{k}'\omega'}}{\omega'}\left(\frac{\mathbf{k}}{\omega} + \frac{\mathbf{k}'}{\omega'} - \frac{\mathbf{k}''}{\omega''}\right)\right)\mathbf{E}_{\mathbf{k}''\omega''} + \left(\mathbf{E}_{\mathbf{k}''\omega''}\,\frac{\mathbf{k}}{\omega}\right)\frac{\mathbf{E}_{\mathbf{k}'\omega'}}{\omega'}\right] \frac{\mathrm{d}\omega'\,\mathrm{d}\omega''\,\mathrm{d}^3k'\,\mathrm{d}^3k''}{(2\pi)^4} \tag{3.5.6}$$

When calculating (3.5.6), we made use of the relations

$$\nabla_v f_0 = -\frac{m}{\mathrm{k}T}\, f_0 \mathbf{v}$$

$$\nabla_v(\mathbf{E}\nabla_v)f_0 = -\frac{m}{\mathrm{k}T}\,\nabla_v(\mathbf{E}\mathbf{v})f_0 = \left(\frac{m}{\mathrm{k}T}\right)^2 (\mathbf{E}\mathbf{v})\mathbf{v}f_0 - \frac{m}{\mathrm{k}T}\,\mathbf{E}f_0$$

Furthermore, we used identity (3.3.8) and also the identity which holds for any vectors $\mathbf{A}$, $\mathbf{B}$ and $\mathbf{C}$:

$$\int \mathbf{v}(\mathbf{A}\mathbf{v})(\mathbf{B}\mathbf{v})(\mathbf{C}\mathbf{v})\, f_0 \mathrm{d}^3v = n\left(\frac{\mathrm{k}T}{m}\right)^2 (\mathbf{A}(\mathbf{B}\mathbf{C}) + \mathbf{B}(\mathbf{A}\mathbf{C}) + \mathbf{C}(\mathbf{A}\mathbf{B}))$$

The proof of this last identity is quite similar to that of (3.3.8). Componentwise, we can write

$$\int v_\alpha v_\beta v_\gamma v_\delta f_0\, \mathrm{d}^3v = n\left(\frac{\mathrm{k}T}{m}\right)^2 (\delta_{\alpha\beta}\delta_{\gamma\delta} + \delta_{\alpha\gamma}\delta_{\beta\delta} + \delta_{\alpha\delta}\delta_{\beta\gamma})$$

Let us consider an electromagnetic field which is a superposition of two plane waves:

$$\mathcal{E}(\mathbf{r}, t) = \frac{1}{2}\,\mathbf{E}_1 \exp(\mathrm{i}\mathbf{k}_1\mathbf{r} - \mathrm{i}\omega_1 t) + \frac{1}{2}\,\mathbf{E}_2 \exp(\mathrm{i}\mathbf{k}_2\mathbf{r} - \mathrm{i}\omega_2 t) + \text{ C.C.}$$

(C.C. stands for complex-conjugate expressions). Then

$$\begin{aligned}\mathbf{E}_{\mathbf{k}\omega} &= \int \mathcal{E}(\mathbf{r}, t) \exp(-\mathrm{i}\mathbf{k}\mathbf{r} + \mathrm{i}\omega t)\, \mathrm{d}t\, \mathrm{d}^3r \\ &= \frac{(2\pi)^4}{2}\,[\mathbf{E}_1\delta(\omega - \omega_1)\delta^3(\mathbf{k} - \mathbf{k}_1) \\ &\quad + \mathbf{E}_2\delta(\omega - \omega_2)\delta^3(\mathbf{k} - \mathbf{k}_2) + \mathbf{E}_1^*\delta(\omega + \omega_1)\delta^3(\mathbf{k} + \mathbf{k}_1) \\ &\quad + \mathbf{E}_2^*\delta(\omega + \omega_2)\delta^3(\mathbf{k} + \mathbf{k}_2)]\end{aligned}$$

Substituting this expression for $\mathbf{E}_{\mathbf{k}\omega}$ into (3.5.6) and integrating over ω', ω'', $\mathbf{k}'$, and $\mathbf{k}''$, we find

$$\mathbf{P}^{(2)}_{\mathbf{k}\omega} = \frac{(2\pi)^4}{2}\,[\mathbf{P}_+\delta(\omega-\omega_1-\omega_2)\delta^3(\mathbf{k}-\mathbf{k}_1-\mathbf{k}_2) + \mathbf{P}^*_+\delta(\omega+\omega_1+\omega_2)\delta^3(\mathbf{k}+\mathbf{k}_1+\mathbf{k}_2) + \mathbf{P}_-\delta(\omega-\omega_1+\omega_2) \times \delta^3(\mathbf{k}-\mathbf{k}_1+\mathbf{k}_2) + \mathbf{P}^*_-\delta(\omega+\omega_1-\omega_2)\delta^3(\mathbf{k}+\mathbf{k}_1-\mathbf{k}_2)]$$

that is,

$$\begin{aligned}\mathcal{P}^{(2)}(\mathbf{r},t) &= \int \mathbf{P}^{(2)}_{\mathbf{k}\omega}\exp(\mathrm{i}\mathbf{k}\mathbf{r}-\mathrm{i}\omega t)\,\frac{\mathrm{d}\omega\,\mathrm{d}^3k}{(2\pi)^4}\\ &= \frac{1}{2}\,(\mathbf{P}_+\exp[\mathrm{i}(\mathbf{k}_1+\mathbf{k}_2)\mathbf{r}-\mathrm{i}(\omega_1+\omega_2)t]\\ &\quad + \mathbf{P}_-\exp[\mathrm{i}(\mathbf{k}_1-\mathbf{k}_2)\mathbf{r}-\mathrm{i}(\omega_1-\omega_2)t]) + \mathrm{C.C.}\end{aligned}$$

In these expressions, the polarization amplitude $\mathbf{P}_+$ at the sum frequency $\omega_1+\omega_2$ is

$$\mathbf{P}_+ = -\frac{\mathrm{i}e^3 n}{2m^2(\omega_1+\omega_2)\omega_1\omega_2} \times \left[(\mathbf{E}_1\mathbf{E}_2)\,\frac{\mathbf{k}_1+\mathbf{k}_2}{\omega_1+\omega_2} + (\mathbf{E}_2\mathbf{k}_2)\,\frac{\mathbf{E}_1}{\omega_2} + (\mathbf{E}_1\mathbf{k}_1)\,\frac{\mathbf{E}_2}{\omega_1}\right] \quad (3.5.7)$$

Expression (3.5.7) is automatically symmetric with respect to $\mathbf{E}_1\mathbf{k}_1\omega_1 \leftrightarrow \mathbf{E}_2\mathbf{k}_2\omega_2$. Componentwise, we can write

$$\begin{aligned}P^{(2)}_{+\alpha} &= \chi^{(2)}_{\alpha\beta\gamma}E_{1\beta}E_{2\gamma}\\ \chi^{(2)}_{\alpha\beta\gamma} &= -\frac{\mathrm{i}e^3 n}{2m^2(\omega_1+\omega_2)\omega_1\omega_2}\left[\frac{(\mathbf{k}_1+\mathbf{k}_2)_\alpha}{\omega_1+\omega_2}\,\delta_{\beta\gamma} + \frac{k_{2\gamma}}{\omega_2}\,\delta_{\alpha\beta} + \frac{k_{1\beta}}{\omega_1}\,\delta_{\alpha\gamma}\right]\end{aligned}$$

If $\mathbf{E}_1$ and $\mathbf{E}_2$ are transverse waves, then $(\mathbf{E}_1\mathbf{k}_1) = (\mathbf{E}_2\mathbf{k}_2) = 0$, so that we have from (3.5.7) that

$$\mathbf{P}_+ = -\frac{\mathrm{i}e^3 n(\mathbf{E}_1\mathbf{E}_2)}{2m^2(\omega_1+\omega_2)\omega_1\omega_2}\,(\mathbf{k}_1+\mathbf{k}_2)$$

This means that the wave vector of the wave generated is $\mathbf{k}_1 + \mathbf{k}_2$ and its polarization is along the wave vector. Two transverse waves thus generate a longitudinal wave. The longitudinal wave exists only in the plasma and is not emitted beyond its boundaries. Its damping mechanism is very specific (the Landau damping), producing the resultant nonlinear damping of transverse waves.

If the vector $\mathbf{E}_1$ is parallel to $\mathbf{k}_1$ and $\mathbf{E}_2$ is perpendicular to $\mathbf{k}_2$ and if also the vectors $\mathbf{k}_1$ and $\mathbf{k}_2$ are parallel, then (3.5.7) implies that by virtue of $(\mathbf{E}_1\mathbf{E}_2) = 0$ and $(\mathbf{E}_2\mathbf{k}_2) = 0$, we have

$$\mathbf{P}_+ = -\frac{ie^3 n(\mathbf{E}_1\mathbf{k}_1)}{2m^2(\omega_1 + \omega_2)\omega_1^2\omega_2}\mathbf{E}_2$$

In this case, a transverse wave at the sum frequency is generated.

Let us evaluate the order of quadratic nonlinearity of the plasma. We have

$$\frac{\mathcal{P}^{(2)}}{\mathcal{P}^{(1)}} \sim \frac{\chi^{(2)}E}{\chi^{(1)}} \sim \frac{e^2nkE}{m^2\omega^4}\Big/\frac{e^2n}{m\omega^2} = \frac{eE}{m\omega}\Big/\frac{\omega}{k} = \frac{v_{\text{osc}}}{v_{\text{ph}}}$$

The relative field-quadratic polarization is of the order of the ratio of the oscillational velocity v_{osc} of electrons in the field to the phase velocity of the wave, v_{ph}. It could be expected that the relative nonlinear polarization equals the ratio of the oscillational velocity of electrons in the field to the thermal electron velocity v_{T}. However, since the linear polarization is a result of spatial dispersion, an additional factor arises, $v_{\text{T}}/v_{\text{ph}}$, which is nothing but the smallness parameter for nonlocality. As a result,

$$\frac{\mathcal{P}^{(2)}}{\mathcal{P}^{(1)}} \sim \frac{v_{\text{osc}}}{v_{\text{T}}}\frac{v_{\text{T}}}{v_{\text{ph}}} = \frac{v_{\text{osc}}}{v_{\text{ph}}}$$

This ratio replaces $\mathcal{E}/\mathcal{E}_{\text{at}}$ which characterizes, as was shown in Chapter 2, the smallness of nonlinear polarization and the possibility of expanding the polarization in powers of field in the case of bound electrons.

3.6. Quantum Description of the Effect of Time-Dependent External Field on Collision Processes

We will briefly consider in terms of quantum mechanics the effects of high-frequency external fields on electron–ion collisions.

We begin with a quantum-mechanical analysis in the framework of the dipole approximation of the simplest problem of interaction between a free electron and the electromagnetic wave field. The amplitude of an electron's oscillations in the wave field, $eE/m\omega^2$, is much less than the radiation wavelength, which justifies the validity of the dipole approximation.

We have mentioned in Section 1.1 that in the dipole approximation, the field can be described in terms of only the scalar potential, so that

$$\varphi(\mathbf{r}, t) = -\mathbf{E}(t)\mathbf{r}$$

or only the vector potential

$$\mathbf{A}(\mathbf{r}, t) = -c\int^{t}\mathbf{E}(t')\,dt'$$

These two representations are related by a gauge transformation, with a function

$$\chi(\mathbf{r},t) = c\mathbf{r}\int^{t} \mathbf{E}(t')\,\mathrm{d}t' = -\mathbf{A}(t)\mathbf{r}$$

It should be pointed out that the wave function is also transformed in going from one representation to another:

$$\psi'(\mathbf{r},t) = \psi(\mathbf{r},t)\exp\left\{\frac{\mathrm{i}e}{\hbar c}\chi(\mathbf{r},t)\right\}$$

It is easier to find the wave function for the harmonic field $\mathbf{E}(t) = \mathbf{E}_0\cos\omega t$ using the Schrödinger equation and the vector potential

$$\mathrm{i}\hbar\frac{\partial\psi}{\partial t} = \frac{1}{2m}\left(-\frac{\mathrm{i}}{\hbar}\nabla - \frac{e}{c}\mathbf{A}(t)\right)^2\psi$$

where $\mathbf{A}(t) = -(c/\omega)E_0\sin\omega t$.

In this particular case, the Hamiltonian remains explicitly independent of the electron's coordinates, so that the spatial dependence of the wave function retains the form $\exp(\mathrm{i}\mathbf{pr}/\hbar)$, which it had for the free electron. The wave function has the form

$$\psi(\mathbf{r},t) = \phi(t)\exp(\frac{\mathrm{i}}{\hbar}\mathbf{pr})$$

where $\phi(t)$ satisfies the equation

$$\mathrm{i}\hbar\frac{\mathrm{d}\phi}{\mathrm{d}t} = \frac{1}{2m}\left(\mathbf{p} - \frac{e}{c}\mathbf{A}(t)\right)^2\phi \tag{3.6.1}$$

that is,

$$\begin{aligned}\phi(t) &= \exp\left\{-\frac{\mathrm{i}}{2m\hbar}\int^{t}\left(\mathbf{p} - \frac{e}{c}\mathbf{A}(t')\right)^2\mathrm{d}t'\right\}\\ &= \exp\left\{-\frac{\mathrm{i}}{\hbar}\left(\frac{p^2}{2m}t - \frac{e\mathbf{p}\mathbf{E}_0}{m\omega^2}\cos\omega t + \frac{e^2E_0^2}{4m\omega^2}\left(t - \frac{\sin 2\omega t}{2\omega}\right)\right)\right\}\end{aligned}$$

Equation (3.6.1) shows that the expression

$$\frac{1}{2m}\left(\mathbf{p} - \frac{e}{c}\mathbf{A}(t)\right)^2 = \frac{p^2}{2m} + \frac{e\mathbf{p}\mathbf{E}_0}{m\omega}\sin\omega t + \frac{e^2E_0^2}{2m\omega^2}\sin^2\omega t$$

gives the electron's energy and that its average value equals the sum of the kinetic energy of the electron's translational motion and the energy of its oscillations in the oscillating field.

In view of all this, and assuming that we use the dipole approximation in the gauge in which the interaction energy is written as $\mathcal{H}' = -e\mathbf{r}\mathbf{E}(t)$ (i.e. the scalar potential is used), the electron wave function in a time-dependent field has the form

$$\psi_{\mathbf{p}}(\mathbf{r},t) = \exp\left\{\frac{\mathrm{i}}{\hbar}\left(\mathbf{p} + e\int^{t}\mathbf{E}(t')\,\mathrm{d}t'\right)\mathbf{r}\right.$$
$$\left. - \frac{\mathrm{i}}{2m\hbar}\int^{t}\mathrm{d}t'\left(\mathbf{p} + e\int^{t'}\mathbf{E}(t'')\,\mathrm{d}t''\right)^{2}\right\}$$
$$= \exp\left\{\frac{\mathrm{i}}{\hbar}\left(\mathbf{p} + \frac{e\mathbf{E}_0}{\omega}\sin\omega t\right)\mathbf{r} - \frac{\mathrm{i}}{2m\hbar}\int^{t}\left(\mathbf{p} + \frac{e\mathbf{E}_0}{\omega}\sin\omega t'\right)^{2}\mathrm{d}t'\right\} \tag{3.6.2}$$

We will now consider transitions between the states of a free electron in an oscillating field, treating the ion potential $U(\mathbf{r})$ as a perturbation (the Born approximation). Let the electron momentum in the initial state be $\mathbf{p}_1$. The amplitude $a(\mathbf{p}_2)$ of a state with momentum $\mathbf{p}_2$ to first order in perturbation theory obeys the equation

$$\mathrm{i}\hbar\frac{\partial a(\mathbf{p}_2,t)}{\partial t} = \int\psi^{*}_{\mathbf{p}_2}(\mathbf{r},t)U(\mathbf{r})\psi_{\mathbf{p}_1}(\mathbf{r},t)\,\mathrm{d}^3r = \exp\left(\frac{\mathrm{i}}{\hbar}\epsilon t\right)Q(\mathbf{q})f(t) \tag{3.6.3}$$

where ϵ is the difference between the kinetic energies of the translational motion in the initial and the final states:

$$\epsilon = \frac{1}{2m}\left(p_2^2 - p_1^2\right)$$

and $Q(\mathbf{q})$ is given by the integral

$$Q(\mathbf{q}) = \int U(\mathbf{r})\exp\left(-\frac{\mathrm{i}\mathbf{q}\mathbf{r}}{\hbar}\right)\mathrm{d}^3r$$

in which $\mathbf{q} = \mathbf{p}_2 - \mathbf{p}_1$ is the change in the electron momentum due to scattering.

The function $f(t)$ is

$$f(t) = \exp\left\{-\frac{\mathrm{i}}{\hbar}\frac{e(\mathbf{p}_2 - \mathbf{p}_1)\mathbf{E}_0}{m\omega^2}\cos\omega t\right\}$$
$$= \exp\left\{-\frac{\mathrm{i}}{\hbar}\frac{e(\mathbf{p}_2 - \mathbf{p}_1)\mathbf{E}_0}{m\omega^2}\sin\left(\omega t + \frac{\pi}{2}\right)\right\}$$

This periodic function can be expanded into a Fourier series

$$f(t) = \sum_{n=-\infty}^{\infty} c_n e^{-in\omega t} \tag{3.6.4}$$

whose expansion coefficients c_n are

$$c_n = \frac{\omega}{2\pi} \int_{-\pi/\omega}^{\pi/\omega} e^{in\omega t} f(t)\, dt$$

In view of the integral representation of the Bessel function,

$$J_n(z) = \frac{1}{2\pi} \int_{-\pi}^{\pi} e^{-inx + iz\sin x}\, dx$$

we obtain the following expressions for the coefficients c_n:

$$c_n = (-i)^n J_n \left(\frac{e\mathbf{E}_0}{m\hbar\omega^2} (\mathbf{p}_2 - \mathbf{p}_1) \right)$$

Substituting (3.6.4) into (3.6.3) and integrating, we obtain

$$a(\mathbf{p}_2, t) = -\frac{i}{\hbar} Q(\mathbf{q}) \int_{t_0}^{t} \sum_n c_n \exp\left\{ \frac{i}{\hbar} (\epsilon - n\hbar\omega) t' \right\} dt' \tag{3.6.5}$$

The rate of transition to an arbitrary final state is

$$\begin{aligned}
w &= \int \left\{ \lim_{t\to\infty} \frac{d}{dt} |a(\mathbf{p}_2, t)|^2 \right\} \frac{d^3 p_2}{(2\pi\hbar)^3} \\
&= \frac{1}{\hbar^2} \int \frac{d^3 p_2}{(2\pi\hbar)^3} |Q(\mathbf{q})|^2 \lim_{t\to\infty} \sum_n |c_n|^2 \\
&\quad \times \left(\exp\left\{ -\frac{i}{\hbar} (\epsilon - n\hbar\omega) t \right\} \int_{t_0}^{t} \exp\left\{ \frac{i}{\hbar} (\epsilon - n\hbar\omega) t' \right\} dt' \right. \\
&\quad \left. + \exp\left\{ \frac{i}{\hbar} (\epsilon - n\hbar\omega) t \right\} \int_{t_0}^{t} \exp\left\{ -\frac{i}{\hbar} (\epsilon - n\hbar\omega) t' \right\} dt' \right) \\
&= \frac{1}{\hbar^2} \int \frac{d^3 p_2}{(2\pi\hbar)^3} |Q(\mathbf{q})|^2 \sum_n |c_n|^2 \int_{-\infty}^{\infty} \exp\left\{ \frac{i}{\hbar} (\epsilon - n\hbar\omega) t' \right\} dt' \\
&= \frac{2\pi}{\hbar} \int |Q(\mathbf{q})|^2 \sum_n |c_n|^2 \delta(\epsilon - n\hbar\omega) \frac{d^3 p_2}{(2\pi\hbar)^3}
\end{aligned} \tag{3.6.6}$$

A double sum over n and n' arises in the calculation of (3.6.6). However, nondiagonal terms with $n \neq n'$ bring oscillating contributions to the transition rate w and vanish under averaging over time.

The coefficient with $\delta(\epsilon - n\hbar\omega)$ in (3.6.6) is the transition rate $w^{(n)}$ when photons are absorbed ($n > 0$) or emitted ($n < 0$). This process of absorption or emission of photons at the frequency ω of the external field when scattering occurs on a static potential is known as *induced bremsstrahlung*.

The transition rate in the continuous spectrum depends on the normalization of wave functions. Hence, it is more convenient, when dealing with such transitions, to operate with the transition cross section (the ratio of transition rate to flux density in the initial state). Deriving a formula, analogous to (3.6.6), for the transition rate $\overline{w}$ in the scattering by a static potential without time-dependent fields, we obtain

$$\begin{aligned}\overline{w} &= \frac{2\pi}{\hbar}\int |Q(\mathbf{q})|^2 \delta(\epsilon)\,\frac{\mathrm{d}^3 p}{(2\pi\hbar)^3} \\ &= \frac{m}{(2\pi\hbar^2)^2}\int |Q(\mathbf{q})|^2 p_2 \delta(p_2^2 - p_1^2)\,\mathrm{d}\Omega\,\mathrm{d}(p_2^2) \\ &= \frac{m p_2}{(2\pi\hbar^2)^2}\int |Q(\mathbf{q})|^2\,\mathrm{d}\Omega\end{aligned}$$

Here $\mathrm{d}\Omega$ is an element of solid angle.

The incident flux density in the normalization selected equals the velocity of electrons, $v = p_1/m$, so that the differential scattering cross section into unit solid angle is

$$\frac{\mathrm{d}\sigma}{\mathrm{d}\Omega} = \frac{\overline{w} m}{p_1} = \frac{m^2}{(2\pi\hbar^2)^2}\,|Q(\mathbf{q})|^2$$

since for elastic scattering we have $p_1 = p_2$.

Correspondingly, we find for the differential scattering cross sections with absorption ($n > 0$) or emission ($n < 0$) of photons (see [17])

$$\frac{\mathrm{d}\sigma^{\pm n}}{\mathrm{d}\Omega} = \frac{\mathrm{d}\sigma}{\mathrm{d}\Omega}\,\frac{p_2}{p_1}\,J_n^2\left(\frac{e\mathbf{E}_0\mathbf{q}}{m\hbar\omega^2}\right) \qquad (3.6.7)$$

Note that owing to δ functions, $p_2^2 = p_1^2 + 2mn\hbar\omega$.

If we use (3.6.7) to calculate the power absorbed by unit volume of plasma, we find that one-photon transitions ($n = \pm 1$) are predominant up to very strong fields; the power absorbed is shown to be a linear function of the incident power, in agreement with the classical theory.

3.7. Ionization in Intense Electromagnetic Field

We will calculate the probability of ionization of an atom in the field of an intense electromagnetic field whose frequency ω is less than $I/\hbar$ (I is the ionization potential). In the limiting case of low frequencies, the expressions to be derived below transform into familiar formulas for tunneling ionization; at high frequencies, these expressions describe processes in which several photons are simultaneously absorbed.

The electromagnetic field affects most strongly the states of the continuum spectrum. As a first step, therefore, we will treat only these effects. To do this, we will use the wave function of the free electron in electric field (3.6.2).

First we calculate the probability of transition from the bound state (the ground state) of the atom to a (3.6.2)-like free electron state. The difference between the procedure used here and the ordinary perturbation theory thus lies only in that we calculate the probability of transition not to the stationary final state but to state (3.6.2) which takes into account exactly the main effect of the electric field: the acceleration imparted to the free electron. It is sufficient to include in the calculation the matrix elements of transition between atomic bound states only to lower orders of perturbation theory, since they are proportional to eE_0a_0; we will soon see that the matrix elements of continuous spectrum which we take into account are proportional to $eE_0a_0\sqrt{I/\hbar\omega}$. We assume that the ratio of the photon energy to the ionization potential is small. For lasers in the optical and infrared ranges, this ratio is about 0.1 or less.

Similarly to (3.6.6), the probability of direct transition from the atomic ground state ψ_0 with energy $-I$ to the continuous spectrum has the form

$$w_0 = \int \left\{ \lim_{t\to\infty} \frac{\mathrm{d}}{\mathrm{d}t} |b(\mathbf{p},t)|^2 \right\} \frac{\mathrm{d}^3p}{(2\pi\hbar)^3}$$

where $b(\mathbf{p},t)$ satisfies the equation

$$\mathrm{i}\hbar \frac{\partial b(\mathbf{p},t)}{\partial t} = \int \psi^*_{\mathbf{p}}(\mathbf{r},t)(-e\mathbf{r}\mathbf{E}_0 \cos\omega t)\psi_0(\mathbf{r})e^{\mathrm{i}It/\hbar}\,\mathrm{d}^3r$$

According to this equation,

$$b(\mathbf{p},t) = \frac{\mathrm{i}}{\hbar} L(\mathbf{p},t)$$

where

$$L(\mathbf{p},t) = \int_{t_0}^{t} \mathrm{d}t' \exp\left\{ \frac{\mathrm{i}}{\hbar} \int_0^{t'} \left[I + \frac{1}{2m}\left(\mathbf{p} + \frac{e\mathbf{E}_0}{\omega}\sin\omega t''\right)^2 \right] \mathrm{d}t'' \right\}$$
$$\times V_0\left(\mathbf{p} + \frac{e\mathbf{E}_0}{\omega}\sin\omega t'\right)$$

and $V_0(\mathbf{p})$ is a matrix element,

$$V_0(\mathbf{p}) = \int \exp\left\{-\frac{\mathrm{i}}{\hbar}\mathbf{p}r\right\} e\mathbf{r}\mathbf{E}_0\psi_0(\mathbf{r})\,\mathrm{d}^3r$$

Expanding $L(\mathbf{p},t)$ into a Fourier series

$$L(\mathbf{p},t) = \sum_n \exp\left\{\frac{\mathrm{i}}{\hbar}\left(I + \frac{p^2}{2m} + \frac{e^2E_0^2}{4m\omega^2} - n\hbar\omega\right)\right\} L_n(\mathbf{p})$$

where

$$L_n(\mathbf{p}) = \frac{1}{2\pi}\int_{-\pi}^{\pi} V_0\left(\mathbf{p} + \frac{e\mathbf{E}_0}{\omega}\sin x\right)\exp\left\{\frac{\mathrm{i}}{\hbar\omega}\left(n\hbar\omega x - \frac{e\mathbf{E}_0\mathbf{p}}{m\omega}(\cos x - 1) - \frac{e^2E_0^2}{8m\omega^2}\sin 2x\right)\right\}\mathrm{d}x \tag{3.7.1}$$

we obtain for the probability w_0 the expression

$$\begin{aligned} w_0 &= \int\left\{\lim_{t\to\infty}\frac{\mathrm{d}}{\mathrm{d}t}|b(\mathbf{p},t)|^2\right\}\frac{\mathrm{d}^3p}{(2\pi\hbar)^3} \\ &= \frac{2\pi}{\hbar}\int\frac{\mathrm{d}^3p}{(2\pi\hbar)^3}\sum_n\frac{1}{4}\left|L_{n+1}(\mathbf{p}) + L_{n-1}(\mathbf{p})\right|^2 \\ &\quad\times\delta\left(I + \frac{p^2}{2m} + \frac{e^2E_0^2}{4m\omega^2} - n\hbar\omega\right) \end{aligned} \tag{3.7.2}$$

Owing to the δ functions in (3.7.2), we can set $n\hbar\omega = I + p^2/(2m) + e^2E_0^2/(4m\omega^2)$ in the expressions for L_{n+1} and L_{n-1}, which are derived from (3.7.1). As a result, instead of (3.7.2), we obtain

$$\begin{aligned} w_0 = \frac{2\pi}{\hbar}\int\frac{\mathrm{d}^3p}{(2\pi\hbar)^3}\sum_n &\left|\frac{1}{2\pi}\int_{-\pi}^{\pi} V_0\left(\mathbf{p} + \frac{e\mathbf{E}_0}{\omega}\sin x\right)\right. \\ &\left.\times\exp\left\{\frac{\mathrm{i}}{\hbar\omega}\int_0^x\left[I + \frac{1}{2m}\left(\mathbf{p} + \frac{e\mathbf{E}_0}{\omega}\sin x'\right)^2\right]\mathrm{d}x'\right\}\cos x\,\mathrm{d}x\right|^2 \\ &\times\delta\left(I + \frac{p^2}{2m} + \frac{e^2E_0^2}{4m\omega^2} - n\hbar\omega\right) \end{aligned}$$

Substituting $u = \sin z$ we transform the integral over z to an integral over a closed contour encircling the segment $(-1, 1)$. This gives

$$w_0 = \frac{2\pi}{\hbar}\int\frac{\mathrm{d}^3p}{(2\pi\hbar)^3}|L(\mathbf{p})|^2\sum_{n=-\infty}^{\infty}\delta\left(I + \frac{p^2}{2m} + \frac{e^2E_0^2}{4m\omega^2} - n\hbar\omega\right) \tag{3.7.3}$$

where

$$L(\mathbf{p}) = \frac{1}{2\pi} \oint V_0\left(\mathbf{p} + \frac{e\mathbf{E}_0}{\omega}u\right) \times \exp\left\{\frac{i}{\hbar\omega}\int_0^u \left[I + \frac{1}{2m}\left(\mathbf{p} + \frac{e\mathbf{E}_0}{\omega}v\right)^2\right]\frac{dv}{\sqrt{1-v^2}}\right\} du \quad (3.7.4)$$

Formula (3.7.3) is a sum of multiphoton processes. The exponential in (3.7.4) is rapidly oscillating, so the integral can be calculated by the steepest descent method. For details of calculations, see [18]. The specific feature here lies in the matrix element $V_0(\mathbf{p} + e\mathbf{E}_0 u/\omega)$ having poles in the saddle points that are found from the condition

$$I + \frac{1}{2m}\left(\mathbf{p} + \frac{e\mathbf{E}_0}{\omega}u_s\right)^2 = 0$$

This result can be verified by using the matrix element $V_0(\mathbf{p})$ for the ground state of the hydrogen atom,

$$\begin{aligned} V_0(\mathbf{p}) &= \frac{e\mathbf{E}_0}{\sqrt{\pi a_0^3}} \int e^{-i\mathbf{p}r/\hbar}\mathbf{r}e^{-r/a_0}\, d^3r \\ &= \frac{ie\hbar\mathbf{E}_0}{\sqrt{\pi a_0^3}} \frac{\partial}{\partial \mathbf{p}} \int e^{-i\mathbf{p}r/\hbar} e^{-r/r_0}\, d^3r \\ &= 8i\sqrt{\pi a_0^3}\, e\hbar\mathbf{E}_0 \frac{\partial}{\partial \mathbf{p}} \frac{1}{(1 + p^2a_0^2/\hbar^2)^2} \end{aligned}$$

and taking into account that the ionization potential I of the hydrogen atom equals one Rydberg, $I = \hbar^2/(2ma_0^2)$. This is not, however, a feature unique for hydrogen; it was shown to hold in the general case and to follow from the analytic properties of the scattering amplitude.

Appropriate transformations which include integration over $\mathbf{p}$ (taking account of the fact that only small $\mathbf{p}$ (such that $p^2 \ll 2mI$) contribute to the total probability) and also introduce a correction factor for the Coulomb interaction in the final "free" state, result in an approximate formula (see [18])

$$w_0 = A\omega\left(\frac{I}{\hbar\omega}\right)^{3/2}\left(\frac{\gamma}{\sqrt{1+\gamma^2}}\right)^{5/2} S\left(\gamma, \frac{\tilde{I}}{\hbar\omega}\right) \times \exp\left\{-\frac{2\tilde{I}}{\hbar\omega}\left[\operatorname{arc\,sinh}\gamma - \gamma\frac{\sqrt{1+\gamma^2}}{1+2\gamma^2}\right]\right\} \quad (3.7.5)$$

where $\gamma = \omega\sqrt{2mI}/eE_0$ is the so-called adiabaticity parameter. As shown below, it is equal to the product of the field frequency by the characteristic time of the escape of an electron from the atom by tunneling in the applied field. The effective ionization potential in (3.7.5),

$$\tilde{I} = I + \frac{e^2E_0^2}{4m\omega^2} = I\left(1 + \frac{1}{2\gamma^2}\right)$$

is greater than the ionization potential I by the kinetic energy of the electron's oscillations in the electromagnetic wave field. It is this quantity that we find in the δ functions that implement the energy conservation law in the general formula (3.7.4).

Furthermore, A in formula (3.7.5) is a numerical factor on the order of unity and $S(\gamma, \tilde{I}/\hbar\omega)$ is a function of frequency and field which varies slowly in comparison with an exponential factor:

$$S(\gamma, x) = \sum_{n=0}^{\infty} \exp\left\{-2[< x+1 > -x + n]\left(\operatorname{arc\,sinh}\gamma - \frac{\gamma}{1+\gamma^2}\right)\right\}$$
$$\times\, \Phi\left\{\left[\frac{2\gamma}{\sqrt{1+\gamma^2}}\right]^{1/2}\right\} \qquad (3.7.6)$$

Here $\langle x\rangle$ stands for the integral part of a number x and the function $\Phi(z)$ is Dawson's integral

$$\Phi(z) = \int_0^z e^{y^2 - z^2}\, dy$$

which is closely connected with the familiar error function $\operatorname{erf}(z)$. The function $S(\gamma, \tilde{I}/\hbar\omega)$ describes the structure of the spectrum which is related to the discrete nature of the number of photons absorbed; it displays typical $\sqrt{\tilde{I} - n\hbar\omega}$-type threshold singularities.

In the case of low frequencies and very high fields, with the adiabaticity parameter $\gamma \ll 1$, large numbers of photons $n \sim \gamma^{-3}$ produce the main contribution in the expression for S. Hence, it is possible to switch in calculating S from summation over n to integration; the result is $S(\gamma, \tilde{I}/\hbar\omega) \approx \sqrt{3\pi}/4\gamma^2$. We then arrive at a formula which transforms for $\omega \to 0$ to the expression for the probability of tunneling ionization of atoms in the electric field. The tunneling autoionization means that electrons tunnel across a barrier when an external field is applied. An essential feature of tunneling is that it is virtually free of inertia: the tunneling probability remains unchanged up to very high frequencies. The reason for this is that the tunneling time is determined essentially by the time of free flight of an electron across the barrier of width $l = I/(eE)$, where E is the instantaneous value of the electric field strength. Since the mean electron velocity is of the order of $\sqrt{I/m}$, the tunneling effect is determined simply by the instantaneous field strength up to frequencies of the order

$$\omega_t = \frac{eE}{\sqrt{2mI}}$$

The frequency dependence of the tunneling probability must be observed only at frequencies $\omega > \omega_t$, since electrons are not given enough time to pierce the barrier during one period. Estimates show that the dispersion can become appreciable at infrared and optical frequencies. Note that the parameter γ is the ratio of frequency to the frequency ω_t:

$$\gamma = \omega / \omega_t$$

In the opposite extreme case of high frequencies and moderately high fields, when the zeroth term is predominant in (3.7.6), formula (3.7.5) describes the probability of simultaneous absorption of several photons.

Formula (3.7.5) can be generalized to a situation in which an atom can transfer to a continuum state not directly from the ground state but via an intermediate excited state.

Experiments with single atoms (e.g., see review [19]) give a qualitative corroboration of switching between the tunneling and multiphoton ionization modes, which is implied by the theory. Constructing a quantitative theory meets with formidable difficulties, since it is necessary to take into account various transitions between the levels of the discrete spectrum and also to calculate more carefully the effect of the Coulomb field on the final state of the electron.

A subject of intense interest in recent years is the multiphoton ionization by microwave field of atoms in high-excitation states (Rydberg states). This process greatly resembles the random walk of atoms among energy levels and can be described in many respects by the classical theory (see reviews [20] and [21]).

4

Atoms and Molecules in Electromagnetic Fields

4.1. Symmetry of Atoms and Molecules. Fundamentals of the Theory of Symmetry

The general theory of interaction between electromagnetic radiation and matter, as presented in Chapter 2, assumed that the wave functions of the stationary states of matter and its energy spectrum were known. The calculation of wave functions and energy levels is a problem which is exactly solved in quantum mechanics only for the simplest systems (a hydrogen atom, an oscillator). If a system is more complex, beginning with the two-electron helium atom and the hydrogen molecule, one can only hope to obtain a more or less accurate — always approximate — calculation of the spectrum and the wave functions.

Nevertheless, an exhaustive and exact classification of the states of atoms and molecules, based on the symmetry of these states, can always be worked out. In terms of quantum mechanics this means that we can find the quantum numbers which characterize various states of the system. The classification in types of symmetry is possible without solving the relevant equations of quantum mechanics. Symmetry properties also tell us much about matrix elements, and thus about selection rules, that is, they indicate when matrix elements vanish and when they are nonzero. In high-symmetry states, many matrix elements are identically zero, so that selection rules constitute a very important type of information about matrix elements.

Group theory, whose concepts and theorems describe symmetry properties, is widely used to study the symmetry of atoms and molecules and also the types of symmetry of their states. We will outline the main definitions of group theory and their corollaries, which we will need in this book.

It will be convenient to introduce the notion of *group* using the symmetry group of a molecule as an example. As a rule, an equilibrium configuration of nuclei in a molecule possesses high symmetry. For example, carbon nuclei of the benzene molecule C_6H_6 are located at the vertices of an equilateral planar hexagon. A methane molecule CH_4 is a regular tetrahedron in which a carbon atom is at the center and hydrogen atoms are at the vertices.

The symmetry of a solid or a molecule is determined by the set of displacements which map the solid (or the molecule) onto itself. Such displacements are known as *symmetry operations*, or *symmetry elements*. If there are two symmetry operations, a consecutive application of these operations generates a new symmetry operation which is called a *product of operations*. In general, the operation product depends on the order in which the constituent operations are applied; this means that an operation product is not necessarily commutative. Each of the possible symmetry operations can be viewed as a product of operations of several basic types: rotation around an axis, mirror reflection in a plane, and translation by a certain distance. This last type of symmetry operation is relevant only to an infinite body, such as a crystal. In finite bodies, and thus in atoms and molecules, only symmetry operations of the first two types and their combinations are possible.

If a rotation by an angle φ is a symmetry operation, then m-fold repetition of this operation results in the rotation by the angle $m\varphi$. Hence, there must exist an n such that $n\varphi = 2\pi$, that is, $\varphi = 2\pi/n$. One then speaks of an n-fold symmetry axis. The corresponding operation is denoted by C_n.

If a body is mapped onto itself by reflection in a plane, one speaks of a plane of symmetry. The operation which consists in the reflection in a plane is denoted by σ.

A symmetry operation product S_n is possible of a rotation and a reflection in a plane orthogonal to the symmetry axis. This symmetry element is known as an n-fold reflection–rotation axis. A body may have this symmetry while the rotation or the reflection in themselves are not its symmetries.

An important case of symmetry is *inversion* I, in which $\mathbf{r} \to -\mathbf{r}$. Obviously, $I = S_2 = C_2\sigma = \sigma C_2$.

The set of all symmetry elements of a given body (or molecule) is called its group of symmetry elements, or simply its *symmetry group*. A set of elements (in our case, a set of operations) is said to be a group if the following axioms are satisfied:

1. A multiplication operation is defined on the set of elements, which puts each pair of elements f and g in correspondence with some other element h of the same set; this is written as $h = fg$. In the general case, $fg \neq gf$;
2. Multiplication is associative: $f(gh) = (fg)h$;
3. There is an identity element e such that for any f, $ef = fe = f$;

4. For any f, there is an inverse element f^{-1} for which $ff^{-1} = f^{-1}f = e$.

If the number of elements in a group is finite, the group is said to be *finite*; otherwise this is an infinite group. The number of elements in a finite group is called its *order*.

If the multiplication in a group is commutative, that is, if $fg = gf$ for any f and g, the group is said to be commutative, or *Abelian*.

An element g is said to be conjugate to an element f if there is an element of the group x such that $xgx^{-1} = f$. If g is adjoint to f, then f is adjoint to g. All mutually adjoint elements form the class of mutually adjoint elements. The class of elements adjoint to e consists of a single element e. Each class of a commutative group consists of a single element because $xgx^{-1} = g$.

A subgroup is defined as a subset of a group if this subset is closed, that is, if it is a group with respect to the same group multiplication.

Two groups are said to be isomorphic if one-to-one correspondence $f \leftrightarrow f'$ can be established between their elements, such that $fg = h$ implies $f'g' = h'$.

Operations forming the symmetry group of a finite body (a molecule) must be such that one point should always stay fixed, that is, all symmetry planes and axes must have a common intersection point. Indeed, rotations of a body around nonintersecting axes or reflections from nonintersecting planes result in a translational motion of the body, which is an operation forbidden for a finite body. Symmetry groups that include only operations with a fixed point are known as *point groups*. There is a complete classification of all finite point groups; they classify into 14 types. These groups are listed and described in textbooks on quantum mechanical applications of group theory.

Here are several examples of molecules and group symmetries they belong to:

- H_2O — group C_{2v}
- CH_3F — group C_{3v}
- BF_3 — group D_{3h}
- C_6H_6 — group D_{6h}
- CH_4 — group T_d (tetrahedral symmetry group)
- CF_6 — group O_h (cubic symmetry group).

The group notation gives the symmetry operations contained in the group. The symbol C_{2v} indicates that there is a twofold symmetry axis and a symmetry plane passing through the axis. The symbol D_{3h} indicates that in addition to a threefold axis, there are twofold axes orthogonal to it and a symmetry plane orthogonal to the main threefold axis (a symmetry plane orthogonal to the axis is indicated by the index h).

The group symmetry describing linear molecules is $C_{\infty v}$, which includes all rotations by arbitrary angles around the symmetry axis and the reflection in the plane passing through the axis. If a diatomic molecule consists of two identical atoms, its symmetry group $D_{\infty h}$ contains, in addition to the infinite-fold symmetry axis, also twofold axes orthogonal to the main axis and a symmetry plane orthogonal to the axis of the molecule.

The last two are examples of infinite groups. Another important example of infinite groups is the group of rotations K in the three-dimensional space. If inversion I is added to the group of rotations, we obtain the group K_h of rotations and reflections in three-dimensional space.

We will now introduce the concept of the *representation* of a group, using as an example operations of electron wave functions of a molecule. Instead of treating symmetry operations as actual rotations and mirror reflections of the object under consideration, it is sometimes more convenient to regard the object as fixed (not moving) and compare its descriptions in different systems of coordinates which differ in rotations about one of the axes or in sign reversal for one of the axes (i.e., a reflection). Note that any symmetry operation applied to a molecule can be treated as a transformation of coordinates or as a corresponding matrix of transformation of coordinates written as

$$x'_\alpha = g_{\alpha\beta} x_\beta$$

where $g_{\alpha\beta}$ is an orthogonal transformation matrix. If two transformations $x'_\alpha = g^{(1)}_{\alpha\beta} x_\beta$ and $x''_\alpha = g^{(2)}_{\alpha\beta} x'_\beta$ are performed consecutively, then $x''_\alpha = g^{(2)}_{\alpha\beta} g^{(1)}_{\beta\gamma} x_\gamma$, that is, the product of symmetry operations corresponds to the product of matrices, $g_{\alpha\gamma} = g^{(2)}_{\alpha\beta} g^{(1)}_{\beta\gamma}$. The group of orthogonal matrices forms a group which is isomorphic to the group of symmetry operations (rotations and reflections).

With this more formal approach, it is also possible to analyze a partial symmetry of the system with respect to the transformation of a particular set of coordinates (of electrons, for instance) with all other coordinates unchanged (e.g., those of nuclei) or assuming that the transformation in question is a symmetry element for the second set of coordinates (e.g., it maps the nuclear "skeleton" of the molecule onto itself). Obviously, each transformation of the coordinates of all electrons, corresponding to a symmetry operation on a molecule, transforms the electron wave function while leaving the configuration of nuclei unaltered. The electron wave function $\psi(\mathbf{r})$ transforms into

$$\psi'(\mathbf{r}) = \psi(g^{-1}\mathbf{r}) = D_g \psi(\mathbf{r})$$

where $g^{-1}\mathbf{r}$ are the "old" coordinates given in terms of the "new," transformed ones. The transformation matrix is thus put in correspondence with a linear operator D_g which is defined by the above equality. This gives

$$D_{g_2}D_{g_1}\psi(\mathbf{r}) = D_{g_2}\psi(g_1^{-1}\mathbf{r}) = \psi(g_1^{-1}g_2^{-1}\mathbf{r}) = \psi((g_2g_1)^{-1}\mathbf{r}) = D_{g_2g_1}\psi(\mathbf{r})$$

The group of operators D_g is thus isomorphic to the group of symmetry operations g or the group of matrices $g_{\alpha\beta}$. Let $\mathscr{H}_{0\mathrm{e}}$ be the Hamiltonian of electrons of a molecule in the field created by its nuclei. In an explicit form, it can be written as

$$\mathscr{H}_{0\mathrm{e}} = -\frac{\hbar}{2m}\sum_i \nabla_i^2 + U(\mathbf{r}, \mathbf{R})$$

where the index i enumerates electrons and U is the potential energy of the Coulomb interaction of electrons between themselves and with nuclei. Whatever an orthogonal transformation g, the operator ∇_i^2 remains unchanged and the same is true for the potential energy U which depends on distances between particles, provided g is a symmetry operation for the molecule.

Therefore, $D_g\mathscr{H}_{0\mathrm{e}}\psi = \mathscr{H}_{0\mathrm{e}}D_g\psi$ and, hence, $\mathscr{H}_{0\mathrm{e}}D_g = D_g\mathscr{H}_{0\mathrm{e}}$, that is,

$$\mathscr{H}_{0\mathrm{e}} = D_g^{-1}\mathscr{H}_{0\mathrm{e}}D_g$$

The Hamiltonian commutes with the operators D_g which form a group isomorphic to the symmetry group g. This is a definition of the invariance of the Hamiltonian with respect to the transformations g; one says in this case that g forms the symmetry group of the Hamiltonian.

Let ψ_n be an eigenfunction of the Hamiltonian, corresponding to the energy value E_n. Hence,

$$\mathscr{H}_{0\mathrm{e}}D_g\psi_n = D_g\mathscr{H}_{0\mathrm{e}}\psi_n = E_nD_g\psi_n$$

that is, $D_g\psi_n$ is an eigenfunction of $\mathscr{H}_{0\mathrm{e}}$, which corresponds to the same value E_n.

If E_n is a nondegenerate energy level, then $D_g\psi_n$ must coincide, up to a factor whose modulus equals unity, with ψ_n. The normalization of the function ψ_n is conserved because D_g is a unitary operator which conserves the integral for any pair of functions ψ_n and ψ_n',

$$\int \psi_{n'}^*\psi_n \,\mathrm{d}^3r = \int (D_g\psi_{n'})^* D_g\psi_n \,\mathrm{d}^3r$$

since D_g is reducible to an orthogonal transformation of coordinates.

Assume now that the level E_n is k-times degenerate. Then $D_g\psi_n^{(r)}$ $(r = 1, \ldots, k)$ is an eigenfunction of $\mathscr{H}_{0\mathrm{e}}$ with the same energy value E_n; hence, this is a linear combination of the functions $\psi_n^{(r)}$:

$$D_g\psi_n^{(r)} = \sum_{r'} A_{r'r}^{(g)}\psi_n^{(r')}$$

Furthermore,

$$D_{g_2g_1}\psi_n^{(r)} = D_{g_2}D_{g_1}\psi_n^{(r)} = \sum_{r'r''} A_{r''r'}^{(g_1)} A_{r'r}^{(g_1)} \psi_n^{(r'')}$$

that is,

$$A_{r''r}^{(g_2g_1)} = \sum_{r'} A_{r''r'}^{(g_2)} A_{r'r}^{(g_1)}$$

In the matrix form, this is

$$A^{(g_2g_1)} = A^{(g_2)}A^{(g_1)}$$

We thus see that the matrices $A^{(g)}$ constitute a group of matrices in which to each g corresponds an $A^{(g)}$ and the product of symmetry operations g_2g_1 corresponds to the product of the matrices. This correspondence is not necessarily one-to-one (the same matrix A may be in correspondence with two or more different g) and is known as the *representation* or the *linear representation* of a symmetry group. The functions $\psi_n^{(r)}$ form the basis of the representation and k is the dimension of the representation. D_g being a unitary operator, all matrices A are unitary matrices ($A^{-1} = A^{+}$) if $\psi_n^{(r)}$ is an orthonormalized basis. If a different basis is chosen, all matrices undergo a similarity transformation

$$A' = S^{-1}AS$$

where S is a unitary matrix of transformation to a new basis.

A representation is said to be *irreducible* if it is not possible to choose such (unitary) matrix S that all A' turn into block matrices, that is, if it is not possible to choose such linear combinations of wave functions $\psi_n^{(r)}$, whose number $k' < k$, that transformations of D_g produced by them give linear combinations of only these same k' new functions (no subspace exists which is invariant under the transformations D_g). A reducible representation can be split into several irreducible ones by using a suitable matrix S.

Only nonequivalent irreducible representations (those not related by the similarity transformation) are important. An arbitrary representation can be decomposed into a sum of irreducible ones by choosing a suitable basis.

The matrices of irreducible representations of a group are completely defined up to a similarity transformation and are independent of the specifics of the problem in hand. For example, electron wave functions of all molecules with a specific symmetry group are the representation bases of this particular group. As a rule, if it is not the case of accidental degeneracy or some other symmetry, each value of energy corresponds to an irreducible representation of the symmetry group.

A particular energy level corresponds to an irreducible representation of a symmetry group, and this representation determines the symmetry type of the wave functions for this level. Levels are classified according to the irreducible representations of the symmetry group.

All nonequivalent irreducible representations have been found for all important groups. Thus, we know them for all point groups, for the group of rotations etc. The following theorems are useful for the determination of irreducible representations of symmetry groups.

1. The number of different irreducible representations of a group equals the number of classes in the group.
2. The sum of squared dimensions of irreducible representations of a group equals the order of this group.
3. Abelian groups have only one-dimensional irreducible representations (since the number of classes in them equals the order of the group).
4. Among group representations, there always is a one-dimensional identity representation which is realized by a function that is invariant under all symmetry operations.

We can conclude from these arguments that the dimensions of irreducible representations equal the possible order of degeneracy of energy levels. They are found without actually solving the problem. If the levels were classified on the basis of symmetry considerations, one knows the multiplicity of degeneracy of each level.

Let us consider this using as an example the electron wave functions of a diatomic molecule. This molecule was put in correspondence with the symmetry groups $C_{\infty v}$ (its symmetry elements are the rotation by an arbitrary angle φ around the axis of a molecule and the reflection in a plane passing through the axis) and $D_{\infty h}$ for a molecule formed of identical atoms (an additional symmetry element is a reflection in a plane orthogonal to the axis and drawn through the midpoint of the axis). We will also find, as we go, all irreducible representations of these groups.

4.2. *Classification of Electron States of a Diatomic Molecule. Selection Rules*

The group of all rotations $C(\varphi)$ by an angle φ around the axis of a molecule — the C_∞ group — is obviously an Abelian group with only one-dimensional irreducible representations. If ψ is the wave function which realizes this representation, the unitarity of the operators D_g implies

$$D_{C(\varphi)}\psi = \exp[\mathrm{i}\mu(\varphi)]\psi$$

where $\mu(\varphi)$ is some real function φ.

The group property $C(\varphi_1)C(\varphi_2) = C(\varphi_1 + \varphi_2)$ implies that $\mu(\varphi_1) + \mu(\varphi_2) = \mu(\varphi_1 + \varphi_2)$. Thus if $\varphi_2 \to 0$, then $d\mu(\varphi)/d\varphi = d\mu(0)/d\varphi$. Hence, $\mu(\varphi) = m\varphi$; note that the single-valuedness of ψ under rotation by 2π implies that $\mu(2\pi) = 2\pi m$, where m is an integer. All irreducible representations of the group C_∞ have been found. They are enumerated by the integers $m = 0, \pm 1, \pm 2, \ldots$ and are realized by the functions ψ_m for which

$$D_{C(\varphi)}\psi_m = \exp(im\varphi)\psi_m$$

In terms of quantum mechanics this means that m is a quantum number for systems possessing cylindrical symmetry C_∞. From the physical standpoint, $m\hbar$ is the projection of angular momentum onto the symmetry axis M_z. Owing to the independence of the energy $\mathscr{H}$ of the angle φ in systems with cylindrical symmetry, M_z is conserved and thus is an integral of motion. In quantum mechanics this signifies that the corresponding operator $\widehat{M_z}$ must commute with the Hamiltonian $\hat{\mathscr{H}}$. The operator $\widehat{M_z}$ coincides with $\partial/\partial\varphi$ up to the constant factor $i\hbar$ and is, therefore, the operator of infinitesimal rotation around the axis, since for any other function

$$\psi(\varphi + \delta\varphi) \simeq (\varphi) + \delta\varphi \frac{\partial}{\partial\varphi}\psi(\varphi) \equiv \psi(\varphi) + \frac{i}{\hbar}\delta\varphi\widehat{M_z}\psi$$

The eigenfunctions of the operator $\widehat{M_z}$ are the functions $e^{im\varphi}$; its eigenvalues are $m\hbar$ — one of the quantum numbers of the system.

For systems whose energy is independent of, for example, the coordinate x, that is, systems in which there are no forces along x (translational symmetry), the literal reproduction of the above arguments shows, after x is substituted for φ, that this symmetry corresponds to the conservation of the projection of momentum p_x. This is the eigenvalue of the operator of infinitesimal translation, $p_x = -i\hbar\partial/\partial x$.

Therefore, the symmetry of dynamic systems is the most general source of the integrals of motion (of quantum numbers). The examples we discuss below are particular cases of this general approach, even though they are not so obviously clear as are momentum or angular momentum.

The group $C_{\infty v}$ is not Abelian any more since it is easy to show that

$$C(\varphi)\sigma = \sigma C(-\varphi)$$

This last equality yields

$$D_{C(\varphi)}D_\sigma\psi_m = D_\sigma D_{C(-\varphi)}\psi_m = e^{-im\varphi}D_\sigma\psi_m$$

As a result, we have

$$D_\sigma\psi_m = \psi_{-m}$$

and the pair of functions ψ_m and ψ_{-m} realize a representation of the group $C_{\infty v}$. This representation is two-dimensional at $\Lambda = |m| \neq 0$. If $m = 0$, we have $D_0\psi_\sigma = \psi_0'$, and in general $\psi_0' \neq \psi_0$. The functions ψ_0 and ψ_0' realize a two-dimensional reducible representation because

$$D_\sigma(\psi_0 + \psi_0') = \psi_0 + \psi_0'$$
$$D_\sigma(\psi_0 - \psi_0') = -(\psi_0 - \psi_0')$$

and the functions $\psi_0^+ = \psi_0 + \psi_0'$ and $\psi_0^- = \psi_0 - \psi_0'$ realize irreducible one-dimensional representations for which

$$D_\sigma\psi_0^+ = \psi_0^- \qquad D_\sigma\psi_0^- = -\psi_0^-$$

The construction of the similar combinations $\psi_m \pm \psi_{-m}$ for $m \neq 0$ does not turn the two-dimensional representation into a reducible one since each of the functions $\psi_m + \psi_{-m}$ and $\psi_m - \psi_{-m}$ transforms under rotations $C(\varphi)$ not into itself but into a certain combination of the two.

The group $C_{\infty v}$ has two one-dimensional representations, denoted by Σ^+ (when $\Lambda = |m| = 0$ and $D_\sigma = +1$) and Σ^- (when $D_\sigma = -1$), and also has two-dimensional representations denoted by Π, Δ, etc.(for $\Lambda = 1, 2, \dots$). The notations have been introduced by analogy with the classification of atomic spectra S, P, D ($L = 0, 1, 2, \dots$), with the Greek letters replacing the Roman ones.

As follows from the above description, these same symbols classify the electron states of a diatomic molecule. If $\Lambda = 0$, these are nondegenerate states, having a certain symmetry under the reflection in the plane passing through the axis of the molecule; if $\Lambda \neq 0$, these states are twice-degenerate. The quantum numbers Λ and m dictate the law of transformation for wave functions under rotations around the axis of the molecule, that is, how the angular momentum is projected onto the axis.

If the nuclei of a diatomic molecule are identical (a homonuclear molecule), its symmetry group is $D_{\infty h}$. It is a product of the $C_{\infty h}$ group and the inversion group which consists of two elements: the inversion I, that is, the sign reversal of all coordinates, $\mathbf{r} \to -\mathbf{r}$, and the identity transformation. The inversion group is Abelian, it has two one-dimensional representations, $D_I = \pm 1$, since an arbitrary function $f(\mathbf{r})$ can be presented as a combination of a part $f(\mathbf{r}) + f(-\mathbf{r})$ symmetric with respect to inversion and a part $f(\mathbf{r}) - f(-\mathbf{r})$ antisymmetric to inversion. These representations are denoted by the letters g (if $D_I = +1$) and u (if $D_I = -1$). Therefore, the number of representations of the group $C_{\infty v}$ is doubled; the representations that arise are one-dimensional ($\Sigma_{g,u}^{\pm}$) and two-dimensional ($\Pi_{g,u}$, $\Delta_{g,u}$ etc.. The electron states of a homonuclear diatomic molecule are classified according to these representations. The g states are not affected under inversion while the sign of the u states is reversed.

Inversion determines the quantum number known as the *parity* of a state: $D_I\psi = P\psi$, with $P = \pm 1$ for the even and odd states g and u, respectively. As a rule, the ground state of a molecule is Σ_g^+.

By analogy to the case of diatomic molecules, the classification of the electron states of polyatomic molecules follows the irreducible representations of the point symmetry group of the molecule. There is no special notation for point group representations which would be analogous to the group representations $C_{\infty v}$ and $D_{\infty v}$. Typically, they are enumerated by $\Gamma_1, \Gamma_2, \ldots$ — as many as irreversible representations. The notation A and B is often used for one-dimensional representations, E for two-dimensional and T for three-dimensional.

An empirical rule was formulated: the wave function of the ground electron term of polyatomic molecules has the maximum symmetry, that is, the wave function of the ground state is invariant with respect to all elements of the symmetry group of the molecule. In other words, the wave function of the ground state belongs to the single, nondegenerate representation of the symmetry group.

The theory of group representations makes it possible to classify states without the exact information on wave functions. It also allows us to determine the multiplicity of level degeneracy and, if we also use perturbation theory, their splitting in external fields (Zeeman and Stark effects, including the splitting in the internal crystal field). Finally, the theory of representations yields selection rules without requiring the complete determination of wave functions. Let us again consider this, using the example of a diatomic molecule.

As we have seen in chapters 1 and 2, the dipole approximation usually provides an adequate description of the interaction of electromagnetic field with atoms and molecules. The reason for this is their small size (of the order of a_0) in comparison with the wavelength. In the dipole approximation, the perturbation operator includes the operator of the dipole moment; its matrix elements are in the expressions for susceptibilities. They also determine the probabilities of the spontaneous emission and absorption. These probabilities are zero in the dipole approximation if the matrix elements of the dipole moment are zeros. The transition is then said to be forbidden. Selection rules indicate which transitions are allowed, that is, for which transitions the matrix elements of the dipole moment are nonzero.

By definition, the transition in a molecule is allowed if the integral

$$I_\alpha = \int \psi_2^* \, \mathrm{d}_\alpha \psi_1 \, \mathrm{d}^3 r \tag{4.2.1}$$

does not vanish; the integration in (4.2.1) is in fact carried out over the coordinates of all electrons. The dipole moment operator $d_\alpha = e\Sigma(\mathbf{r}_i)_\alpha$ is transformed under rotations of the coordinate system as the vector $\mathbf{r}_i$. It is convenient to work with the components of the vector $d_0 = d_z$, $d_\pm = d_x \pm \mathrm{i}d_y$.

Rotations around the axis z, pointing along the axis of the molecule, transform d_0 as z and keep it invariant while $d_\pm$ is transformed as $x \pm \mathrm{i}y$. The law of transformation of these combinations is easily found:

$$D_{C(\varphi)}f(x, y, z) = f(x\cos\varphi - y\sin\varphi, x\sin\varphi + y\cos\varphi, z)$$
$$D_{C(\varphi)}(x \pm \mathrm{i}y) = \mathrm{e}^{\pm\mathrm{i}\varphi}(x \pm \mathrm{i}y)$$

The law of transformation under reflections is found in a similar manner:

$$D_{\sigma_x} z = z$$
$$D_{\sigma_x}(x \pm \mathrm{i}y) = x \mp \mathrm{i}y$$

These expressions show that d_0 is transformed in accord with the representation Σ^+ of the group $C_{\infty v}$ and $d_\pm$, in accord with the representation Π.

Whatever the orthogonal transformation of coordinates, integral (4.2.1) must remain unaltered since for this integral, an orthogonal transformation only means a replacement of integration variables.

Let ψ_1 correspond to a representation Λ_1 and to a quantum number m_1 ($|m_1| = \Lambda_1$) and let ψ_2 correspond to a representation Λ_2 and to a quantum number m_2. Then integrals (4.2.1), $I_\pm = I_x \pm \mathrm{i}I_y$, are multiplied under rotations around the axis by $\exp[-\mathrm{i}(m_2 - m_1 \pm 1)\varphi]$. If $I_\pm \neq 0$, this factor must equal unity, so that $m_2 - m_1 \pm 1 = 0$, that is,

$$|\Lambda_2 - \Lambda_1| = 1$$

Likewise, $I_0 = I_z \neq 0$ is unaltered under rotations only if $m_2 = m_1$, that is,

$$|\Lambda_1 - \Lambda_2| = 0$$

We have thus obtained selection rules for the quantum number Λ:

$$|\Lambda_2 - \Lambda_1| = 0, 1$$

Likewise, if we apply the reflection operator D_σ, we obtain $I_\pm = 0$ for the Σ states and $I_0 \neq 0$ only for the transitions $\Sigma^+ \to \Sigma^+$ and $\Sigma^- \to \Sigma^-$.

This line of reasoning can be extended to the case of homonuclear molecules for which the electron wave functions ψ are transformed according to the representations of the group $D_{\infty h}$. Applying the inversion operator and taking into account that under inversion, $d_\alpha \to -d_\alpha$ (i.e., d_0 is transformed under representation Σ_u^+ and $d_\pm$, under the representation Π_u of the group $D_{\infty h}$), we find that $I_\alpha \neq 0$ only if

$$P_2 P_1 = -1$$

This is the selection rule for parity. It means that only the transitions $g \to u$ and $u \to g$ are allowed.

In the general case of a polyatomic molecule, integrals like (4.2.1) are not zero if the integrand is a function which transforms either as a maximum-symmetry (identity) representation of the symmetry group or as a reducible representation whose decomposition into irreducible representations contains an identity representation. It is therefore possible to find selection rules because d transforms as a representation (not necessarily irreducible) of the symmetry group, while ψ_2 and ψ_1 transform as some irreducible representation. The entire integrand transforms as a symmetry group representation which is generally reducible; the integral may not equal zero if this reducible representation contains an identity irreducible one. To find the selection rules it is thus necessary to have tables of decomposition of reducible representations into irreducible ones. Such tables can be found in handbooks and reference sources for all groups important for quantum mechanics.

4.3. Atomic Spectra. Selection Rules for Atoms

In the reference frame of the atomic nucleus, the symmetry group of the atom's Hamiltonian is K_h, that is, the groups of rotations and reflections. The states of the atom (or ion) are classified in accordance with the irreducible representations of this group.

The atom's Hamiltonian commutes with any operator from the symmetry group K_h, that is, with any rotation, including an infinitesimal rotation around any axis. Infinitely small rotations correspond to Hermitian operators of the angular momentum $\hbar\mathbf{J}$. Therefore, the momentum of an atom is conserved by virtue of the fundamental notions of quantum mechanics.

Irreducible representations of the group K_h are determined by an integer J (if the spin is taken into account, J may be integral or half-integral). The dimension of this representation is $2J+1$, and the functions that realize this representation are enumerated by the number $m = -J, -J+1, \ldots, J$. They are eigenvectors of the operator $\mathbf{J}^2$, whose eigenvalues are $J(J+1)$ and the eigenfunctions are J_z, with the eigenvalue m.

Any rotation from K_h is given by Euler's angles α, β, and γ. Therefore, if there is a representation of the group K_h, the functions $D^J_{m'm}(\alpha, \beta, \gamma)$ (that is, the elements of the matrix which correspond to the rotations by Euler's angles in the representation J) can be found.

Euler's angles for a specific rotation are assigned in a familiar manner. Let x, y, z be the original system of coordinates. A rotation around the axis z by an angle α transforms it to the system $\tilde{x}$, $\tilde{y}$, $\tilde{z}$ (see Figure 4.1). Now we execute a rotation by an angle β around the axis $\tilde{y}$, which brings the axis z to the position z'. Finally, we rotate the system by an angle γ

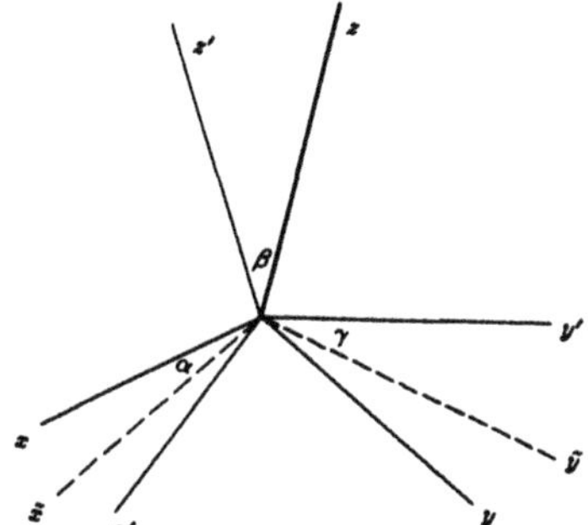

Figure 4.1. Euler's angles for an arbitrary rotation.

around the axis z', thus obtaining the new axes x', y', z' which determine the rotation by Euler's angles α, β and γ. The functions $D^J_{m'm}$ are known as Wigner functions, or generalized spherical functions; they can be written as

$$D^J_{m'm}(\alpha,\beta,\gamma) = \exp(\mathrm{i}m'\gamma)d^J_{m'm}(\beta)\exp(\mathrm{i}m\alpha)$$

where $d^J_{m'm}$ are functions of the angle β, both tabulated, with a known general expression. If J is integral and $m' = 0$, the Wigner functions transform to the ordinary spherical functions:

$$Y_{lm}(\beta,\alpha) = \mathrm{i}^l\sqrt{\frac{2l+1}{4\pi}}\,D^l_{0m}(\alpha,\beta,\gamma) \tag{4.3.1}$$

In the nonrelativistic approximation, the atomic Hamiltonian does not include the spin. The wave functions then depend only on the coordinates of electrons, and the orbital angular momentum **L** is conserved. The number L plays the role of J but this is always an integer. In addition to the rotational symmetry, an atom with several electrons also has a permutational symmetry; the Hamiltonian is invariant with respect to permutations of the electrons' coordinates. The group of permutations is also a symmetry group of the Hamiltonian. Hence, in addition to the number L (the orbital angular momentum which determines the representation of the group of rotations), the wave functions are also characterized by an irreducible representation of the group of permutations of N elements (if N is the number of electrons in the atom).

If the spin is taken into account, wave functions become dependent on spin variables; these functions must be antisymmetric since electrons are fermions. In the nonrelativistic approximation, with the Hamiltonian having no terms corresponding to the spin–orbital interaction, wave functions can be written as products of the coordinate and spin wave functions. The spin wave function has the following important property: its permutational symmetry determines the rotational symmetry. In other words, S is an integral or half-integral number which determines the representation under which the spin wave function is transformed and which also fixes its permutational symmetry. Since the permutational symmetry of the spin wave function determines the permutational symmetry of the coordinate wave function (so as to make their product antisymmetric under permutations), the number S determines the permutational symmetry of the coordinate wave function.

This is the reason why the atomic energy levels are classified by the quantum numbers L and S, the former determining the representation of the group of rotation and the latter the representation of the group of permutations under which the coordinate wave function is transformed.

If the relativistic spin–orbital interaction is taken into account, only $\mathbf{J} = \mathbf{L} + \mathbf{S}$ is conserved, that is, the state is characterized by the quantum number $\mathbf{J}$, already mentioned above. This is the fine structure of atomic levels, with the rough structure being characterized by the nonrelativistic Hamiltonian and the numbers L and S. Selection rules are determined by the rule of multiplication of irreducible representations of the rotations group. The (reducible) product of the representations J_1 and J_2 equals the sum of representations from $|J_1 - J_2|$ to $|J_1 + J_2|$. A clear illustration of this is one vector model in quantum mechanics. Since $\mathbf{d}$ is transformed under the vector representation $J = 1$, an integral of the type of (4.2.1) is nonzero, as follows from Section 4.2, that is, transitions are allowed only if

$$|J_1 - J_2| \leq 1 \leq J_1 + J_2$$

Furthermore, since $\mathbf{d}$ reverses sign under inversion,

$$P_1 P_2 = -1$$

where P_1 and P_2 are the state parities.

If the spin–orbital interaction is taken into account, similar arguments show that the rough-structure level corresponding to specific L and S splits to fine-structure levels characterized by the numbers J, where

$$|L - S| \leq J \leq L + S$$

If $L \geq S$, the number of fine-structure components equals $(2S + 1)$, so that the number S determines the multiplicity of the term. The degree of degeneracy of a term with a given J is $(2J + 1)$.

4.4. Adiabatic Approximation

When an atom is treated quantum-mechanically, converting to a system of coordinates connected with the center of inertia of the atom, the problem is reduced to the Schrödinger equation for electrons in the field of the nucleus. The motion of the nucleus is then automatically taken into account.

The situation is quite different when several nuclei are present, as in a molecule or a solid. Since atomic nuclei are much heavier than electrons, they move much slower; as a zero approximation, nuclei can be regarded as fixed and electrons as moving in the field of stationary nuclei. The motion of nuclei can be taken into account as the next stage. This idea is basic for the adiabatic approximation, which is the foundation of the theory of molecules and solids.

The possibility of this approach stems from the existence of a small parameter in the problem of a molecule, m/M, where m is the electron mass and M is the mean mass of the nucleus. In fact, the role of the small parameter is played, as will be seen below, by $\mu = (m/M)^{1/4}$. The value of this parameter is of the order of 0.1 or less since $m/M \sim 10^{-5}$–10^{-4}. According to expression (2.4.5), the nonrelativistic Hamiltonian of a molecule or crystal can be written as

$$\mathscr{H}_0 = \sum_i \frac{\mathbf{P}_i^2}{2m} + \sum_k \frac{\mathbf{P}_k^2}{2M_k} + \frac{1}{2}\sum_{i\neq i'} \frac{e^2}{|\mathbf{r}_i - \mathbf{r}_{i'}|} + \frac{1}{2}\sum_{k\neq k'} \frac{Z_k Z_{k'} e^2}{|\mathbf{R}_k - \mathbf{R}_{k'}|} - \sum_{i,k} \frac{Z_k e^2}{|\mathbf{r}_i - \mathbf{R}_k|} \tag{4.4.1}$$

where the subscript i enumerates electrons and k enumerates nuclei, $\mathbf{r}_i$ are electrons' coordinates and $\mathbf{R}_k$ are those of nuclei, $\mathbf{p}_i = -i\hbar\partial/\partial\mathbf{r}_i$ and $\mathbf{P}_i = -i\hbar\partial/\partial\mathbf{R}_k$ are the momentum operators of electrons and nuclei. The first term in (4.4.1) is the operator of the kinetic energy of electrons, the second term is that of nuclei, the third term is the Coulomb interaction of electrons, the fourth term is that of nuclei, and the fifth term is the interaction between electrons and nuclei.

Since $m \ll M_k$, the operator of the kinetic energy of nuclei in (4.4.1) is small and can be treated as a perturbation in calculating the spectrum of $\mathscr{H}_0$. Hence, $\mathscr{H}_0$ can be written as

$$\mathscr{H}_0 = \mathscr{H}_{0e} + T$$

where

$$T = \sum_k \frac{\mathbf{P}_k^2}{2M_k}$$

The operator $\mathscr{H}_{0e}$ depends on coordinates of nuclei as parameters and does not contain derivatives with respect to nuclear coordinates. It can be treated as the electron Hamiltonian at fixed positions of nuclei. Assume that the eigenfunctions and eigenvalues of the electron Hamiltonian have been found,

$$\mathscr{H}_{0e}\varphi_n(\mathbf{r},\mathbf{R}) = U_n(\mathbf{R})\,\varphi_n(\mathbf{r},\mathbf{R}) \tag{4.4.2}$$

They depend on the coordinates of nuclei as on parameters (here and below, $\mathbf{r}$, are the coordinates of all electrons and $\mathbf{R}$ are the coordinates of all nuclei).

As will be clear from further arguments below, the amplitude of oscillations of nuclei is of the order of μa_0, so the energy of oscillations is of second order in the small parameter μ. The rotational energy and also the terms related to the interaction between electron motion and rotation add to T a contribution of the order of μ^4.

A systematic solution of the Schrödinger equation with the Hamiltonian $\mathscr{H}_0$, using perturbation theory, makes it possible to find the energy levels E of the molecule, taking into account the oscillational and rotational motion and the interaction between them. It is found that up to terms of orders in μ above four, the wave function can be written as

$$\psi_n(\mathbf{r},\mathbf{R}) = \varphi_n(\mathbf{r},\mathbf{R})\,\Phi_n(\mathbf{R}) \tag{4.4.3}$$

where the function $\Phi_n(\mathbf{R})$ satisfies the Schrödinger equation for the nuclear motion,

$$(T + U_n(\mathbf{R}))\,\Phi_n(\mathbf{R}) = E\Phi_n(\mathbf{R})$$

with the effective potential energy $U_n(\mathbf{R})$ defined in (4.4.2). By virtue of its physical meaning, $U_n(\mathbf{R})$ is the total electron energy for fixed nuclear coordinates $\mathbf{R}_k$ plus the energy of the Coulomb interaction between nuclei. Obviously, $U_n(\mathbf{R})$ is a function of the electron state; this is indicated by the subscript n.

The wave function (4.4.3) has the form of the product of the function $\Phi_n(\mathbf{R})$, which describes the motion of nuclei, by the electron wave function $\varphi_n(\mathbf{r},\mathbf{R})$ which indicates that while nuclei move, electrons move as if nuclei were frozen in their instantaneous positions. It is said that electrons adiabatically follow the motion of nuclei; furthermore, in this motion electrons do not jump from one electron state (enumerated by n) to another.

The adiabaticity is lost and the wave function cannot be represented any more by (4.4.3) if we take into account in the expressions for E the terms of orders in μ higher than four.

We assume that the translational degrees of freedom are excluded by the transformation to the reference frame of the center of inertia of the molecule or its nuclei. The rotational degrees of freedom give the rotational energy which is of the order of μ^4; it is thus very small in heavy molecules.

In the case of macroscopic solids, the rotation need not be considered at all, so that we neglect rotation for the time being and consider only oscillations of nuclei.

Let us expand the wave function $\psi(\mathbf{r}, \mathbf{R}, t)$, which satisfies the time-dependent Schrödinger equation

$$\mathrm{i}\hbar \frac{\partial \psi}{\partial t} = \mathscr{H}_0 \psi \tag{4.4.4}$$

in the eigenfunctions of electron Hamiltonian (4.4.2):

$$\psi(\mathbf{r}, \mathbf{R}, t) = \sum_n \Phi_n(\mathbf{R}, t)\, \varphi_n(\mathbf{r}, \mathbf{R}) \tag{4.4.5}$$

This can always be done since the Hamiltonian $\mathscr{H}_{0\mathrm{e}}$ for fixed $\mathbf{R}$ is a Hermitian operator acting on the electrons' coordinates. It has a complete set of eigenfunctions $\varphi_n(\mathbf{r}, \mathbf{R})$ which depend on $\mathbf{R}$ as on parameters. These eigenfunctions can be chosen orthogonal and normalized:

$$\int \varphi_n^*(\mathbf{r}, \mathbf{R})\, \varphi_{n'}(\mathbf{r}, \mathbf{R})\, \mathrm{d}^3 r = \delta_{nn'}$$

Let us substitute expansion (4.4.5) into equation (4.4.4), multiply it by $\varphi_n^*(\mathbf{r}, \mathbf{R})$, and integrate over all electron coordinates. In view of the orthogonality of the functions φ_n, we obtain

$$\mathrm{i}\hbar \frac{\partial \Phi_n}{\partial t} = (T + U_n(\mathbf{R}))\Phi_n + \sum_{k,n'} \left(A_{nn'}^{(k)} \frac{\partial \Phi_{n'}}{\partial \mathbf{R}_k} + B_{nn'}^{(k)} \Phi_{n'} \right) \tag{4.4.6}$$

where

$$A_{nn'}^{(k)} = -\frac{\hbar^2}{M_k} \int \varphi_n^* \frac{\partial \varphi_{n'}}{\partial \mathbf{R}_k}\, \mathrm{d}^3 r$$

$$B_{nn'}^{(k)} = -\frac{\hbar^2}{2M_k} \int \varphi_n^* \frac{\partial^2 \varphi_{n'}}{\partial \mathbf{R}_k^2}\, \mathrm{d}^3 r$$

If $A_{nn'}^{(k)}$ and $B_{nn'}^{(k)}$ were diagonal matrices, the equations for the functions Φ_n would separate. No approximations were introduced so far. Now we will make use of the smallness of $\mu = (m/M)^{1/4}$ and show that the terms with A and B are small and can be neglected as long as μ is small. This is what we know as the adiabatic approximation; the terms with A and B are nonadiabatic terms which can be found using perturbation theory.

The coefficients A and B include M_k^{-1} and it is therefore easy to see why nonadiabatic terms are small in comparison with the electron energy $U_n(\mathbf{R})$. However, although T also contains the factor M_k^{-1}, nonadiabatic terms are still small in comparison with T. We will show this by evaluating the terms containing T and those containing A and B.

The order of magnitude of the effective potential energy U_n is obvious. By the order of magnitude, the electron energy is

$$I_0 = e^2/2a_0 = \hbar^2/2ma_0^2 = me^4/2\hbar^2 = 13.6\,\mathrm{eV}$$

The value of the energy of the Coulomb interaction between nuclei,

$$\sum_{k \neq k'} \frac{Z_k Z_{k'} e^2}{|\mathbf{R}_k - \mathbf{R}_{k'}|}$$

is of the same order of magnitude since $|\mathbf{R}_k - \mathbf{R}_{k'}| \sim a_0$; it is contained in U_n.

Nuclei oscillate around the equilibrium configuration $\mathbf{R}_0$ in which $\partial U_n(\mathbf{R}_0)/\partial \mathbf{R} = 0$. The value assumed by $U_n(\mathbf{R}_0)$ in this configuration, U_{n0}, is also of the order of I_0. For the purpose of estimation near equilibrium, we can retain only quadratic terms in the expansion of U_n in powers of $\delta\mathbf{R} = \mathbf{R} - \mathbf{R}_0$:

$$U_n(\mathbf{R}) = U_{n0} + \frac{1}{2} \frac{\partial^2 U_n(\mathbf{R}_0)}{\partial \mathbf{R}^2} \, \delta\mathbf{R}^2 = U_{n0} + \frac{1}{2} U''_{n0} \, \delta\mathbf{R}^2$$

The scale of the oscillation amplitude $\delta\mathbf{R}$ can be found using the following arguments. In harmonic oscillations, the kinetic and potential energies are equal (the virial theorem). Hence,

$$\frac{P^2}{2M} \simeq \frac{1}{2} U''_{n0} \, \delta R^2$$

On the other hand, the uncertainty relation implies that $P \gtrsim \hbar \delta R^{-1}$, whence

$$U'' \, \delta R^2 \simeq \frac{P^2}{2M} \geq \frac{\hbar^2}{2M \, \delta R^2}$$

that is,

$$\delta R \sim \sqrt[4]{\frac{\hbar^2}{MU''}} \sim \sqrt[4]{\frac{\hbar^2 a_0^2}{M I_0}} \sim \sqrt[4]{\frac{m}{M}} \, a_0 \sqrt[4]{\frac{\hbar^2}{m a_0^2 I_0}} \sim \sqrt[4]{\frac{m}{M}} \, a_0 = \mu a_0$$

We have taken into account here that the characteristic scale of change of $U_n(\mathbf{R})$ is the Bohr radius a_0. As a result, $U'' \sim I_0/a_0^2$.

We thus find that in agreement with earlier statements, for lower oscillational levels we have $\delta R \sim \mu a_0$. Note also that in these states,

$$P \sim \frac{\hbar}{\delta R} \sim \mu^{-1} \frac{\hbar}{a_0}$$

that is, the oscillational momentum of nuclei is much greater than the electron momentum, which is of the order of $\hbar/a_0$. However, the oscillational energy

$$E_o \sim \frac{P^2}{2M} \sim \frac{\hbar^2}{2Ma_0^2\mu^2} = \frac{\hbar^2}{2ma_0^2}\frac{1}{\mu^2} = \mu^2 I_0$$

is much smaller than the characteristic electron energy I_0. The velocity of nuclei is of the order of

$$\frac{P}{M} \sim \frac{\hbar}{\mu a_0 M} = \frac{\hbar}{a_0 m}\frac{m}{M\mu} = \mu^3 \frac{\hbar}{a_0 m}$$

that is, much smaller than the electron velocity $\hbar/a_0 m = e^2/\hbar$.

The angular momentum of nuclei is of the same order of magnitude as the electron momentum because the characteristic scale of change of the rotational wave function is 2π (by angle of arc) or a_0 (on the circle). Hence, the rotational kinetic energy

$$E_r \sim \frac{1}{2M}\left(\frac{\hbar}{a_0}\right)^2 \sim \frac{m}{M}\frac{\hbar^2}{2ma_0^2} \sim \mu^4 I_0$$

is much smaller than both the electron and the oscillational energies.

Now we can evaluate the nonadiabatic terms:

$$A \sim \frac{\hbar^2}{M}\frac{1}{a_0} = \frac{\hbar^2}{ma_0^2}\frac{m}{M}a_0 \sim \mu^4 I_0 a_0$$

$$B \sim \frac{\hbar^2}{Ma_0^2} \sim \mu^4 I_0$$

We have taken into account here that $\varphi_{n'}$ mostly depends on the difference between the electron and nuclear coordinates. For this reason, the scale of variation of this function along $\mathbf{R}$ is the same as along $\mathbf{r}$, that is, it equals a_0. Hence,

$$\frac{\partial\varphi_{n'}}{\partial R} \sim \frac{\varphi_{n'}}{a_0} \qquad \frac{\partial^2\varphi_{n'}}{\partial R^2} \sim \frac{\varphi_{n'}}{a_0^2}$$

The functions $\varphi_{n'}$ are normalized, so that

$$\int \varphi_n^* \frac{\partial\varphi_{n'}}{\partial R}\,\mathrm{d}^3r \sim a_0^{-1} \qquad \int \varphi_n^* \frac{\partial^2\varphi_{n'}}{\partial R^2}\,\mathrm{d}^3r \sim a_0^{-2}$$

If now we take into account that, since the scale of variation of the oscillational wave function $\Phi_{n'}$ is of the order of μa_0,

$$\frac{\partial\Phi_{n'}}{\partial R} \sim \frac{\Phi_{n'}}{a_0\mu}$$

we find that nonadiabatic terms with $\mathbf{A}$ and B are of the order of $\mu^3 I_0$ and $\mu^4 I_0$, respectively, and can be neglected in comparison with the kinetic energy of the oscillational motion T, which is of the order of $\mu^2 I_0$.

Strictly speaking, the estimates are valid, as we have already indicated, only for the lower oscillational levels. However, as the oscillational energy grows, the ratio of the adiabatic term T in (4.4.6) to nonadiabatic terms only increases since $T \sim P^2$, $A\partial\Phi/\partial R \sim P$ and the order of magnitude of the term $B\Phi$ does not change at all.

If now we drop the terms with $\mathbf{A}$ and B in (4.4.6), we obtain

$$i\hbar\frac{\partial\Phi_n}{\partial t} = \Big(\sum_k \frac{\mathbf{P}_k^2}{2M_k} + U_n(\mathbf{R})\Big)\Phi_n \tag{4.4.7}$$

In the adiabatic approximation, the equations of nuclear motion for different electron states enumerated by the subscript n have separated. In this approximation, transitions between different electron states do not occur (in the absence of external fields). The total wave function has the form (4.4.5) but noncoupled equations exist for the wave functions $\Phi_n(\mathbf{R}, t)$. The effective potential energy $U_n(\mathbf{R})$ depends on the subscript n and is different for different electron states. We can use perturbation theory to take into account nonadiabatic terms in (4.4.6) which contain $\mathbf{A}$ and B. The first perturbation-theory correction for nondegenerate electron states consists in retaining in (4.4.6) only those terms with the coefficients $\mathbf{A}_{nn}$ and B_{nn} which are diagonal in n. Since the electron Hamiltonian $\mathscr{H}_{0e}$ is real, the wave functions φ_n can be chosen real; this gives

$$\mathbf{A}_{nn}^{(k)} = -\frac{\hbar}{M_k}\int \varphi_n \frac{\partial\varphi_n}{\partial\mathbf{R}_k}\,\mathrm{d}^3r = -\frac{\hbar^2}{2M_k}\frac{\partial}{\partial\mathbf{R}_k}\int \varphi_n\varphi_n\,\mathrm{d}^3r = 0$$

The term with the functions $B_{nn}^k(\mathbf{R})$ then simply gives a correction to the adiabatic energy:

$$U_n(\mathbf{R}) \to U_n(\mathbf{R}) + \sum_k B_{nn}^{(k)}(\mathbf{R})$$

The correction to the adiabatic energy is of order four in μ, so that its dependence on $\mathbf{R}$ can be ignored by setting $\mathbf{R} = \mathbf{R}_0$ ($\mathbf{R}_0$ is an equilibrium configuration of nuclei, corresponding to the minimum of $U_n(\mathbf{R})$). After adding this correction, (4.4.7)-type equations become valid up to terms of order μ^4, that is, they give a correct description of the rotational motion as well.

In view of the adiabatic-energy correction, equation (4.4.7) for the nuclear motion can also be written as a Schrödinger equation

$$i\hbar\frac{\partial\Phi_n}{\partial t} = \mathscr{H}_n\Phi_n$$

where

$$\mathscr{H}_n = T + \langle \varphi_n | \mathscr{H}_{0e} | \varphi_n \rangle = \sum_k \frac{\mathbf{P}_k^2}{2M_k} + \int \varphi_n^* \mathscr{H}_{0e} \varphi_n \, d^3 r \tag{4.4.8}$$

is the Hamiltonian of the molecule, averaged over the electron wave function.

For degenerate or nearly degenerate electron states in equations (4.4.6), we need to retain terms with $\mathbf{A}$ and B which relate electron states with equal or almost equal energies.

4.5. Types of Bonding in Molecules and Solids

The behavior of nuclei of a molecule or a solid in the adiabatic approximation is determined by the effective potential energy $U_n(\mathbf{R})$ which includes the Coulomb interaction energy of nuclei and the electron energy. Note that the electron energy includes, in its turn, the kinetic energy of electrons and the potential energy, which also depends on the Coulomb interaction between electrons and nuclei.

The direct Coulomb interaction between nuclei always produces repulsion. If the electron energy is also taken into account, the function $U_n(\mathbf{R})$ may reverse the sign of its derivative with respect to $\mathbf{R}$, that is, it may correspond to attraction. The derivative, $\partial U_n/\partial \mathbf{R}_k$, can be interpreted as the force of effective interaction between the atoms of a molecule (the force acting on the kth atom or its nucleus in an arbitrary configuration of nuclei).

In an equilibrium configuration, all these forces are zero and there is a local minimum of the energy, $U_n(\mathbf{R})$. This minimum signifies the chemical bonding between atoms.

The characteristic features of the effective potential energy $U_n(\mathbf{R})$ make it possible to speak of different types of bonding. The decisive factor here is the spatial distribution of the electron density of a given electron state.

As we see from Hamiltonian (4.4.1), the true nature of the interaction is always the same: the Coulomb interaction among electrons and nuclei. Depending on the specific type of the wave function φ_n in (4.4.2), however, the form of the effective interaction $U_n(\mathbf{R})$ can be very different even for different electron states of the same molecule.

Hamiltonian (4.4.1) also ignores magnetic interactions. However, they are weaker than the Coulomb interaction by four orders of magnitude, since v/c for electrons in atoms is of the order of $e^2/\hbar c = \alpha = 1/137$ (the fine-structure constant). As a result, magnetic interactions can produce subtle effects but do not determine the stability of molecules and solids as such.

Bonding is roughly classified into the following types, or classes: ionic, covalent, metallic, hydrogen and van der Waals. None of these is ever found in the pure form, so a given substance is subsumed under the class according to the predominant type of bonding.

We shall briefly describe these types of bonding.

1. **Ionic bonding** is characteristic, for example, of alkali halides NaCl, KBr, CsI, etc. that is, of compounds in which one element is an alkali and the other comes from group VII of Mendeleyev's periodic table ($A_I B_{VII}$ compounds). The ionic bonding is a result of the Coulomb interaction between oppositely charged ions. The situation is not reducible to this interaction, however.

When ions are brought closer, repulsive forces arise which are not produced by the electrostatic interaction between ions. These forces, of quantum-mechanical nature, are implied by the Pauli principle: filled electron shells of ions cannot penetrate into each other. Of course, repulsion between nuclei also gives a contribution but the repulsion of electron shells has a considerably steeper dependence on distance. Two ions cannot be separated by a distance less than the sum of radii of their electron shells. An experimental confirmation of this is the fact that the lattice period of an alkali halide crystal can be predicted with considerable accuracy as the sum of radii of its ions, and such that the radii of, for example, the Na ion are the same in different such compounds. The electron energy grows steeply because the constraint of the Pauli principle has to be satisfied: when two filled shells are "squeezed" into each other, some electrons have to transfer to states of higher energy than in isolated atoms and ions.

This can be qualitatively illustrated by the simplest model of a multielectron atom: the Thomas–Fermi model. If the number of electrons is large, the electron shell can be subdivided into smaller regions δV for which the local electron density $n(\mathbf{r})$ is introduced. Within each region, the Pauli principle demands that the Fermi distribution occur for the electrons: they occupy levels up to the limiting energy $\varepsilon_F = p_F^2/2m$, where p_F is the electron momentum at the Fermi level. The density of states in the momentum space is $\delta V/(2\pi\hbar)^3$, so that with the number of states doubled owing to the spin, we have

$$n = \frac{4\pi p_F^3}{3}\,\frac{2}{(2\pi\hbar)^2} = \frac{p_F^3}{3\pi^2\hbar^3}$$

The Fermi energy is therefore connected with the electron density $n(\mathbf{r})$ by the relation

$$\varepsilon_F = \frac{p_F^2}{2m} = \frac{\hbar^2(3\pi^2 n)^{2/3}}{2m}$$

The total energy of electrons with energy from 0 to ε_F per unit volume is

$$\frac{\delta E}{\delta V} = 2 \int\limits_{|p|<p_F} \frac{p^2}{2m} \frac{d^3p}{(2\pi\hbar)^3} = (10\,m)^{-1}\hbar^2\pi^{4/3}3^{5/3}n^{5/3}$$

When filled electron shells overlap, the local value increases to $n_a + n_b$, where n_a and n_b are the local electron densities for the overlapping parts of the shell of the ions *a* and *b*. Note that $\delta E/\delta V \sim (n_a + n_b)^{5/3} > n_a^{5/3} + n_b^{5/3}$. If, for example, $n_a = n_b$ then $2^{5/3} > 2$ and the difference $2^{5/3}n^{5/3} - 2n^{5/3} = (2^{5/3} - 2)n^{5/3} > 0$ describes the increase in energy owing to shell overlapping.

So far we assumed that electron shells remain practically undeformed. In fact, this is true only for totally filled shells. This is not true, however, for partially occupied valence shells: electrons are substantially redistributed among the shells of the interacting atoms and thus the ion or covalent bonding arises.

Generally, the states (wave functions) of valence electrons in molecules differ essentially from their states in isolated atoms. The restructuring is determined by the maximum energy release and thus ensures that atoms stay bound as a molecule. For a qualitative analysis, however, and even for crude quantitative estimates, it is convenient to approximately model the wave function of electrons in the molecule as a superposition of electron states of the original individual atoms. We do not necessarily mean the same states that were filled by electrons in the isolated atoms. Quite often it is more favorable (energywise) that one or more valence electrons transfer to states which were not filled at all in the isolated atom (to excited states). A partial loss in energy is amply compensated for by the formation of a chemical bond which is stronger than the one produced without this restructuring of valence shells.

An extreme case of this restructuring is the formation of the ionic bond, when an electron from the valence shell of one of the interacting atoms is transferred to the shell of the other atom and thus forms a pair of ions. The formation of an ionic molecule or a crystal of NaCl can thus be described in the following manner. The energy required to remove an electron from a Na atom (its ionization potential) is 5.2 eV; the energy released when a Cl atom gains an electron is 3.8 eV. The transfer of an electron from a Na atom to a Cl atom thus consumes $5.2 - 3.8 = 1.4$ eV. This energy deficit is compensated for since the ions produced move close together and the negative Coulomb energy of attraction arises, $-e^2/R$, which balances out 1.4 eV at $R \sim 10^{-7}$cm and even leaves a gain. An excess of negative energy determines the energy of bonding.

When electron states of a molecule are described in terms of atomic states, the superposition of different electron configurations is especially important, as we have already mentioned above. In other words, the external electron shell of each atom is not necessarily in its ground state or

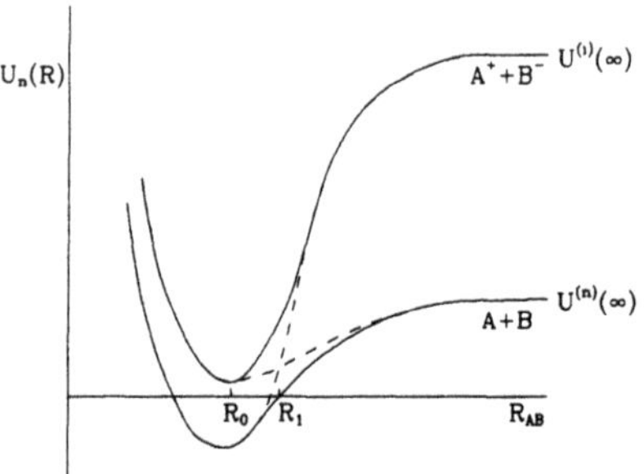

Figure 4.2. Effective energy of interaction between ions.

in a specific excited state. Its wave function can be an arbitrary linear combination of these states and this means that electrons can be in either of these states with, in general, unequal probabilities. Which linear combination exists in reality, that is, what the probabilities are of finding an electron in a specific state, is again determined only by the gain in energy.

We will illustrate this phenomenon by an analysis of formation of the ionic bond, which is more realistic than the scenario given above. Figure 4.2 schematically plots the potential curves for two different electron states of a molecule composed of two atoms A and B forming the ionic bond. The curve $U^{(\mathrm{n})}$ corresponds to the electron state which describes a pair of neutral atoms separated by a large distance R_{AB}, and the curve $U^{(\mathrm{i})}$ corresponds to the state of a pair of ions A^+ and B^-. This situation corresponds qualitatively to the example with NaCl outlined above: as $R_{\mathrm{AB}} \to \infty$, $U^{(\mathrm{i})} > U^{(\mathrm{n})}$, that is, the neutral configuration is favored. As the distance R_{AB} decreases for reasons explained above, $U^{(\mathrm{i})}$ begins to fall off. If one of the states remained purely ionic, $\mathrm{A}^+ + \mathrm{B}^-$, and the other purely neutral, their energies plotted by broken curves in Figure 4.2 would become equal at a point $R_{\mathrm{AB}} = R_1$, after which the ionic configuration would become more favorable. This is the moment at which an electron should be transferred from A to B.

The actual situation is somewhat different. The lower state ceases to be purely neutral even as $R_{\mathrm{AB}} \gg R_1$: the neutral state is superposed on a small admixture of the ionic state. As R_{AB} decreases, this admixture grows and reaches 50% at $R_{\mathrm{AB}} \simeq R_1$, i.e., the state becomes half-ionic, half-neutral. As R_{AB} decreases further, this admixture keeps growing and the state becomes predominantly ionic.

There is, therefore, a continuous transition from a pair of neutral atoms to a pair of ions. The ionic component of the wave function is dominant in the neighborhood of the equilibrium point R_0. A fraction of the neutral state still survives, however. Typically, it is not less than 10–15% even in strongly ionic compounds, so that the idea of a purely ionic bond is a considerable idealization of the reality. Similarly, in the case of the purely ionic state at $R_{\mathrm{AB}} \to \infty$, an admixture of the neutral configuration grows gradually as R_{AB} decreases, and becomes dominant at $R_{AB} < R_1$. By virtue of this mixing of states, their energies change, as shown by solid curves in Figure 4.2. The curves are as if repulsed from each other and do not cross at the point R_1. The energy of the lower state remains at all times lower than the energies of both the purely neutral and the purely ionic configurations. This is an additional stabilization of the chemical bond. This effect is especially strong close to R_1 where the energies of both initial configurations coincide (the case of resonance).

The interatomic distance in ionic crystals and molecules is of the order of 2–3 Å, and the bonding energies are about 8–10 eV per molecule.

The compounds formed by the ionic type of bonding are those in the $A_I B_{VII}$ and partly $A_{II} B_{VI}$ groups. Compounds and crystals formed of elements of groups III and V of the periodic table have even less pronounced ionic nature, and elements of group IV form purely covalent crystals.

2. Covalent bonding. Its typical representative is the bond forming the hydrogen molecule. It is studied in quantum mechanics textbooks (the Heitler–London theory of the hydrogen molecule), so only the main results will be outlined here.

The main role in the formation of the bond in the hydrogen molecule is played by the exchange interaction, which is a specific feature of quantum mechanics. It arises from the familiar Coulomb interaction after the Pauli principle has been taken into account. As we have already mentioned in connection with atoms, the nonrelativistic Hamiltonian does not cover spins so that the wave function can be written as a product of the coordinate wave function and the spin wave function. The Pauli principle, which is equivalent to stipulating that the total wave function be antisymmetric under permutations of electrons, results in a close relationship of the permutational symmetry of the coordinate and spin wave functions.

Owing to the invariance of the electron Hamiltonian of the hydrogen molecule under permutations of electrons(at fixed positions of the nuclei), the coordinate wave function (it is the eigenfunction of the electron Hamiltonian) must correspond to one of the representations of the group of permutations of two electrons. This group has only two irreducible representations, the symmetric and the antisymmetric. The wave functions of the first kind do not change under the permutation, while those of the second kind have their sign reversed.

As a result of the antisymmetry of the total wave function, the symmetric coordinate wave function corresponds to the antisymmetric spin function, and vice versa. The symmetric spin wave function corresponds to the total spin $S = 1$; the antisymmetric wave function corresponds to $S = 0$. The spin thus determines the permutational symmetry of the coordinate wave function, and $S = 0$ corresponds to the symmetric function.

When different symmetric and antisymmetric coordinate wave functions are used for an approximate determination of the eigenvalues of the

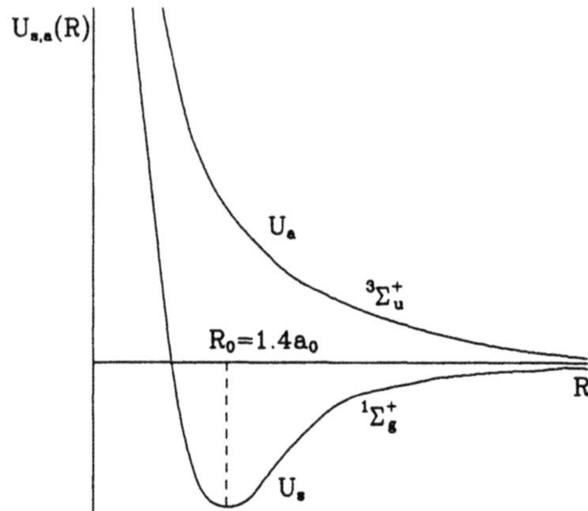

Figure 4.3. Schematic diagram of the electron terms of the hydrogen molecule.

electron Hamiltonian, it is found that the symmetric wave function gives lower energy than the antisymmetric one. In the Heitler–London method, approximate wave functions are constructed of atomic wave functions, which for the hydrogen atom are known. A symmetric and an antisymmetric wave function can be constructed out of the atomic wave functions corresponding to the ground state of the hydrogen atom. The former energy then has an energy U_s less than the sum of the atomic energies and the energies of interaction between nuclei, and the latter energy U_a is higher than this sum. Figure 4.3 plots U_s and U_a as functions of the internuclear spacing R.

In accordance with the classification of electron states of a diatomic molecule, the level U_s obviously corresponds to the state ${}^1\sigma_g^+$ (the upper index gives, as usual, the multiplicity of the term $2S+1$) and the level U_a to the state ${}^3\sigma_u^+$.

As we see from Figure 4.3, only the singlet state, with spin $S=0$ and a symmetric coordinate wave function, corresponds to a stable state of the hydrogen molecule. The triplet state with $S=1$ has no minimum on the potential curve and thus does not bind hydrogen atoms into a molecule. In such an electron state, a molecule dissociates into atoms.

If we now consider the distribution of the electron density for $S=0$ and $S=1$, the electron density at the midpoint between hydrogen nuclei (for $S=0$ and a symmetric wave function) is higher than for individual atoms, owing to the overlapping and interference of the wave functions of the two atoms; if $S=1$, the midpoint density is lower than for individual atoms.

The symmetric state ($S=0$) is preferable energywise because electrons located between the nuclei interact with the nuclei and bind them together. An increased electron density in the space between atomic nuclei is a characteristic feature of the covalent bonding.

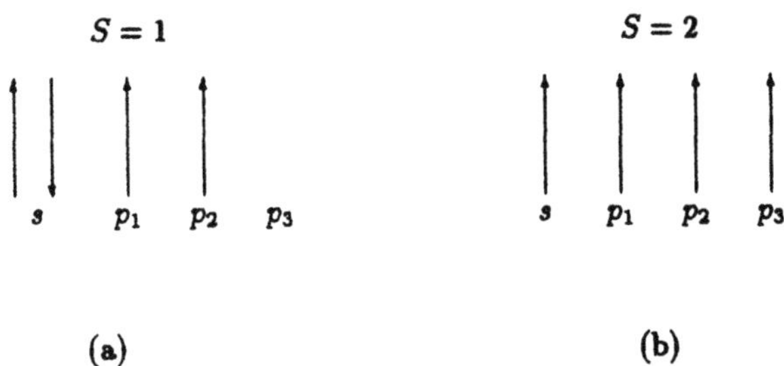

Figure 4.4. (a) Ground state and (b) excited state of the carbon atom.

A covalent bond arises if $S = 0$, that is, if the spins of the two electrons compensate for each other. As the molecule is formed, electrons in the outer shells of the atoms are rearranged so as to saturate the atomic valences. The saturation of the valences consists in the mutual compensation of spins of the pairs of valence electrons. The chemical valence must be defined as the number of electrons in the outer shell with noncompensated (unpaired) spins (hence, the valence equals $2S$).

For this reason, noble gases cannot form valence bonds. Their electron shells are filled and the electron spins are compensated.

A classical example of covalent crystals is found in the semiconductors Si and Ge, and also in C and β-Sn; these are all elements of group IV of the periodic table. The carbon atom C has two electrons in the s state and two in the p state. The spins of the last two are not compensated, which could correspond to valence 2 (see Figure 4.4(a)). However, the energies of the s and p electrons differ very slightly, so that as the atoms move closer, it is energetically favorable for the electron shell of the atom to rearrange itself so as to correspond to the excited state with $S = 2$ (see Figure 4.4(b)) with four electrons whose spins are not compensated. These are the four electrons that form four valence bonds.

If there are several bonds, they are arranged at specific angles to one another. In group IV semiconductors, this gives a symmetric tetrahedral structure of the nearest surroundings of the atom. As a result of the formation of the superposition (mixing) of s and p states, all four bonds are equivalent.

With covalent bonding, the interaction energy decreases exponentially with distance. The bonding energy is of the order of 5–10 eV, as it is for the ionic bond.

Ionic crystals are typically dielectric while typical covalent crystals are semiconductors; this observation agrees with the partial delocalization of charge in covalent bonds.

The properties of $A_{III}B_{V}$ compounds (crystal structure etc.) are quite close to those of covalent Ge and Si crystals. The bonding here is intermediate: the charge is partly transferred from the group III atom to the group V atom. At the same time, electron exchange arises and the charge density, localized between the nuclei, generates covalent bonding. Among such compounds we find GaAs, InSb and others. CdS- and ZnO-type semiconductors $A_{II}B_{VI}$ are compounds in which the bonding is intermediate between ionic and covalent but the ionic character of the bonding manifests itself stronger than for $A_{III}B_{V}$ compounds.

The ionic and covalent bonds are the extreme cases of the same valence bond.

3. Metallic bonds exist in solids (in metals) and can be treated as the limiting case of the covalent bonding. It is a result of collectivization of conduction electrons in the metal. The electron density is not concentrated any more in the gap between a pair of atoms but fills the entire space between atoms.

The metallic bonding can also be treated as a limiting case of the ionic bonding. Indeed, we can assume that the sodium atom is a Na^{+} ion plus an electron e^{-}. The energy of interaction between particles includes the energy of Coulomb interaction of the ion and the electron, which is of the order of $-e^2/a$ per unit cell, and the energy of degenerate electron gas. This latter energy is positive and, according to item 1 of this section, is proportional to $n^{5/3}$, that is, it is of the order of $\hbar^2/ma^2$ per unit cell (since $n \sim a^{-3}$).

The total energy

$$U \sim -\frac{e^2}{a} + \frac{\hbar}{ma^2}$$

has a minimum at $a = 2\hbar^2/e^2m$:

$$U_{\min} = -\frac{e^2m}{4\hbar^2} = -\frac{I_0}{2}$$

This is a quite reasonable estimate. The actual bonding energy in metallic sodium is about 1 eV per atom.

4. Hydrogen bonding. Hydrogen is a special case as far as bonding is concerned. The reason for this is that, on one hand, a hydrogen atom can surrender its electron to an electrically negative atom, such as oxygen, and on the other hand, the hydrogen ion is the proton (which is pointlike on the atomic scale), has no inner electron shells and can therefore come very close to other atoms (or ions).

Under certain conditions, a hydrogen atom is bound by strong attractive forces to two neighboring atoms, which form the so-called hydrogen bond. A typical example of the hydrogen bond is found in ice crystals. In these crystals, water molecules are bound by hydrogen bonds into a certain structure. A proton is located close to one oxygen atom on the line which connects it with the neighboring oxygen ion. It binds two oxygen ions to one another, thus producing a complex, fairly regular ice structure.

The energy of the hydrogen bond, by the order of magnitude, is 0.1 eV.

5. Van der Waals or molecular bonding arises when neither ionic nor covalent bonds are formed. The nature of the van der Waals interaction is completely universal. It arises between any atoms, molecules, and even macroscopic bodies. It is there even if the valence bonding is present but is much weaker than it at interatomic distances at which the covalent bond is appreciable. However, the energy of the valence bond decreases with distance exponentially, so that the van der Waals attractive forces become important at a certain distance between nuclei; their energy depends on distance as R^{-6}.

As a result, there is a very shallow potential well at a distance of about several a_0, even on repulsion-type potential curves (as the curve U_a in Figure 4.3). The depth of this well is of the order of 10^{-2} eV, which is why it could not be shown on the scale of Figure 4.3.

A similar shallow potential well arises owing to the van der Waals interaction of noble gas atoms or molecules with filled electron shells, where the valence bond cannot arise. At small distances, filled electron shells are repulsed but at large distances the van der Waals attraction is stronger. A combination of these factors creates a potential well about 10^{-2} eV deep at distances of several a_0.

The van der Waals interaction connects crystals of noble gases and molecular crystals at low temperatures. The low melting points of such crystals are the result of the weak van der Waals interaction. Examples of molecular crystals are solid H_2 and N_2 and many organic crystals. Elementary building blocks of molecular crystals are molecules at distances which are greater than the distances between atoms in each molecule.

The classical explanation of the nature of the van der Waals interaction is that atoms (molecules) interact via their fluctuating dipole moments. Assume that we have two noble gas atoms spaced by a distance R. On average, the space charge distribution in each atom is spherically symmetric but at each moment of time an atom may have a nonzero dipole moment (whose time average is zero). If $\mathbf{d}_1$ is the instantaneous value assumed by the dipole moment of atom 1, then it produces an electric field $\mathcal{E}$ of the order of d_1/R^3 at the distance R from the atom. We can use the dipole approximation since R is much larger than the atomic size.

This electric field induces in the second atom a dipole moment of the order of $d_2 \sim \varkappa\mathcal{E} \sim \varkappa d_1/R^3$, where $\varkappa$ is the polarizability of the atom. Finally, the interaction energy of two dipoles is $d_1 d_2/R^3 \sim \varkappa d_1^2/R^6$.

It is essential that $\overline{d_1^2} \neq 0$ even if $\bar{d}_1 = 0$. Furthermore, the energy of interaction is very low and falls off with distance as R^{-6}.

These arguments can be translated into the language of quantum mechanics. We will show that there is a weak interaction at distances of about several a_0 where electron shells do not overlap.

In the adiabatic approximation, the electron Hamiltonian of two atoms at fixed positions of the two atoms is

$$\mathscr{H}_{0e} = \mathscr{H}_{0e}^{(1)} + \mathscr{H}_{0e}^{(2)} + V \tag{4.5.1}$$

where $\mathscr{H}_{0e}^{(1)}$ and $\mathscr{H}_{0e}^{(2)}$ are the electron Hamiltonians of isolated atoms and V is the interaction energy. The Coulomb interaction of atoms at large distances can be expanded into a series in multipoles; since the total charge of each atom is zero, the first term is the dipole–dipole interaction

$$V = -\frac{1}{R^3}\left(\mathbf{d}_1\mathbf{d}_2 - 3(\mathbf{d}_1\mathbf{n})(\mathbf{d}_2\mathbf{n})\right) \tag{4.5.2}$$

where $\mathbf{n}$ is the unit vector along the vector R connecting the nuclei of the atoms.

The average value of the operator of the dipole moment, $\mathbf{d}_1$ or $\mathbf{d}_2$, in the stationary state of the atom is zero.

Let us find the electron energy U, that is, the eigenvalue of Hamiltonian (4.5.1) using perturbation theory. The energy U can be written as

$$U_{nm} = E_n^{(1)} + E_m^{(2)} + \Delta E_{nm}$$

where $E_n^{(1)}$ and $E_m^{(2)}$ are the energies of the atoms, that is, the eigenvalues of $\mathscr{H}_{0e}^{(1)}$ and $\mathscr{H}_{0e}^{(2)}$, and ΔE is the perturbation-theory correction to energy.

The first-order correction ΔE to the energy of the ground state $\varphi_0^{(1)}\varphi_0^{(2)}$ is zero, since the average value of the perturbation V for this state is zero. This is implied by the fact that the average value of the product of dipole moments equals the product of average values (which equal zero) for factorized states of the type of $\varphi_0^{(1)}\varphi_0^{(2)}$.

The second-order perturbation-theory correction to the ground state $(n = m = 0)$ is

$$\Delta E_{00} = \sum_{nm} \frac{|V_{00,nm}|^2}{E_0^{(1)} + E_0^{(2)} - E_n^{(1)} - E_m^{(2)}}$$

A specific feature of the second order in perturbation theory is that for the ground state we always have $\Delta E_{00} < 0$, that is, the interaction between atoms is attractive. Making use of the explicit expression (4.5.2) for, V, we obtain

$$\Delta E_{00} = -\frac{1}{R^6}\sum_{nm} \frac{(d_{1\alpha})_{0n}(d_{1\beta})_{n0}(d_{2\alpha})_{0m}(d_{2\beta})_{m0}}{E_n^{(1)} + E_m^{(2)} - E_0^{(1)} - E_0^{(2)}} + \dots \tag{4.5.3}$$

In (4.5.3), α and β are polarization indices over which summation is carried. We have written above only the first term which emerges from the scalar product $\mathbf{d}_1\mathbf{d}_2$ in (4.5.2). Other terms have a similar structure.

The double sum in (4.5.3) can be factorized into a product of single sums over n and m by the following method. We use the identity

$$\frac{1}{E_n^{(1)} + E_m^{(2)} - E_0^{(1)} - E_0^{(2)}} = \int \frac{d(\hbar\omega_1)\, d(\hbar\omega_2)}{\hbar\omega_1 + \hbar\omega_2} \delta(E_n^{(1)} - E_0^{(1)} - \hbar\omega_1)\, \delta(E_n^{(2)} - E_0^{(2)} - \hbar\omega_2)$$

Equality (4.5.3) then changes to

$$\Delta E_{00} = -\frac{1}{R^6\hbar} \int \frac{d\omega_1\, d\omega_2}{\omega_1 + \omega_2} \varkappa_{\alpha\beta}^{(1)''}(\omega_1)\, \varkappa_{\alpha\beta}^{(2)''}(\omega_2) + \dots \tag{4.5.4}$$

where $\varkappa_{\alpha\beta}^{(1)''}$ and $\varkappa_{\alpha\beta}^{(2)''}$ are the imaginary parts of polarizability of the ground state of the atoms 1 and 2. According to Chapter 2, these polarizabilities are found in terms of the matrix elements of dipole moments:

$$\varkappa_{\alpha\beta}^{(1)''}(\omega_1) = \hbar \sum_n (d_{1\alpha})_{0n} (d_{1\beta})_{n0}\, \delta(E_n^{(1)} - E_0^{(1)} - \hbar\omega_1)$$

A similar formula holds for $\varkappa_{\alpha\beta}^{(2)''}$.

For an atom or an isotropic molecule, we have

$$\varkappa_{\alpha\beta} = \varkappa\delta_{\alpha\beta}$$

whence

$$\varkappa_{\alpha\beta}^{(1)''}\, \varkappa_{\alpha\beta}^{(2)''} = 3\varkappa^{(1)''}\, \varkappa^{(2)''}$$

The remaining terms of (4.5.4) have the form $9\varkappa_{xx}^{(1)''}(\omega_1)\varkappa_{xx}^{(2)''}(\omega_2)$ (the axis x is directed along $\mathbf{R}$) or $-6\varkappa_{\alpha x}^{(1)''}(\omega_1)\varkappa_{\alpha x}^{(2)''}(\omega_2)$. In the isotropic case, they finally give the formulas $9\varkappa^{(1)''}(\omega_1)\varkappa^{(2)''}(\omega_2)$ and $-6\varkappa^{(1)''}(\omega_1)\varkappa^{(2)''}(\omega_2)$. Collecting all the terms, we arrive at the London formula,

$$\Delta E_{00} = -\frac{A}{R^6}$$

$$A = \frac{6}{\hbar} \int \frac{\varkappa^{(1)''}(\omega_1)\, \varkappa^{(2)''}(\omega_2)}{\omega_1 + \omega_2} d\omega_1\, d\omega_2$$

which relate the van der Waals interaction energy with atomic polarizabilities.

It is not difficult to estimate ΔE_{00} using (4.5.3). Since $d \sim ea_0$ and $E_n^{(1)} + E_m^{(2)} - E_0^{(1)} - E_0^{(2)} \sim I_0$, we obtain

$$\Delta E_{00} \sim \frac{(ea_0)^4}{R^6 I_0} \sim \frac{e^4 a_0^4 a_0}{R^6 e^2} \sim \frac{e^2}{a_0}\left(\frac{a_0}{R}\right)^6 \sim I_0\left(\frac{a_0}{R}\right)^6$$

If $R \sim (3\text{–}4)a_0$, this gives a reasonable estimate $\Delta E \sim 10^{-2}$ eV. In the case of hydrogen, it is possible to find the values of all matrix elements of dipole moments and numerically sum up (4.5.3). This gives

$$\Delta E_{00} = -6.47\,\frac{e^2 a_0^5}{R^6}$$

which is in accord with the estimate given above.

4.6. Oscillational and Rotational Energy of Molecules

It has been shown in Section 4.4 that the wave function of a molecule in the adiabatic approximation has the form

$$\psi_n(\mathbf{r}, \mathbf{R}, t) = \Phi_n(\mathbf{r}, t)\varphi_n(\mathbf{r}, \mathbf{R})$$

where $\Phi_n(\mathbf{r}, \mathbf{R})$ satisfies the equation

$$i\hbar\,\frac{\partial \Phi_n}{\partial t} = \mathcal{H}_n \Phi_n$$

in which $\mathcal{H}_n$ is the vibrational–rotational Hamiltonian, including the kinetic energy of the nuclei, T, and the effective potential energy for the nth electron state $U_n(\mathbf{R})$.

The translational energy of the molecule as a whole is $\mathbf{P}^2/2M$, where $\mathbf{P}$ is the total momentum of the molecule and M is its mass; this energy is assumed to be singled out by changing to the center-of-mass reference frame. The next natural step is to single out the energy of rotation of the molecule as a whole.

The vibrational–rotational Hamiltonian can be written under certain conditions as

$$\mathcal{H}_n = T + U_n = U_n(\mathbf{R}_0) + \mathcal{H}_v + \mathcal{H}_r + \mathcal{H}_{vr}$$

where $U_n(\mathbf{R}_0)$ is the value of $U_n(\mathbf{R})$ in the equilibrium configuration of nuclei, $\mathbf{R}_0$, $\mathcal{H}_v$ is the oscillational (vibrational) Hamiltonian, $\mathcal{H}_r$ is the rotational Hamiltonian, and $\mathcal{H}_{vr}$ describes the interaction between vibrations and rotations.

The effective potential energy $U_n(\mathbf{R})$ does not change if a molecule is rotated as a whole, that is, when the relative distances between nuclei remain constant. It is assumed that $U_n(\mathbf{R}_0)$ corresponds to a stable state of the molecule, in other words, $U_n(\mathbf{R})$ has a (local) minimum at a certain equilibrium configuration $\mathbf{R}_0$. Owing to this, the configuration $\mathbf{R}_0$ is defined up to rotations of the molecule as a whole.

The rotational degrees of freedom can be singled out by converting to a system of coordinates fixed to the molecule. The orientation of this system of coordinates relative to the laboratory frame of reference is given by Euler's angles α, β, and γ (Figure 4.1).

The rotational Hamiltonian $\mathscr{H}_r$ takes the form

$$\mathscr{H}_r = \frac{\hbar^2}{2}\left(\frac{J_{x'}^2}{I_{x'}} + \frac{J_{y'}^2}{I_{y'}} + \frac{J_{z'}^2}{I_{z'}}\right) \tag{4.6.1}$$

where $J_{x'}$, $J_{y'}$, $J_{z'}$ are the components of the angular momentum of the molecule in the rotating reference frame x', y', z' with the orientation of axes given by Euler's angle, and $I_{x'}$, $I_{y'}$, $I_{z'}$ are the principal moments of inertia of the molecule in the equilibrium configuration.

The operators $J_{x'}$, $J_{y'}$, $J_{z'}$ satisfy the commutation relations of the type

$$[J_{x'}, J_{y'}] = -\mathrm{i}J_{z'}$$

while the operators J_x, J_y, J_z of the angular momentum components in the laboratory system of coordinates satisfy commutation relations of the same type but with the sign in the righthand side reversed. The operators $J_{x'}$, $J_{y'}$, $J_{z'}$, as well as J_x, J_y, J_z, are operators which act on angular variables, i.e., Euler's angles, which determine the vibrational–rotational wave function Φ_n.

The potential energy $U_n(\mathbf{R})$ can be expanded in a series in powers of the deviation of the nuclear configuration $\mathbf{R}$ from the equilibrium configuration $\mathbf{R}_0$. Since there is a minimum in the configuration at $\mathbf{R}_0$, there can be no linear terms and the expansion begins with quadratic terms. If the number of atoms (nuclei) in the molecule is N, the function U_n depends on $3N-6$ variables since $3N$ is the number of degrees of freedom of nuclei out of which 3 degrees of freedom are the eliminated translational degrees and 3 are the rotational degrees of freedom, which now correspond to the coordinates α, β, and γ. The function U_n is independent of α, β, γ so that all in all there are $3N-6$ vibrational coordinates (the number of such coordinates for a linear molecule, whose orientation is determined by only two angular variables α and β — the angles in the spherical system of coordinates, is $3N-5$).

We can diagonalize the quadratic form, $U_n - U_n(\mathbf{R}_0)$, using the linear transformation of coordinates and converting to the so-called normal coordinates; this gives the expression which is exact up to terms of higher than second order in displacement $(\mathbf{R}-\mathbf{R}_0)$. These higher-order terms (the anharmonic terms) can also be written via normal coordinates and treated as perturbation by merging them with $\mathcal{H}_{vr}$. The vibrational Hamiltonian $\mathcal{H}_v$ then has the form of a sum of Hamiltonians of the oscillators each of which corresponds to a specific normal coordinate Q_l $(l = 1, \ldots, 3N - 6)$:

$$\mathcal{H}_v = \frac{1}{2}\sum_l (P_l^2 + \omega_l^2 Q_l)^2 \tag{4.6.2}$$

In (4.6.2), ω_l is the frequency of normal oscillations and

$$P_l = -i\hbar \frac{\partial}{\partial Q_l}$$

In addition to anharmonic terms, the perturbation operator $\mathcal{H}_{vr}$ includes terms which describe the interaction of vibrations and rotation. The nature of these terms is the dependence of moments of inertia I on the vibrational coordinates (centrifugal distortion) and also the Coriolis coupling which manifests itself in that the energy of the rotational motion is a function of vibrational velocities (vibrational momenta) for a given, conserved moment $\mathbf{J}$. Ignoring for the moment the operator $\mathcal{H}_{vr}$, which can be taken into account by perturbation theory, we write the eigenfunction of the vibrational–rotational Hamiltonian $\mathcal{H}_n$ in the form

$$\Phi_{nvJ} = \Phi_J(\alpha, \beta, \gamma)\Phi_v(Q)$$

where J are the quantum numbers of Hamiltonian (4.6.1) and v are the quantum numbers of the states of the ensemble of oscillators which are described by Hamiltonian (4.6.2). We know from quantum mechanics that v are positive integers $v_1, \ldots, v_{3N-6}$, corresponding to the vibrational energy

$$E_v = \sum_l \hbar\omega_l(v_l + 1/2)$$

It is rather easy to find the eigenvalues of the rotational Hamiltonian (4.6.1) for molecules of the type of the symmetric top, with $I_{x'} = I_{y'}$. If a molecule has a threefold symmetry axis or is of a higher order of symmetry, two moments of inertia are always equal. Hamiltonian (4.6.1) then becomes

$$\mathcal{H}_r = \frac{\hbar^2}{2}\left[\frac{\mathbf{J}^2}{I_{x'}} + \left(\frac{1}{I_{z'}} - \frac{1}{I_{x'}}\right) J_{z'}^2\right]$$

It commutes with J^2, J_z, and $J_{z'}$, which commute among themselves, so that the eigenfunction of this operator coincides with the eigenfunction of the operators J^2, J_z, and $J_{z'}$:

$$\mathbf{J}^2\Phi_{JMK} = J(J+1)\Phi_{JMK}$$
$$J_z\Phi_{JMK} = M\Phi_{JMK}$$
$$J_{z'}\Phi_{JMK} = K\Phi_{JMK}$$
$$\mathscr{H}_r\Phi_{JMK} = \frac{\hbar^2}{2}\left[\frac{J(J+1)}{I_{x'}} + \left(\frac{1}{I_{z'}} - \frac{1}{I_{x'}}\right)K^2\right]\Phi_{JMK}$$

The wave function $\Phi_{JMK}(\alpha, \beta, \gamma)$, which is the eigenfunction of $\mathbf{J}^2$, J_z, and $J_{z'}$, coincides up to a normalizing factor with the Wigner function $D^J_{KM}(\alpha, \beta, \gamma)$ mentioned in Section 4.3:

$$\Phi_{JMK}(\alpha, \beta, \gamma) = \sqrt{2J+1}\, D^J_{KM}(\alpha, \beta, \gamma) \tag{4.6.3}$$

In the case of symmetric-top-type molecules, therefore, the rotational energy is

$$E_J = \frac{\hbar}{2}\left[\frac{J(J+1)}{I_{x'}} + \left(\frac{1}{I_{z'}} - \frac{1}{I_{x'}}\right)K^2\right] \tag{4.6.4}$$

The energy of the stationary state of the molecule, however, splits into the electron energy $E_n = U_n(R_0)$, the vibrational energy E_v and the rotational energy E_J:

$$E_{nvJ} = E_n + E_v + E_J$$

We have already mentioned that for small quantum numbers v and J, $E_J \ll E_v \ll E_n$, and that by the order of magnitude,

$$E_n \sim I_0 \quad E_v \sim \mu^2 I_0 \quad E_J \sim \mu^4 I_0 \quad \mu = (m/M)^{1/4}$$

In a symmetric general-type molecule, in which $I_{x'} \neq I_{y'} \neq I_{z'}$, an eigenfunction of operator $\mathscr{H}_r$ (4.6.1) is a linear combination of the functions D^J_{KM} for a given value of J and M and for arbitrary K ($|K| \leq J$). It corresponds to a specific value of the angular momentum J of the molecule and to its projection M onto an axis fixed in space. If $I_{x'}$, $I_{y'}$, and $I_{z'}$ are known, the values of the rotational energy can be found by solving the eigenvalue problem for the matrix of rank $(2J+1)$ which represents operator (4.6.1) in the basis composed of functions $\Phi_{JMK}(\alpha, \beta, \gamma)$.

Let us consider in more detail the separation of the vibrational and rotational motions for a diatomic molecule. In the electron Σ-state, for example, in the ground state of a diatomic molecule, the energy $U_n(\mathbf{R})$ is a function of only the distance R between nuclei, independent of the direction of the vector $\mathbf{R}$. The separation of the vibrational and rotational motions is then achieved immediately by converting to spherical coordinates.

In the center-of-mass reference frame of nuclei we have

$$\mathscr{H}_n = -\frac{\hbar}{2M}\nabla^2 + U_n(R) \tag{4.6.5}$$

where M is the reduced mass of the nuclei or atoms of the molecule, $M^{-1} = M_1^{-1} + M_2^{-1}$ and $\nabla^2 = \partial^2/\partial\mathbf{R}^2$, where $\mathbf{R}$ is the vector connecting the nuclei of the molecule.

Operator (4.6.5) can be written in the spherical coordinates R, θ and φ as

$$\mathscr{H}_n = \frac{\hbar^2}{2M}\left[\frac{\mathbf{J}^2}{R^2} - \frac{1}{R^2}\frac{\partial}{\partial R}R^2\frac{\partial}{\partial R}\right] + U_n(R)$$

where $\mathbf{J}^2$ is the operator of squared angular momentum which acts on the angular variables θ and φ.

The vibrational variable R and the rotational variables θ and φ allow separation,

$$\Phi_n(R, \Theta, \varphi) = \Phi_r(\Theta, \varphi)\,\Phi_v(R)$$

so that the rotational wave functions Φ_r coincide with the spherical functions Y_{JM}, which are the eigenfunctions of $\mathbf{J}^2$ with the eigenvalue $J(J+1)$ and of J_z with the eigenvalue M.

For the vibrational wave function Φ_r we obtain the equation

$$\left[-\frac{\hbar^2}{2MR^2}\left(\frac{\partial}{\partial R}R^2\frac{\partial}{\partial R} - J(J+1)\right) + U_n(R) - E\right]\Phi_v(R) = 0$$

Assuming $\Phi_v = \widetilde{\Phi}_v/R$, we rewrite the above equation as

$$\left[-\frac{\hbar^2}{2M}\frac{\mathrm{d}^2}{\mathrm{d}R^2} + \frac{J(J+1)}{2MR^2} + U_n(R) - E\right]\widetilde{\Phi}_v(R) = 0$$

Let the function $U_n(R)$ have a minimum at a point R_0; we then find for small $\rho = R - R_0$ that

$$U_n(R) \simeq U_n(R_0) + \frac{1}{2}\frac{\mathrm{d}^2U_n}{\mathrm{d}R^2}\rho^2 = E_n + \frac{1}{2}M\omega_n^2\rho^2 \tag{4.6.6}$$

This relation determines the vibrational frequencies ω_n.

In the harmonic approximation (4.6.6), the equation for $\widetilde{\Phi}_v$ is the linear oscillator equation, whose solution is

$$\widetilde{\Phi}_v = \mathrm{const}\cdot\exp\left(\frac{M\omega_n^2\rho^2}{\hbar}\right)H_v\left(\sqrt{\frac{M\omega_n}{\hbar}}\,\rho\right) \tag{4.6.7}$$

where H_v are Hermite polynomials, and

$$E = E_n + B_n J(J+1) + \hbar\omega_n(v + 1/2) \tag{4.6.8}$$

Formula (4.6.8) is the decomposition of the energy into the electron, rotational, and vibrational components. The rotational constant is $B_n = \hbar^2/2MR_0^2$. By comparing (4.6.8) and (4.6.4), we see that the diatomic molecule is a particular case of the symmetric-top molecule, with $I_{z'} = 0$ and $K = 0$. As $I_{z'} \to 0$, one rotational degree of freedom and one rotational coordinate (the Euler angle γ) vanish. We need to assume here that $K = 0$, so that by virtue of (4.3.1) the wave functions Φ_{JMK} transform into spherical functions Y_{JM}. For the electron states of a diatomic molecule with $\Lambda \neq 0$ (the Π, Δ, etc. states), the rotational wave function in the adiabatic approximation mostly coincides with the rotational wave function Φ_{JMK} of the symmetric top molecule, (4.6.3), with $K = \pm\Lambda$, the only difference being the factor $\sqrt{2\pi}$. This factor arises because we need not introduce the third Euler angle γ for a diatomic molecule but can set $\gamma = 0$ in Φ_{JMK}; also, we need not integrate over γ under conditions of normalization. Note that $I \geq |K| = \Lambda$: the total momentum of the molecule cannot be less than its electron momentum.

In the center-of-mass reference frame, the operator of kinetic energy of the nuclei of a diatomic molecule can be written as

$$T = \frac{\hbar^2}{2MR^2}\Big(-\frac{\partial}{\partial R}\Big(R^2\frac{\partial}{\partial R}\Big) + (\mathbf{J}-\mathbf{L})^2\Big)$$

where $\mathbf{J}$ is the operator of the total (orbital) momentum and $\mathbf{L}$ is the operator of the electron angular momentum. The wave function ψ of the molecule must satisfy the condition $(J_{z'} - L_{z'})\psi = 0$, since the angular momentum of nuclei along the axis of the molecule z' is zero.

After averaging over the electron state with the quantum number $K = \pm\Lambda$, we obtain, as implied by (4.4.8),

$$\mathcal{H}_K = \frac{\hbar^2}{2MR^2}\Big(-\frac{\partial}{\partial R}\Big(R^2\frac{\partial}{\partial R}\Big) + \mathbf{J}^2 + \overline{\mathbf{L}^2} - 2\Lambda^2\Big) + U_K(R)$$

since the mean electron momentum $\bar{\mathbf{L}} = \int \varphi_k^* \mathbf{L} \varphi_k \, \mathrm{d}^3 r$ is directed along the axis of the molecule. Therefore, if $\Lambda \neq 0$, the potential energy $U_K(R)$ is replaced by

$$U_K'(R) = U_K(R) + \frac{\hbar^2}{2MR^2}(\overline{\mathbf{L}^2} - 2\Lambda^2)$$

In the adiabatic approximation, the energy levels remain twice degenerate, in correspondence with two signs of $K = \pm\Lambda$. If we use perturbation theory to take into account the matrix elements of the operator $(\mathbf{J} - \mathbf{L})^2$, which is nondiagonal in electron states (i.e., if we consider nonadiabatic terms of $\mathscr{H}_0$), and if $\Lambda = 1$, we find that the splitting of the levels with $K = \pm\Lambda$ is very small, of the order of B_K^2/I_0 (B_K is the rotational constant). This is the so-called Λ doubling. If $\Lambda > 1$, this splitting is so small that it is negligible in all cases.

Here, as before, we have neglected the electron and nuclear spins. This corresponds to the nonrelativistic Hamiltonian (4.4.1) and to neglecting the spin–orbital and spin–spin interactions. If the spin–orbital interaction becomes of the order of the interval between the neighboring rotational levels, then this interaction must be taken into account if the total spin $S \neq 0$. Also, the total orbital momentum $\mathbf{J}+\mathbf{S}$ plays the role of the orbital moment $\mathbf{J}$. Since in all sufficiently heavy molecules the energy of the spin–orbital interaction is greater than the interval of the rotational structure (if $\Lambda \neq 0$), this remark is often relevant; our analysis is mostly valid for singlet electron terms ($S = 0$).

If we take into account higher-order (anharmonic) terms in the expansion of U_n in powers of ρ and also expand the term with rotational energy $\hbar^2 J(J+1)/2M(R_0+\rho)^2$ in the vibrational Hamiltonian in powers of ρ, we obtain the perturbation-theory corrections to energies (4.6.8). These corrections express the interaction of rotation and vibration:

$$E = E_n + B_{nv}J(J+1) + \hbar\omega_n(v+1/2) - \xi(v+1/2)^2 w - \nu J^2(J+1)^2$$

It is essential that the rotational constant $B_{nv} = B_n - \eta(v+\frac{1}{2})$ now depends on the vibrational quantum number (note that ξ, ν and η are small).

In polyatomic linear molecules (e.g., carbon dioxide molecule CO_2), a molecule may have a vibrational momentum at the twice-degenerate frequency of deformational vibrations; this momentum assumes the values v, $v-2$, ..., $-v$ ($v+1$ values in all). If vibrations are harmonic, the energy is a function of v and is independent of the vibrational momentum. In the case of anharmonic vibrations, the degeneracy in the vibrational momentum is removed. A similar situation may be found for symmetric-top-type molecules.

The classification of vibrations in molecules is introduced by using the notions of the molecular symmetry group. The theory of representations is used here to an even greater extent than for the electron states of molecules. It makes it possible to find the number of normal vibrations with certain symmetry and the degeneracy of these vibrations once the structure of the molecule (or of the unit cell of the crystal) is known. The theory of representations offers efficient methods of decomposing representations into irreducible ones, of determining the number of irreducible representations, etc. The same theory helps to determine the selection rules: which oscillations can be active in molecular spectra. As an example, we will now consider the water molecule.

The equilibrium configuration of the H_2O molecule is shown in Figure 4.5(a). The symmetry group of the molecule is C_{2v}; the symmetry

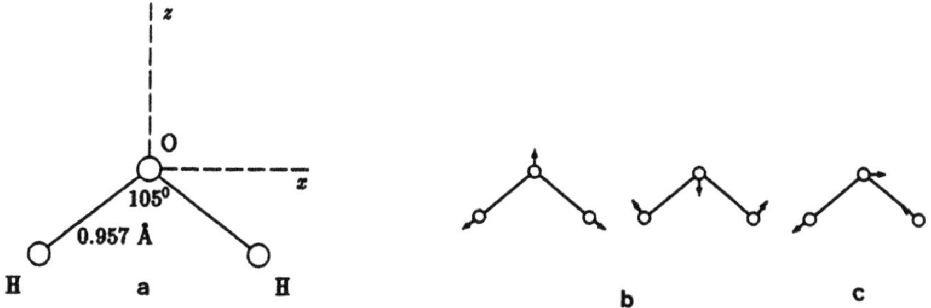

Figure 4.5. Vibrations of the water molecule: (*a*) equilibrium configuration, (*b*) fully symmetric vibrations, (*c*) partly symmetric vibrations.

transformations are the identity transformation E, rotation C_2 around the z axis by the angle π, reflection σ_v in the xz plane, and reflection $\sigma_{v'}$ in the yz plane. This symmetry group is commutative (Abelian) and has four one-dimensional irreducible representations: A_1, A_2, B_1 and B_2. Each symmetry transformation corresponds to a one-dimensional matrix, that is, ± 1. The table of these one-dimensional representations is

A_1	1	1	1	1
A_2	1	1	−1	−1
B_1	1	−1	−1	1
B_2	1	−1	1	−1

Displacements of three atoms of the molecule correspond to nine degrees of freedom and are transformed under a reducible representation which contains three fully symmetric (identity) representations A_1, two representations A_2, two representations B_1, and two representations B_2. However, we need to remove from this list the translational displacements and rotations that correspond to one representation A_1, one representation A_2, two representations B_1, and two representations B_2.

The vibrational degrees of freedom take up two representations A_1 and one representation B_2. The representation A_1 corresponds to two fully symmetric vibrations shown in Figure 4.5(*b*), the stretching of valence bonds and the deformation (bending); the representation B_2 corresponds to the partly symmetric valence vibration shown in Figure 4.5(*c*).

As an example of a linear molecule, we can consider the CO_2 molecule with $3 \times 3 - 5 = 4$ degrees of freedom. Two vibrational degrees of freedom correspond to nondegenerate symmetric and antisymmetric valence vibrations along the axis of the molecule, while the other two correspond to the degenerate deformational vibrations in which atoms are displaced perpendicularly to the axis of the molecule.

The interaction of vibrations with the electron motion (i.e., the nona-

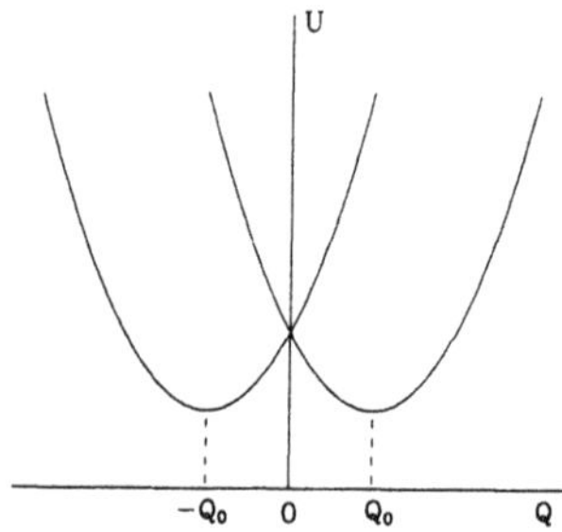

Figure 4.6. Adiabatic energies $U(Q)$ in the case of the Jahn–Teller effect.

diabatic terms in equation (4.4.6)) may play an important role in the degeneracy of electron states. In such cases of degenerate electron states in symmetric molecular configurations, the Jahn–Teller and Renner effects take place. The Jahn–Teller effect consists in the symmetric configuration of the molecule becoming unstable if it is possible to use the adiabatic approach in the case of degenerate electron state of a nonlinear molecule. The dependence of adiabatic energies $U(Q)$ on the vibrational coordinate Q for the electron state which is twice-degenerate at $Q = 0$ has the form shown in Figure 4.6. The symmetric state $Q = 0$ which corresponds to degenerate electron energy $U(Q)$ becomes unstable while the distorted configuration becomes stable.

The Renner effect occurs in polyatomic linear molecules, such as CO_2, BO_2 etc. Deformational vibrations of the molecule split the twice-degenerate electron term Π; vibrations in this electron state are very unconventional, which manifests itself in spectral characteristics.

4.7. Optical Spectra of Molecules

Once the energies of the stationary states of a molecule and their wave functions are known, we can find the frequencies and transition probabilities. In the dipole approximation, the transition probabilities are determined by the matrix elements of the operator of the dipole moment $\mathbf{d}$. The same matrix elements are in the expressions for the susceptibilities derived in Chapter 2 in the dipole approximation.

The probability of transition from a state n, v, J to a state n', v', J' is determined by the familiar formula of perturbation theory in quantum mechanics:

$$w_{n'v'J',nvJ} = \frac{2\pi}{\hbar} |(\mathcal{E}\mathbf{d})_{n'v'J',nvJ}|^2 - \delta(E_{n'v'J'} - E_{nvJ} - \hbar\omega)$$

The calculation reduces to finding the matrix element $(d_\alpha)_{n'v'J',nvJ}$.

According to the previous section, the wave function ψ_{nvJ} in the adiabatic approximation can be rewritten in the form

$$\psi_{nvJ} = \Phi_J(\alpha, \beta, \gamma)\Phi_v(Q)\varphi_n(\mathbf{r}', Q)$$

where both the electron wave function φ_n and the vibrational wave function Φ_v are assumed to be in the reference frame x', y', z' fixed to the molecule. The electron wave function depends on the coordinates $\mathbf{r}'$ of electrons in this system of coordinates and on the coordinates of nuclei (i.e., on the vibrational coordinates Q) as on parameters. It is independent of the rotational coordinates α, β, γ. The adiabatic energy $U_n(Q)$ is also a function of only the vibrational coordinates.

Instead of the components d_α, it is more convenient to consider their linear combinations d_m, where $m = 0, \pm 1$: $d_0 = d_z$, $d_{\pm 1} = d_x \pm \mathrm{i} d_y$. Likewise, $d_{m'}$ is introduced in the reference frame rotating together with the molecule: $d_{0'} = d_{z'}$, $d_{\pm 1'} = d_{x'} \pm \mathrm{i} d_{y'}$. The components d_m of the vector are transformed under the vector representation of the group of rotations, that is,

$$d_m = \sum_{m'} D^1_{m'm}(\alpha, \beta, \gamma) d_{m'}$$

The matrix element of the operator d_m is calculated by integrating over the electron coordinates $\mathbf{r}$, vibrational coordinates Q, and rotational coordinates α, β, γ:

$$\begin{aligned}(d_m)_{n'v'J',nvJ} &= \int \sin\beta \, \mathrm{d}\alpha \, \mathrm{d}\beta \, \mathrm{d}\gamma \\ &\times \int \mathrm{d}Q \int \mathrm{d}^3 r' \Phi^*_{J'}(\alpha, \beta, \gamma)\Phi^*_{v'}(Q)\varphi^*_{n'}(\mathbf{r}', Q) \\ &\times \sum_{m'} D^1_{m'm}(\alpha, \beta, \gamma) \, d_{m'} \Phi_J(\alpha, \beta, \gamma)\Phi_v(Q)\varphi_n(\mathbf{r}', Q) \quad (4.7.1)\end{aligned}$$

Integration over the angles α, β, γ can be carried out immediately if the rotational wave functions $\Phi_{J'}$ and Φ_J are known. We do know them for symmetric-top-type and linear molecules, and, by virtue of (4.6.3), they coincide with the Wigner functions. Integrals of products of Wigner functions $D^{J'*}D^1D^J$ are calculated in the general form and can be written as known coefficients $\varkappa_{m'm}$ which depend on the quantum numbers of rotational states. This implies

$$(d_m)_{n'v'J',nvJ} = \sum_{m'} \varkappa_{m'm} \int \Phi^*_{v'}(Q) \mathrm{d}^{m'}_{n'n}(Q)\Phi_v(Q)\mathrm{d}Q \qquad (4.7.2)$$

where

$$d^{m'}_{n'n}(Q) = \int \varphi^*_{n'}(\mathbf{r}', Q) \mathrm{d}_{m'} \varphi_n(\mathbf{r}', Q) \, \mathrm{d}^3 r'$$

is the dipole moment of the electron transition which depends parametrically on the vibrational coordinates Q.

The quantities $|\varkappa_{m'm}|^2$ determine the relative intensities of the rotational components of the molecular spectrum and are known as the Hönl–London factors. It is essential that they are functions of the rotational quantum numbers of the states and yield, among other results, the familiar selection rule for the rotational quantum number:

$$|J - J'| \leq 1 \leq J + J'$$

For simplicity, we consider only diatomic molecules. The generalization to polyatomic molecules is obvious in most cases. The vibrational coordinate for diatomics is the distance R between nuclei.

First we consider transitions within one (ground) electron state. In this case $n' = n$, $m' = 0$ because the dipole moment of the electron transition in the reference frame fixed to the molecule is directed along the axis of the molecule ($d_{0'} \neq 0$, $d_{\pm 1'} = 0$). We will use the notation $d_n(R) = d^0_{nn}(R)$.

For small $\rho = R - R_0$ (R_0 is the equilibrium internuclear distance) we have

$$d_n(R) \simeq d^0_n + q_n\rho$$

where $d^0_n = d_n(R_0)$ and $q_n = (\partial d_n/\partial R)_{R=R_0}$.

The dimension of the quantity q_n is that of charge; it is known as the effective charge. Estimates for the dipole moment and the effective charge are obvious: $d^0_n \sim d_n \sim ea_0$, $q_n \sim e$.

If we forget about the Hönl–London factors, the transition probability for a diatomic molecule is determined by the matrix element

$$\begin{aligned} d_{v'v} &= \int \widetilde{\Phi}^*_{v'}(\rho)(d^0_n + q_n\rho)\widetilde{\Phi}_v(\rho)\mathrm{d}\rho \\ &= d^0_n \int \widetilde{\Phi}^*_{v'}(\rho)\widetilde{\Phi}_v(\rho)\,\mathrm{d}\rho + q_n \int \widetilde{\Phi}^*_{v'}(\rho)\rho\widetilde{\Phi}_v(\rho)\,\mathrm{d}\rho \end{aligned}$$

The first integral in $d_{v'v}$ equals $d^0_n\delta_{v'v}$ as a result of orthonormalization of the functions $\widetilde{\Phi}_v(\rho)$ in (4.6.7), and is nonzero only if $v' = v$; this is a purely rotational transition. The second integral differs from zero only when $v' = v \pm 1$ for the wave functions of the harmonic oscillator. It corresponds to the vibrational transition. The probability of a purely rotational transition is proportional to the squared constant dipole moment $|d^0_n|^2$ of the molecule. For homonuclear molecules, the dipole moment $d_n(R) = 0$ since the electron wave functions of homonuclear diatomic molecules possess certain parity, which is denoted, as we have mentioned earlier, by the symbols g and u. As the selection rules were derived for the electron spectra of diatomic molecules, it was shown that $d_n(R) \neq 0$ only in the transitions $g \to u$ and $u \to g$. If the electron state remains unchanged, then either

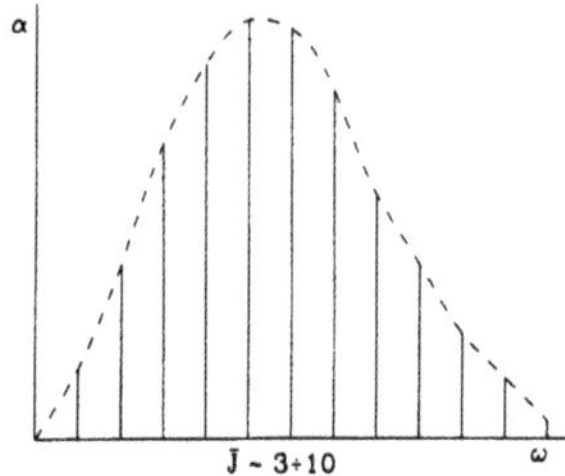

Figure 4.7. Schematic diagram of a purely rotational absorption spectrum of a diatomic molecule, neglecting the finite linewidths.

the electron state remains unchanged, then either $u \to u$ or $g \to g$, and $d_n(R) = 0$. Homonuclear molecules (such as the hydrogen molecule) do not have a purely rotational or purely vibrational spectrum (of absorption or emission).

The frequency of purely rotational transition is

$$\omega = \frac{1}{\hbar} B_{nv}[J'(J'+1) - J(J+1)] = 2\hbar^{-1} B_{nv}(J+1)$$

since the selection rules for J imply $J' = J + 1$. Therefore, the rotational spectrum is formed by equidistant lines at an interval determined by the rotational constant B_{nv}. Generally, the rotational constant is a function of the electron state and the vibrational quantum number v.

The line intensity in the absorption spectrum depends on the populations of the rotational levels, which are proportional to the factor $\exp[-B_{nv}J(J+1)/\mathrm{k}T]$ and to the statistical weight of the rotational level $(2J+1)$, owing to the degeneracy with respect to the quantum number M.

The mean value of J in the absorption spectrum is of the order of $\bar{J} \sim$ 3–10, since

$$\overline{J}^2 \sim \frac{\mathrm{k}T}{B_{nv}} \sim \frac{\mathrm{k}TMR_0^2}{\hbar^2} \sim \frac{M}{m}\frac{\mathrm{k}T}{I_0} \sim 10\text{–}100$$

A typical shape of a purely rotational spectrum is shown in Figure 4.7. The rotational spectrum usually lies in the far infrared range.

For a vibrational–rotational transition we have $v' = v + 1$. For wave functions (4.6.7),

$$d_{v+1,v} = q_n \int \widetilde{\Phi}^*_{v+1} \rho \, \widetilde{\Phi}_v \, \mathrm{d}\rho = q_n \sqrt{\frac{\hbar}{2M\omega_n}} \sqrt{v+1}$$

Taking into account the selection rules $J' - J = 0, \pm 1$, we obtain from (4.6.8) frequencies for such a transition:

$$\omega = \omega_n + \hbar^{-1} B_{nv+1} \begin{cases} 2(J+1) \\ 0 \\ -2J \end{cases} + \hbar^{-1}(B_{nv+1} - B_{nv})J(J+1) \quad (4.7.3)$$

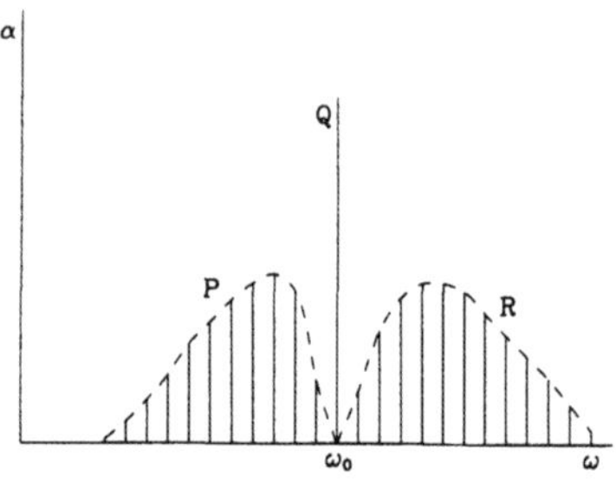

Figure 4.8. Vibrational–rotational absorption spectrum of a diatomic molecule.

The frequencies of the vibrational–rotational spectrum lie in the infrared range. Three values of the frequency correspond to $J' = J + 1$, $J' = J$, $J' = J - 1$ and refer to the R, Q and P branches of the oscillational spectrum (Figure 4.8). If the interaction of rotations and vibrations is taken into account, the rotational constant B_{nv} becomes a function of the quantum number v, as we have mentioned in Section 4.6. The Q-line splits into a band only if this dependence is included in the calculations.

With the next order of expansion terms in the series of $d_n(R)$ in powers of ρ, we would have

$$d_n(R) \simeq d_n^0 + q_n \rho + \frac{1}{2} \frac{\mathrm{d}^2 d_n}{\mathrm{d}R^2} \rho^2$$

The quadratic term is of the order of $e\rho^2/a_0$; it would produce matrix elements of the order of $e\hbar/M\omega_n a_0$. The ratio of the probability of transitions in which the quantum number v changes by 2 ($v' = v + 2$) to the probability of transitions with $v' = v + 1$ is of the order of

$$\frac{w_{v+2,v}}{w_{v+1,v}} \sim \frac{\hbar^2}{a_0^2 M^2 \omega_n^2} \frac{M\omega_n}{\hbar} = \frac{\hbar}{a_0^2 M \omega_n} \sim \sqrt{\frac{m}{M}} = \mu^2 \ll 1$$

Indeed,

$$\hbar\omega_n \sim \mu^2 I_0 \sim \sqrt{\frac{m}{M}} \frac{\hbar^2}{m a_0^2}$$

This last estimate also follows from $\omega_n \sim \sqrt{U''/M} \sim \sqrt{I_0/M a_0^2} \sim (I_0/\hbar)\sqrt{m/M}$.

Assume now that $n' \neq n$. The electron state of the molecule changes and we find an electron transition (in fact, an electron–vibrational–rotational transition). The problem of calculating transition probabilities reduces to calculating the integrals

$$d_{n'v',nv} = \int \widetilde{\Phi}^*_{n'v'}(R)\, d_{n'n}(R) \widetilde{\Phi}_{nv}(R)\, \mathrm{d}R \tag{4.7.4}$$

For such integrals, there are no selection rules for the quantum number v since the vibrational wave functions $\widetilde{\Phi}_{n'v'}$ and $\widetilde{\Phi}_{nv}$ refer to different adiabatic energies $U_n(R)$ and, being eigenfunctions of different Hamiltonians, are not orthogonal. The function $d_{n'n}(R)$ depends on R slower than $\widetilde{\Phi}_{nv}(R)$. The scale of change of the function $d_{n'n}(R)$ is of the order of the molecular size, that is, a_0, while the scale of change of $\Phi_{nv}(R)$ is, as was shown in Section 4.4, of the order of $\mu a_0 \ll a_0$ ($\mu = (m/M)^{1/4}$ is the adiabaticity parameter). If the dependence of $d_{n'n}$ on R is neglected and we set $d_{n'n}(R) \simeq d_{n'n} = \text{const}$, then

$$d_{n'v',nv} = d_{n'n} \int \widetilde{\Phi}^*_{n'v'}(R)\widetilde{\Phi}_{nv}(R)\,\mathrm{d}R$$

and the transition probability is proportional to squared integral of the overlap of the vibrational wave functions:

$$q_{n'v',nv} = \left| \int \widetilde{\Phi}^*_{n'v'}(R)\widetilde{\Phi}_{nv}(R)\,\mathrm{d}R \right|^2$$

The factors $q_{n'v',nv}$ in the transition probability are known as Franck–Condon factors, and the approximation in which the dependence of $d_{n'n}$ on the internuclear distance is ignored is known as the Condon approximation. Since $\Phi_{nv}(R)$ and $\Phi_{n'v'}(R)$ are normalized, the sum rule holds:

$$\sum_{v'} q_{v',v} = \sum_{v} q_{v',v} = 1$$

The Franck–Condon factors are calculated for different models of potentials $U_n(R)$. This is easiest to do for parabolic potentials when $\Phi_{nv}(R)$ are the wave functions of the linear oscillator, even if R_0 and ω_n differ for the electron states n and n'.

Integral (4.7.4), determining the dipole moment of the electron-vibrational transition $d_{n'v',nv}$, can be evaluated in the quasiclassical approximation which is valid at high values of the vibrational quantum number v.

In the quasiclassical approximation, the wave function $\widetilde{\Phi}_{nv}$ is a standing wave

$$\widetilde{\Phi}_{nv}(R) \sim \frac{1}{\sqrt{P_n}} \exp\left(\frac{\mathrm{i}}{\hbar} \int^{R} P_n(R')\,\mathrm{d}R' \right) + \text{ C.C.}$$

Here $P_n = P_n(R)$ is the classical momentum for a given R and a known energy E_{nv}:

$$\frac{P_n^2}{2M} + U_n(R) = E_{nv}$$

The overlap integral (4.7.4) reduces to a sum of integrals of the type

$$\int \frac{1}{\sqrt{P_n P_{n'}}} \exp\left(\frac{\mathrm{i}}{\hbar} \int^R (P_n - P_{n'})\,\mathrm{d}R'\right) d_{n'n}(R)\,\mathrm{d}R$$

The integral contains a function $\exp\left[\mathrm{i}\int(P_n - P_{n'})\hbar^{-1}\,\mathrm{d}R'\right]$ which oscillates rapidly when the quasiclassical approximation is valid. An appreciable contribution to the integral is made only by the points at which the oscillation frequency vanishes (the stationary-phase method).

This is an integral of the type

$$\int \varphi(x) \exp(\mathrm{i}\lambda f(x))\,\mathrm{d}x$$

with a large parameter λ. The stationary-phase method consists in expanding $f(x)$ in the neighborhood of a point x_s where $f'(x_s) = 0$.

We have

$$f(x) \simeq f(x_s) + 1/2 f''(x_s)(x - x_s)^2$$

$$\int \varphi(x) \exp(\mathrm{i}\lambda f(x))\,\mathrm{d}x \sim \varphi(x_s) \exp\left(\mathrm{i}\lambda f(x_s)\right) \sqrt{\frac{\pi}{\lambda |f''(x_s)|}}$$

Making use of these relations in our case, we find

$$\frac{d_{n'n}(R_s)}{\sqrt{P_n(R_s) P_{n'}(R_s)}} \exp\left(\frac{\mathrm{i}}{\hbar} \int^R (P_n - P_{n'})\mathrm{d}R'\right) \sqrt{\frac{\pi\hbar}{\left|\frac{\mathrm{d}}{\mathrm{d}R_s}(P_n - P_{n'})\right|}}$$

The point R_s is found from the condition

$$P_n(R_s) - P_{n'}(R_s) = 0$$

This is the point at which the classical momenta at the same coordinate are equal, so that the transition occurs without changes in the velocities of nuclei.

Differentiating the equation

$$\frac{1}{2M}(P_n^2 - P_{n'}^2) + (U_n - U_{n'}) = E_{nv} - E_{n'v'}$$

with respect to R and assuming $R = R_s$, we obtain

$$\frac{P_n(R_s)}{M} \frac{\mathrm{d}}{\mathrm{d}R_s}(P_n - P_{n'}) = -\frac{\mathrm{d}}{\mathrm{d}R_s}(U_n - U_{n'})$$

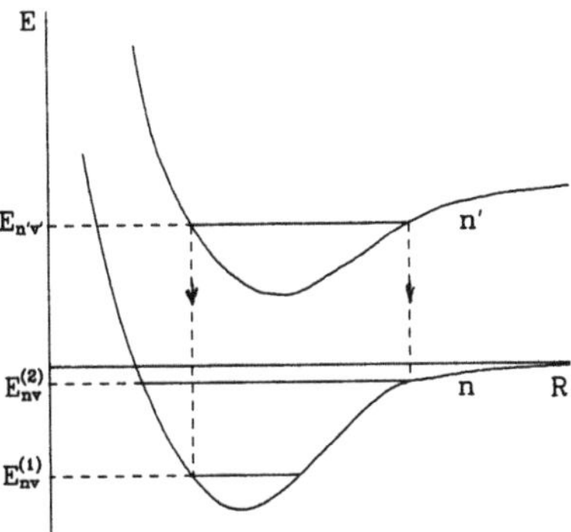

Figure 4.9. Franck–Condon principle.

As a result, the transition probability is proportional to

$$w \sim |d_{n'n}(R_s)|^2 / P_n(R_s) \left| \frac{\mathrm{d}}{\mathrm{d}R_s} (U_n - U_{n'}) \right|$$

This expression yields the Franck–Condon principle: the transition probability is maximal at the point R_s where the momenta $P_n(R_s)$ and $P_{n'}(R_s)$ are equal and small ($P_n(R_s) \simeq 0$). This is the turning point at which $U_n(R_s) = E_{nv}$ and $U_{n'}(R_s) = E_{n'v'}$. The transition occurs without changes in coordinates and momenta and mostly at the turning points, where the velocity is small and the residence time is large (Figure 4.9).

We now have that if $P \simeq 0$, then $w \propto P^{-1} \to \infty$; this result obtains because the quasiclassical approximation is not valid close to the points of turning (stopping) of nuclei. In fact, the probability w is high but finite. The transition is most probable at a point where nuclei stop and where they consequently spend a considerable time.

The set of lines produced by transitions between rotational components of two electron-vibration levels compose a band. The lines in the band are packed quite closely since rotational intervals are small. Their frequencies are given by the expression

$$w_{J'J} = \mathrm{const} + B_{n'v'}J'(J'+1) - B_{nv}J(J+1)$$

In view of the selection rules for the rotational quantum number $J' = J,\ J \pm 1$, this formula is shown by three branches (parabolas) whose points determine the values of frequencies for integral values of J. Figure 4.10 shows the picture for $B_{n'v'} > B_{nv}$. If $B_{n'v'} < B_{nv}$, the parabolas are concave toward lower frequencies, with the R branch ($J' = J + 1$) being the upper one.

The branch that folds back on itself (the P branch in Figure 4.10) causes lines to "crowd" toward a certain frequency, producing a "fringe."

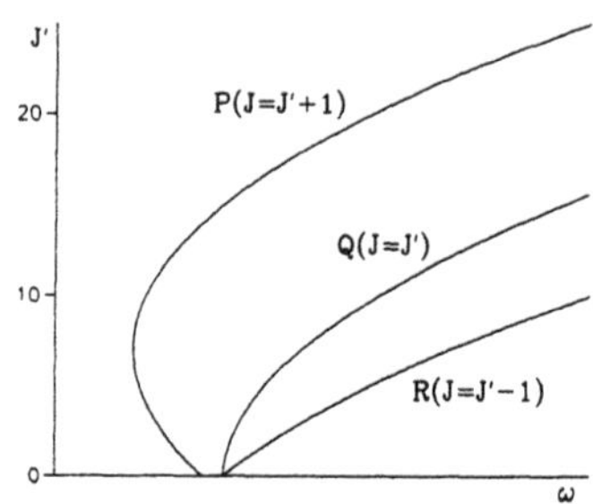

Figure 4.10. Bands of electron vibration spectrum. Transition frequencies correspond to integral values of J'.

4.8. The Placzek Approximation

With the adiabatic approximation, we can write the expression for polarizability tensor of a molecule and for the Raman scattering tensor (2.12.8) and (2.12.9) in terms of electron polarizabilities.

The physical meaning of this approximation (known as the Placzek approximation) is that for small $\mu = (m/M)^{1/4}$ the electron wave function is constantly adjusting to the instantaneous configuration of nuclei (following it adiabatically) and that the polarizability of a molecule is its electron polarizability for a specific nuclear configuration, averaged over the probabilities of various configurations. The Raman scattering tensor (more often referred to as the "scattering tensor") is, correspondingly, the matrix element of the electron polarizability for two vibrational–rotational states.

According to (2.12.8), the polarizability tensor of a molecule in a state

$$\psi_{nvJ} = \Phi_J(\alpha, \beta, \gamma)\Phi_v(Q)\varphi_n(\mathbf{r}', Q)$$

equals

$$\begin{aligned}\varkappa_{\alpha\beta}(\omega) = \sum_{n'v'J'} & (E_{n'v'J'} - E_{nvJ} - \hbar\omega)^{-1}(d_\alpha)_{nvJ,n'v'J'}(d_\beta)_{n'v'J',nvJ} \\ & + (E_{n'v'J'} - E_{nvJ} + \hbar\omega)^{-1}(d_\beta)_{nvJ,n'v'J'}(d_\alpha)_{n'v'J',nvJ} \qquad (4.8.1)\end{aligned}$$

The electron states of the molecule are separated by energy intervals of the order of $I_0 \gg kT$, so that only the lowest, ground electron state $n = 0$ is of interest. We will assume it to be nondegenerate and will calculate the polarizability and the scattering tensor for states ψ_{0vJ} which belong to these electron states.

Let us single out in $\varkappa_{\alpha\beta}(\omega)$ the contribution connected with the transition inside the electron ground state ($n' = 0$):

$$\tilde{\varkappa}_{\alpha\beta}(\omega) = \sum_{v'J'} \frac{(d_\alpha)_{0vJ,0v'J'}(d_\beta)_{0v'J',0vJ}}{E_{0v'J'} - E_{0vJ} - \hbar\omega} + \frac{(d_\beta)_{0vJ,0v'J'}(d_\alpha)_{0v'J',0vJ}}{E_{0v'J'} - E_{0vJ} + \hbar\omega} \qquad (4.8.2)$$

By analogy to Section 4.7, we can now express the components of the operator d_α in terms of the operators $d_{\alpha'}$ of the components of $\mathbf{d}$ in the system of coordinates fixed to the molecule,

$$d_\alpha = \sum_{\alpha'} D^1_{\alpha'\alpha} d_{\alpha'}$$

where $D^1_{\alpha'\alpha}$ are functions of Euler's angles, which describe the orientation of the molecule.

The energy differences $E_{0v'J'} - E_{0vJ}$ are the differences between the vibrational–rotational energies of the ground electron state $E_{v'J'} - E_{vJ}$.

The matrix element $(d_\alpha)_{0vJ,0v'J'}$ can be rewritten, by analogy to what we did in Section 4.7, as a product of matrix elements,

$$\sum_{\alpha'} (D^1_{\alpha'\alpha})_{J,J'} (d^{\alpha'}_{00})_{v,v'}$$

Here $D^1_{\alpha'\alpha}$ are functions of Euler's angles and $d^{\alpha'}_{00}$ is the molecular dipole moment, which is a function of vibrational coordinates, averaged over the electron ground state.

The matrix elements $(D^1_{\alpha'\alpha})_{J,J'}$ can again be calculated in the general form for linear molecules or symmetric-top-type molecules, for which the rotational wave functions are known.

The dipole moment $d^{\alpha'}_{00}$ of a molecule, averaged over the electron state, is nonzero only for molecules with ionic bonding. For example, it is zero for homonuclear diatomic molecules. In polar molecules, the ion contribution to polarizability $\tilde{\varkappa}_{\alpha\beta}$ is of the same order of magnitude as the electron contribution at the lowest frequencies (the static polarizability). Therefore, if $\hbar\omega \gg E_{v'J'} - E_{vJ}$, the ion contribution $\tilde{\varkappa}_{\alpha\beta}$ is negligibly small. This last inequality is always assumed satisfied in the Raman scattering of light (ω is the frequency of light) but in the infrared range, where $\hbar\omega \sim E_{v'J'} - E_{vJ}$ (of the order of intervals of the vibrational–rotational energy), the electron and ion parts of polarizability are of comparable magnitudes and must be treated jointly.

The adiabatic part of polarizability is obtained by averaging the adiabatic polarizability of a molecule over the vibrational–rotational function for fixed positions of nuclei. Obviously, the electron polarizability at fixed values of the nuclei's coordinates is

$$\varkappa_{\alpha'\beta'}(\omega, Q) = \sum_{n\neq 0} \frac{d^{\alpha'}_{0n}(Q)\, d^{\beta'}_{n0}(Q)}{U_n(Q) - U_0(Q) - \hbar\omega} + \frac{d^{\beta'}_{0n}(Q)\, d^{\alpha'}_{n0}(Q)}{U_n(Q) - U_0(Q) + \hbar\omega} \quad (4.8.3)$$

We assume here that ω is not close to any of the frequencies of the electron transition $(U_n - U_0)/\hbar$.

If this is correct, then the electron part is given by

$$\varkappa_{\alpha\beta}(\omega) = \sum_{\alpha'\beta'} (\varkappa_{\alpha'\beta'}(\omega, Q) D^1_{\alpha'\alpha} D^1_{\beta'\beta})_{vJ,v'J'}$$
$$= \sum_{\alpha'\beta'} (D^1_{\alpha'\alpha} D^1_{\beta'\beta})_{JJ} (\varkappa_{\alpha'\beta'}(\omega, Q))_{v,v'}$$

For nonoriented molecules, this expression must be averaged over orientations of molecules, that is, we need to find the sum over the values of the quantum number M and divide it by $(2J + 1)$. As a result, $\varkappa_{\alpha\beta}$ is obviously an isotropic tensor,

$$\varkappa_{\alpha\beta}(\omega) = \varkappa(\omega)\delta_{\alpha\beta}$$

where $\varkappa(\omega)$ coincides with the electron polarizability, averaged over vibrations and then over orientations:

$$\varkappa(\omega) = \frac{1}{3} \sum_{\alpha'} (\varkappa_{\alpha'\alpha'}(\omega, Q))_{v,v'}$$

If molecules are oriented (say, by an external field), the tensor $\varkappa_{\alpha\beta}$ is generally anisotropic.

All this equally holds for the Raman scattering of light but in order to calculate $\varkappa_{\alpha\beta}$ we need the nondiagonal matrix element of the electron polarizability tensor,

$$\varkappa^{(21)}_{\alpha\beta}(\omega) = \sum_{\alpha'\beta'} (\varkappa_{\alpha'\beta'}(\omega, Q) D^1_{\alpha'\alpha} D^1_{\beta'\beta})_{v_2 J_2, v_1 J_1}$$
$$= \sum_{\alpha'\beta'} (D^1_{\alpha'\alpha} D^1_{\beta'\beta})_{J_2 J_1} (\varkappa_{\alpha'\beta'}(\omega, Q))_{v_2, v_1}$$

The matrix element $(D^1_{\alpha'\alpha} D^1_{\beta'\beta})_{J_2 J_1}$ over the rotational wave functions can be calculated in complete form for linear and symmetric-top-type molecules, for which we know the rotational wave functions, and the calculation of matrix elements reduces to integration of products of Wigner functions.

Even if the rotational wave functions are unknown (molecules of the type of asymmetric top), the rotational matrix element gives selection rules in rotational quantum numbers:

$$|J_2 - J_1| \leq 2 \leq J_1 + J_2$$

Therefore, in Raman scattering we have $J' = J - 2$, $J - 1$, J, $J + 1$, $J + 2$, which corresponds to the O, P, Q, R, S branches of the Raman scattering of light.

The electron polarizability tensor $\varkappa_{\alpha'\beta'}(\omega, Q)$ (4.8.3) is symmetric ($\varkappa_{\alpha'\beta'} = \varkappa_{\alpha\beta}$). This can be shown in the same manner when we have proved the symmetry of linear susceptibility in the dipole approximation in Chapter 2. The matrix element $(\varkappa_{\alpha'\beta'}(\omega, Q))_{v_2 v_1}$ is therefore also symmetric under a permutation of α' and β'. In reduced axes, this symmetric tensor can be rewritten in a diagonal form:

$$\|\varkappa\| = \begin{pmatrix} \varkappa_1 & 0 & 0 \\ 0 & \varkappa_2 & 0 \\ 0 & 0 & \varkappa_3 \end{pmatrix}$$

For linear and symmetric-top-type molecules, two principal values of the tensor $\varkappa$ coincide, so that

$$\|\varkappa\| = \begin{pmatrix} \varkappa_\perp & 0 & 0 \\ 0 & \varkappa_\perp & 0 \\ 0 & 0 & \varkappa_\parallel \end{pmatrix}$$

where $\varkappa_\parallel$ is the polarizability along the symmetry axis of the molecule (symmetry higher than twofold) and $\varkappa_\perp$ is the polarizability perpendicular to the axis.

The purely rotational Raman scattering is determined by the degree of anisotropy of the tensor $\varkappa$, namely, by the difference $\varkappa_\parallel - \varkappa_\perp$. If $\varkappa_\parallel = \varkappa_\perp$, there is no rotational Raman scattering of light. The intensity of rotational Raman scattering of light and the amplification coefficient for the stimulated rotational Raman scattering are proportional to $(\varkappa_\parallel - \varkappa_\perp)^2$ and the dependence on the rotational quantum numbers can be obtained for symmetric-top-type molecules, as we have already mentioned in connection with the general case.

The selection rules of the Raman scattering of light in a vibrational–rotational transition are obtained in the same manner as for absorption in allowed transitions. Namely, we expand $\varkappa_{\alpha'\beta'}(\omega, Q)$ in powers of coordinates:

$$\varkappa_{\alpha'\beta'}(\omega, Q) = \varkappa_{\alpha'\beta'}(\omega, 0) + \frac{\partial \varkappa_{\alpha'\beta'}}{\partial Q} Q + \dots$$

The first term gives transitions in which only the rotational state changes (the rotational Raman scattering). The second term has nonzero matrix elements of transitions between the vibrational states v_1 and v_2 in the harmonic approximation if

$$v_2 = v_1 \pm 1$$

The generation of overtones, that is, transitions for which $v_2 = v_1 \pm 2$ etc. and also of combination frequencies, is connected with anharmonism (with the difference between vibrational wave functions and harmonic-oscillator wave functions) and with quadratic and other nonlinear terms in the expansion of $\varkappa_{\alpha'\beta'}$ in powers of Q.

Ignoring rotations, as we did in Section 4.4, we write the Schrödinger equations in the adiabatic approximation for a molecule interacting with the electromagnetic field $\boldsymbol{\mathcal{E}}$. They are derived by analogy to equations (4.4.7) (we only need to take the interaction with the field, $\mathbf{d}\boldsymbol{\mathcal{E}}$, into account in the Hamiltonian) and have the form

$$\begin{aligned} i\hbar\frac{\partial\Phi_n}{\partial t} + \frac{\hbar^2}{2M}\nabla^2\Phi_n - (U_n(R) - \boldsymbol{\mathcal{E}}\mathbf{d}_{nn}(R))\Phi_n \\ = -\sum_{n'\neq n}\boldsymbol{\mathcal{E}}\mathbf{d}_{nn'}(R)\Phi_{n'}(R,t) \end{aligned} \tag{4.8.4}$$

We have restricted the analysis to the case of a diatomic molecule, with $U_n(R)$ being the adiabatic energy for the nth electron state and $\mathbf{d}_{nn'}(R)$ a matrix element of the dipole moment operator of the molecule for two electron states, for a given R:

$$\mathbf{d}_{nn'}(R) = \int \varphi_n^* \mathbf{d}\varphi_{n'}\,\mathrm{d}^3r$$

Let a molecule be in a state Φ_0 ($n = 0$) at the time $t \to -\infty$ when there was no field $\boldsymbol{\mathcal{E}}$, with all remaining $\Phi_n = 0$ ($n \neq 0$). Perturbation theory for $n \neq 0$ then gives

$$i\hbar\frac{\partial\Phi_n}{\partial t} + \frac{\hbar^2}{2M}\nabla^2\Phi_n - (U_n(R) - \boldsymbol{\mathcal{E}}\mathbf{d}_{nn})\Phi_n = -\boldsymbol{\mathcal{E}}\mathbf{d}_{n0}\Phi_0$$

Retaining only terms linear in $\boldsymbol{\mathcal{E}}$ and taking into account that $\Phi_n \sim \boldsymbol{\mathcal{E}}$ ($n \neq 0$), we formally write, for a field $\boldsymbol{\mathcal{E}}$ of the type

$$\boldsymbol{\mathcal{E}} = \sum_{\omega}\mathbf{E}_\omega e^{-i\omega t}$$

the equality

$$\Phi_n = -\sum_{\omega'\beta}\left[i\hbar\frac{\partial}{\partial t} + \frac{\hbar^2}{2M}\nabla^2 - U_n\right]^{-1} E_{\omega'\beta}e^{-i\omega' t}(d_\beta)_{n0}\Phi_0$$

Obviously, we can write

$$\begin{aligned} &\left(i\hbar\frac{\partial}{\partial t} + \frac{\hbar^2}{2M}\nabla^2 - U_n\right)\exp(-i\omega' t)\frac{(d_\beta)_{n0}}{U_0 - U_n + \hbar\omega'}\Phi_0 \\ &= \exp(-i\omega' t)\left(i\hbar\frac{\partial}{\partial t} + \frac{\hbar^2}{2M}\nabla^2 - U_n + \hbar\omega'\right) \\ &\times \frac{(d_\beta)_{n0}}{U_0 - U_n + \hbar\omega'}\Phi_0 \end{aligned} \tag{4.8.5}$$

The scale of change of U, U_0 and $(d_\beta)_{n0}$ being of the order of a_0 and that of Φ_0 of the order of $\mu a_0 = (m/M)^{1/4} a_0$, we can write in the adiabatic approximation that (4.8.5) equals

$$\exp(-\mathrm{i}\omega' t)\frac{(d_\beta)_{n0}}{U_0 - U_n + \hbar\omega'}\Big[\mathrm{i}\hbar\frac{\partial}{\partial t} + \frac{\hbar^2}{2M}\nabla^2 - U_0 + (U_0 - U_n) + \hbar\omega'\Big]\Phi_0 = \exp(-\mathrm{i}\omega' t)(d_\beta)_{n0}\Phi_0$$

We have taken into account here that Φ_0 satisfies the equation

$$\Big(\mathrm{i}\hbar\frac{\partial}{\partial t} + \frac{\hbar^2}{2M}\nabla^2 - U_0\Big)\Phi_0 = 0$$

up to terms proportional to $\boldsymbol{\mathcal{E}}$.

We therefore obtain

$$\Big[\mathrm{i}\hbar\frac{\partial}{\partial t} + \frac{\hbar^2}{2M}\nabla^2 - U_n\Big]^{-1}\exp(-\mathrm{i}\omega' t)(d_\beta)_{n0}\Phi_0 = \exp(-\mathrm{i}\omega' t)\frac{(d_\beta)_{n0}}{U_0 - U_n + \hbar\omega'}\Phi_0$$

and hence

$$\Phi_n = -\sum_{\omega'\beta} E_{\omega'\beta}\exp(-\mathrm{i}\omega' t)\frac{(d_\beta)_{n0}}{U_0 - U_n + \hbar\omega'}\Phi_0$$

The function Φ_n is expressed in terms of Φ_0. Substituting this expression into the equation for Φ_0, we arrive at a closed equation for this function:

$$\mathrm{i}\hbar\frac{\partial\Phi_0}{\partial t} + \frac{\hbar^2}{2M}\nabla^2\Phi_0 - [U_0 - \boldsymbol{\mathcal{E}}\mathbf{d}_{00}(R)]\Phi_0 = \sum_{\omega'\beta\omega\alpha} E_{\omega\alpha}E_{\omega'\beta}\exp(-\mathrm{i}(\omega+\omega')t)\sum_{n\neq 0}\frac{(d_\alpha(R))_{0n}(d_\beta(R))_{n0}}{U_0 - U_n + \hbar\omega'}$$

Let us rewrite this equation in a somewhat different form. First we symmetrize the righthand side with respect to $\omega\alpha \leftrightarrow \omega'\beta$,

$$-\frac{1}{2}\sum E_{\omega\alpha}E_{\omega'\beta}\exp(-\mathrm{i}(\omega+\omega')t) \times \sum_{n\neq 0}\left\{\frac{(d_\alpha)_{0n}(d_\beta)_{n0}}{U_n - U_0 - \hbar\omega'} + \frac{(d_\beta)_{0n}(d_\alpha)_{n0}}{U_n - U_0 - \hbar\omega}\right\}$$

and then perform the substitution $\omega \to -\omega$ in the sum over ω, taking into account that $E_{-\omega\alpha} = E^*_{\omega\alpha}$. The equation for Φ_0 now takes the form

$$\begin{aligned} &i\hbar \frac{\partial \Phi_0}{\partial t} + \frac{\hbar^2}{2M} \nabla^2 \Phi_0 - [U_0(R) - \boldsymbol{\mathcal{E}} \mathbf{d}_{00}(R)]\Phi_0 \\ &= -\frac{1}{2} \sum E^*_{\omega\alpha} E_{\omega'\beta} \exp(i(\omega - \omega')t) \\ &\times \sum_{n\neq 0} \left\{ \frac{(d_\alpha)_{0n}(d_\beta)_{n0}}{U_n - U_0 - \hbar\omega'} + \frac{(d_\beta)_{0n}(d_\alpha)_{n0}}{U_n - U_0 + \hbar\omega} \right\} \end{aligned} \tag{4.8.6}$$

The sum over $n \neq 0$ in (4.8.6) is the scattering tensor which depends on R as on a parameter, since $(d_\alpha)_{n0}$, $(d_\beta)_{0n}$, U_n, and U_0 are functions of the distance R between the nuclei of a diatomic molecule. If we assume that $|\omega - \omega'| \ll (U_n - U_0)/\hbar$, then this tensor becomes the R-dependent electron polarizability tensor of the molecule, $\varkappa_{\alpha\beta}$ for the electron ground state.

Finally,

$$\begin{aligned} &i\hbar \frac{\partial \Phi_0}{\partial t} + \frac{\hbar^2}{2M} \nabla^2 \Phi_0 - [U_0(R) - \boldsymbol{\mathcal{E}} \mathbf{d}_{00}(R)]\Phi_0 \\ &= -\frac{1}{2} \sum_{\omega\alpha\omega'\beta} E^*_{\omega\alpha} E_{\omega'\beta} \exp(i(\omega - \omega')t) \varkappa_{\alpha\beta}(R)\Phi_0 \end{aligned} \tag{4.8.7}$$

The result is a Schrödinger-type equation for the function $\Phi_0(R, t)$, and the coefficient with Φ_0 in the righthand side of equation (4.8.7) stands for the energy of interaction between the field and the induced dipole moment.

Solving equation (4.8.7), we can consider induced Raman scattering of light for the transition between vibration states which belong to the electron ground state. This is equivalent to the results presented above.

4.9. Kerr Effect

As we have indicated in the preceding section, if the probability for a molecule to be in all $(2J + 1)$ states with different values of the quantum number M is the same, the resulting susceptibility of the medium is isotropic after averaging over M even if the molecule possesses anisotropic polarizability tensor.

If the electric field is sufficiently strong, the state with a specific value of J, degenerate in M, undergoes splitting and the resulting states have, in general, unequal energies. This is the Stark effect; in high-frequency fields, it is known as the dynamic Stark effect.

As a result of relaxation, the populations of these energy-split states become different (in the sense of the Boltzmann distribution). The mean polarizability of a molecule and, as a consequence, the linear susceptibility of the medium are thus anisotropic. This is the Kerr effect, or the dynamic Kerr effect in high-frequency fields.

The concept of a state with a definite J is meaningful for molecules only in a sufficiently rarefied gas. In a liquid, strong intermolecular interaction makes the quantum number J meaningless, so that different orientations of a molecule must be considered instead of states with different M.

In view of the conclusions formulated at the end of the preceding section, the energy of a molecule in a liquid is a function of the field and the molecular polarizability. For an anisotropic polarizability tensor, the energy is a function of the orientation of a molecule relative to the field. In accordance with the main principles of statistical physics (the Gibbs distribution), this situation results in a nonuniform orientational distribution of molecules and thus a field-induced anisotropy of the medium.

Let us consider the Kerr effect in a liquid composed of symmetric-top-type molecules with a higher-than-twofold symmetry axis. Such molecules have two principal values of polarizability: along the axis of the molecule, $\varkappa_{\parallel}$, and perpendicular to the axis, $\varkappa_{\perp}$.

Let the electric field $\mathcal{E}$ be linearly polarized along the axis z. The dipole moment $\mathbf{d}$ of the molecule then has a component $\mathbf{d}_{\parallel}$ along the axis z' of the molecule which coincides with the symmetry axis, and a component $\mathbf{d}_{\perp}$ which is perpendicular to this axis, and such that

$$d_{\parallel} = \varkappa_{\parallel}\mathcal{E}\cos\Theta \qquad d_{\perp} = \varkappa_{\perp}\mathcal{E}\sin\Theta$$

where Θ is the angle between the axis of the molecule and the electric field.

The energy of a molecule in an electric field is

$$\begin{aligned}\varepsilon &= -\frac{1}{2}\varkappa_{\alpha\beta}\mathcal{E}_{\alpha}\mathcal{E}_{\beta} = -\frac{1}{2}d_{\alpha}\mathcal{E}_{\alpha} = -\frac{1}{2}(d_{\parallel}\mathcal{E}_{\parallel} + d_{\perp}\mathcal{E}_{\perp})\\ &= -\frac{1}{2}(\varkappa_{\parallel}\cos^2\Theta + \varkappa_{\perp}\sin^2\Theta)\mathcal{E}^2 = -\frac{1}{2}[(\varkappa_{\parallel} - \varkappa_{\perp})\cos^2\Theta + \varkappa_{\perp}]\mathcal{E}^2\end{aligned}$$

The orientational distribution of molecules thus has the form

$$\begin{aligned}w(\Theta) &= \exp\left(-\frac{\varepsilon(\Theta)}{kT}\right)\left(\int \exp\left(-\frac{\varepsilon(\Theta)}{kT}\right)\sin\Theta\, d\Theta\right)^{-1}\\ &= \exp\left(\frac{(\varkappa_{\parallel} - \varkappa_{\perp})\cos^2\Theta\bar{\mathcal{E}}^2}{2kT}\right)\\ &\times\left(\int \exp\left(\frac{(\varkappa_{\parallel} - \varkappa_{\perp})\cos^2\Theta\bar{\mathcal{E}}^2}{2kT}\right)\sin\Theta\, d\Theta\right)^{-1}\end{aligned}$$

The polarization of the medium with a number density of n molecules in a unit volume, referred to one unit volume, is

$$\mathcal{P} = \mathcal{P}_z = n \int (d_{\parallel} \cos\Theta + d_{\perp} \sin\Theta) w(\Theta) \sin\Theta \, d\Theta$$
$$= n\mathcal{E} \int (\varkappa_{\parallel} \cos^2\Theta + \varkappa_{\perp} \sin^2\Theta) w(\Theta) \sin\Theta \, d\Theta$$

We introduce now a notation for the dimensionless energy of a molecule in the field:

$$\beta = (\varkappa_{\parallel} - \varkappa_{\perp})\overline{\mathcal{E}}^2/2kT$$

Then (with $\xi = \cos\Theta$)

$$\mathcal{P} = n\mathcal{E}\left[\varkappa_{\perp} + \int_{-1}^{1} (\varkappa_{\parallel} - \varkappa_{\perp}) \exp(\beta\xi^2)\xi^2 \, d\xi \left(\int_{-1}^{1} \exp(\beta\xi^2) \, d\xi\right)^{-1}\right]$$

The integral which contains a factor ξ^2 in addition to $\exp(\beta\xi^2)$ can be integrated by parts; this gives

$$\mathcal{P} = n\mathcal{E}\left[\varkappa_{\perp} + \frac{1}{2\beta}(\varkappa_{\parallel} - \varkappa_{\perp})\left(2e^{\beta}\left(\int_{-1}^{1} \exp(\beta\xi^2) \, d\xi\right)^{-1} - 1\right)\right]$$

Making use of expansions for small β,

$$\int_{-1}^{1} \exp(\beta\xi^2) \, d\xi \simeq 2\left(1 + \frac{\beta}{3} + \frac{\beta^2}{10} + \ldots\right)$$

we find

$$\mathcal{P} = n\mathcal{E}\left[\frac{1}{3}(\varkappa_{\parallel} + 2\varkappa_{\perp}) + \frac{4}{45}(\varkappa_{\parallel} - \varkappa_{\perp})\beta + \ldots\right] \tag{4.9.1}$$

In a weak field, when $\beta \to 0$, the polarization is dictated by the polarizability $(\varkappa_{\parallel} + 2\varkappa_{\perp})/3$ averaged over all orientations. The field contributes to the polarization, so that an additional term proportional to β (i.e., to the field intensity $\overline{\mathcal{E}^2}$) appears in the medium's susceptibility.

The field which induces molecules to align and determines the parameter β, may differ from the probe field $\mathcal{E}$ for which the susceptibility is determined. These fields may have different frequencies, intensities, and polarization. As a result, one field may cause a change in the susceptibility of molecular medium for another field. This is the classical Kerr effect, and formula (4.9.1) gives an expression for the Kerr constant in terms of the anisotropy of the molecular polarizability. If the field is very strong ($\beta \gg 1$), all molecules align in such a way that the axis of maximum $\varkappa$ be oriented along the field (this $\varkappa$ is the greatest of $\varkappa_{\parallel}$ and $\varkappa_{\perp}$). The susceptibility of the medium is then determined by this maximum value of polarizability, not by the average value $(\varkappa_{\parallel} + 2\varkappa_{\perp})/3$ as in the case of weak fields.

5

Interaction of Electromagnetic Radiation with Crystals

5.1. Model of a Crystal

A crystal is essentially a very large molecule; theoretically, one can even consider infinitely large crystals, thereby ignoring surface effects.

Solids are classified by the type of chemical bonding. Types of chemical bonds have already been presented in Chapter 4. We will recall that five types of bonding are distinguished: ionic, covalent, metallic, hydrogen and van der Waals bonding. None of the five are found in nature in its pure form; a substance is subsumed under a particular class according to the prevalent bonding type.

Ionic crystals are dielectrics with high bonding energy and high melting temperature. As a rule, they are transparent in the visual band, since their lattice vibrations correspond to the infrared band and the interband transitions lie in the ultraviolet band.

The number of purely covalent crystals is not large (covalent molecules are much more numerous). However, silicon and germanium, which are so important for electronics technologies, belong to this group, as do diamond and β-Sn (gray tin). These are mostly semiconductors (diamond is a dielectric). Diamond is transparent down to the visible range and semiconductors are transparent down to the near-infrared range. Bonding energies of covalent crystals are high.

The metallic bonding is typical for metals. Metals are opaque, possess high electric and thermal conductivity. The bonding energy is slightly lower than in ionic and covalent crystals but is much higher than in molecular crystals.

The hydrogen bond is a weak bond but it determines the structure of such molecular crystals as ice.

The van der Waals bonding is the most universal and the weakest type of bonding. It does not involve either overlapping of wave functions or charge transfer. Bonding of this type is responsible for the formation of molecular crystals and liquids. These crystals are dielectrics (their electrons are localized in molecules) with low melting point.

As with molecules, the solid state theory is based on separating the electron and nuclear motions. The possibility of this separation stems from the smallness of the ratio m/M and is justified in the adiabatic approximation that we were discussing earlier.

In the adiabatic approximation, the motion of nuclei can be considered in terms of the effective energy $U(\mathbf{R})$, which is a function of coordinates and includes the electron energy (including the electron kinetic energy of electrons). We will drop the index which enumerates the electron state.

The minimum of the function $U(\mathbf{R})$ is reached for an ordered periodic arrangement of nuclei, which determines the periodic structure of crystals. We have already mentioned in Chapter 1 that the structure of a crystal is prescribed by its Bravais lattice

$$\mathbf{R_n} = \mathbf{a}_1 n_1 + \mathbf{a}_2 n_2 + \mathbf{a}_3 n_3$$

where n_1, n_2 and n_3 are integral numbers, and by the basis, that is the structure of a unit cell of the lattice. The basis is prescribed by the radii $\boldsymbol{\rho}_s$ of atoms in the unit cell, where s enumerates atoms of one unit cell so that the radius vector of the sth atom in the nth unit cell is

$$\mathbf{R}_{\mathbf{n}s} = \mathbf{R_n} + \boldsymbol{\rho}_s$$

For instance, $s = 1$ signifies the Na atom for a NaCl crystal, and $s = 2$ stands for the Cl atom. The $\mathbf{R}_{\mathbf{n}s}$ vectors describe the equilibrium configuration of nuclei. In fact, atoms vibrate around these positions, so that their radius vectors are $\mathbf{R}_{\mathbf{n}s} + \mathbf{u}_{\mathbf{n}s}$, where $|\mathbf{u}_{\mathbf{n}s}|$ are small in comparison with a_0 or with the interatomic distances $|\mathbf{a}_1|$, $|\mathbf{a}_2|$, $|\mathbf{a}_3|$.

It has already been demonstrated in Chapter 4, in the discussion of the adiabatic approximation, that the energy of zero (quantum) vibrations of nuclei is small in comparison with the characteristic electron energy I_0, and the vibration amplitudes are small, by a factor of $(m/M)^{1/4}$, in comparison with a_0. The thermal vibrational energy $\mathrm{k}T$ is not greater than $\mathrm{k}T_{\mathrm{melt}}$, where T_{melt} is the melting temperature of the order of 10^3 K or less. Hence, $\mathrm{k}T_{\mathrm{melt}} \sim 10^{-16} \times 10^3$ erg $\sim 10^{-13}$ erg, which is much less than $I_0 \sim 10^{-11}$ erg.

The amplitude of thermal vibrations $\mathbf{u}$ can be evaluated if we equate the potential energy of vibrations (by virtue of the virial theorem, the kinetic energy of vibrations in harmonic oscillations equals the potential energy) to the thermal energy $\mathrm{k}T_{\mathrm{melt}}$. This gives us the estimate

$$\mathrm{k}T_{\mathrm{melt}} \sim U''\mathbf{u}^2 \sim \frac{I_0}{a_0^2}\mathbf{u}^2$$

$$\frac{u}{a_0} \sim \sqrt{\frac{\mathrm{k}T_{\mathrm{melt}}}{I_0}} \sim 0.1$$

This result is in agreement with experimental data: crystals melt when atomic displacements reach about 0.1 of the interatomic distance.

Since the vibration amplitude is small in comparison with interatomic distances, it is possible to expand the energy $U(\mathbf{R}_{\mathbf{n}s} + \mathbf{u}_{\mathbf{n}s}) \equiv U(\mathbf{R} + \mathbf{u})$ in powers of $\mathbf{u}$. We recall that $U(\mathbf{R} + \mathbf{u})$ varies on the scale of a_0. This gives

$$U(\mathbf{R} + \mathbf{u}) = U(\mathbf{R}) + \frac{1}{2} \sum_{\substack{\mathbf{n}s\alpha \\ \mathbf{n}'s'\alpha'}} A_{\mathbf{n}\mathbf{n}'}^{\substack{ss' \\ \alpha\alpha'}} u_{\mathbf{n}s\alpha} u_{\mathbf{n}'s'\alpha'} + \dots \tag{5.1.1}$$

where $U(\mathbf{R})$ is the minimal (equilibrium) value of $U(\{\mathbf{R}_{\mathbf{n}s}\})$; there are no terms linear in $u_{\mathbf{n}s}$ (the expansion is made in the neighborhood of the minimum of energy) and α is the index of displacement polarization ($\alpha = x, y, z$). For the time being, we omit anharmonic terms with the third and higher powers of $u_{\mathbf{n}s\alpha}$.

Very important among symmetry transformations, typical for crystals, are translations $T_{\mathbf{n}}$, namely, operations in which the coordinates of all particles in a crystal are incremented by $\mathbf{R}_{\mathbf{n}} = \mathbf{a}_1 n_1 + \mathbf{a}_2 n_2 + \mathbf{a}_3 n_3$. The translations $T_{\mathbf{n}}$ form a group in which all operations commute:

$$T_{\mathbf{n}_1} T_{\mathbf{n}_2} = T_{\mathbf{n}_2} T_{\mathbf{n}_1} = T_{\mathbf{n}_1 + \mathbf{n}_2} \tag{5.1.2}$$

The Abelian group of translations $T_{\mathbf{n}}$ has only one-dimensional irreducible representations. For any vector or wave function which are transformed under translations according to some irreducible representation of the translation group, the translation $T_{\mathbf{n}}$ results in multiplying the vector or function by a number $t_{\mathbf{n}}$. Owing to the group property (5.1.2), $t_{\mathbf{n}_1} t_{\mathbf{n}_2} = t_{\mathbf{n}_1 + \mathbf{n}_2}$, or $\ln t_{\mathbf{n}_1} + \ln t_{\mathbf{n}_2} = \ln t_{\mathbf{n}_1 + \mathbf{n}_2}$. This equation has only linear solutions $\ln t_{\mathbf{n}} = c_1 n_1 + c_2 n_2 + c_3 n_3 = \mathrm{i}\mathbf{k}\mathbf{R}_{\mathbf{n}}$, where the vector $\mathbf{k}$ is given by the equations $\mathrm{i}(\mathbf{k}\mathbf{a}_1) = c_1$, $\mathrm{i}(\mathbf{k}\mathbf{a}_2) = c_2$, $\mathrm{i}(\mathbf{k}\mathbf{a}_3) = c_3$. Here $\mathbf{k}$ is a real vector, otherwise $|t_{\mathbf{n}}| = |\exp(\mathrm{i}\mathbf{k}\mathbf{R}_{\mathbf{n}})| \neq 1$ and the norm of the vector, regardless of its interpretation, would be changed under translation, which is not allowed.

These very general arguments explain why all possible solutions of the equations for crystals appear as traveling waves $\mathrm{e}^{\mathrm{i}\mathbf{k}\mathbf{R}_{\mathbf{n}}}$. We shall often come across such cases later in the book.

In addition to translations, the Bravais lattice may also have, as symmetry transformations, various point transformations that are typical for molecules and were discussed in the previous chapter. Inversion is always among these transformations but in the general case of nonrectangular lattices, the group of inversion (it comprises inversion and identity transformation) gives the complete point symmetry. Not all point symmetry groups are compatible with the Bravais lattice, that is, with the translational symmetry. It can thus be shown that only two-, three-, four- and sixfold symmetry axes are allowed. In general, there are only seven point groups which are compatible with the symmetry of the Bravais lattice. They define the familiar seven crystallographic systems: cubic, tetragonal, rhombohedral, monoclinic, triclinic, trigonal (rhombohedral), and hexagonal.

If translations are also taken into account, we have 14 space groups which include point transformations and translations, characterizing all possible symmetries of Bravais lattices (14 Bravais lattices).

If, however, we take the entire crystal structure, that is, the Bravais lattice and the basis, we arrive at 32 crystallographic point groups and 230 space groups.

As in the case of point groups, products of the main types of transformations can also be symmetry transformations, even if each of these types may not be a symmetry transformation. Therefore, in addition to translations by $\mathbf{R_n}$, rotations, reflections and their combinations, reflection glide planes (reflection combined with translation by a certain fraction of the basic lattice period) and screw axes (translation by half the lattice period combined with rotation) are possible.

A formal expression for the invariance of the energy $U(\mathbf{R}+\mathbf{u})$ under translations is the requirement that the matrix $A_{\mathbf{nn'}}$ commute with the translations $T_{\mathbf{n}}$ which are represented by the matrices

$$(T_{\mathbf{m}})_{\mathbf{nn'}} = \delta_{\mathbf{n+m,n'}}$$

The commutativity condition $A = T_{\mathbf{m}}^{-1} A T_{\mathbf{m}}$ gives $A_{\mathbf{nn'}} = A_{\mathbf{n-m,n'-m}}$, and this signifies that the matrix A is in fact a function of the difference $\mathbf{n}-\mathbf{n'}$, and we shall take this into account in later sections. Physically, this is obvious: the interaction between two lattice atoms depends only on their relative locations.

5.2. Lattice Vibrations

Using expression (5.1.1) for the effective potential energy of lattice atoms, we can write the Hamiltonian and the Schrödinger equation for the motion of atoms in the lattice. One typically starts with the corresponding classical problem — not the Hamiltonian is considered but Hamilton's function with potential energy (5.1.1). After the classical equations have been solved and the normal modes of vibration have been found, the transition to quantum picture is immediate since the quantization of an ensemble of independent oscillators is a familiar model problem in quantum mechanics. In the general case, U contains not only terms quadratic in displacements but higher-order terms as well. However, even if anharmonic effects are considered later, it is advisable to begin with diagonalizing that part of energy which is quadratic in displacements, that is, to transform to normal coordinates which are determined by the harmonic component of the potential U. It is expedient also to express the anharmonic terms of (5.1.1) via normal coordinates. It becomes possible then to take into account the interaction between normal modes via anharmonic terms as a perturbation, and operate with an ensemble of independent, noninteracting linear oscillators in the nonperturbed problem.

In the harmonic approximation, Hamilton's function for the lattice is

$$\mathscr{H} = \sum_{\mathbf{n}s\alpha} \frac{P^2_{\mathbf{n}s\alpha}}{2M_s} + \frac{1}{2} \sum_{\substack{\mathbf{n}s\alpha \\ \mathbf{n}'s'\alpha'}} A^{ss'}_{\substack{\alpha\alpha' \\ \mathbf{n}-\mathbf{n}'}} u_{\mathbf{n}s\alpha} u_{\mathbf{n}'s'\alpha'} \tag{5.2.1}$$

where $P_{\mathbf{n}s\alpha}$ are the components of atom's momentum, with indices $\mathbf{n}$, s. Mathematically, the transition to normal coordinates is equivalent to simultaneous diagonalization of two quadratic forms: the kinetic and the potential energies.

Using the linear canonical transformation

$$u_{\mathbf{n}s\alpha} = \tilde{u}_{\mathbf{n}s\alpha} \sqrt{\frac{M}{M_s}} \qquad P_{\mathbf{n}s\alpha} = \widetilde{P}_{\mathbf{n}s\alpha} \sqrt{\frac{M_s}{M}} \tag{5.2.2}$$

where $M = \sum_s M_s$ is the sum of the masses of atoms (nuclei) in the unit cell, we can transform (5.2.1) to a form in which the kinetic energy is written as a quadratic form with a matrix which is a multiple of the identity matrix:

$$\mathscr{H} = \frac{1}{2M} \sum_{\mathbf{n}s\alpha} \widetilde{P}^2_{\mathbf{n}s\alpha} + \frac{1}{2} \sum_{\substack{\mathbf{n}s\alpha \\ \mathbf{n}'s'\alpha'}} \widetilde{A}^{ss'}_{\substack{\alpha\alpha' \\ \mathbf{n}-\mathbf{n}'}} \tilde{u}_{\mathbf{n}s\alpha} \tilde{u}_{\mathbf{n}'s'\alpha'}$$

$$\widetilde{A}^{ss'}_{\substack{\alpha\alpha' \\ \mathbf{n}-\mathbf{n}'}} = \frac{M}{\sqrt{M_s M_{s'}}} A^{ss'}_{\substack{\alpha\alpha' \\ \mathbf{n}-\mathbf{n}'}}$$

The quadratic form $(2M)^{-1} \sum \widetilde{P}^2_{\mathbf{n}s\alpha}$ remains unchanged and diagonal under arbitrary orthogonal transformations $\tilde{u}_{\mathbf{n}s\alpha}$. As we know from algebra, a quadratic form with a symmetric matrix $\widetilde{A}$ can be diagonalized by an orthogonal transformation. The diagonalization of the matrix $\widetilde{A}$ is achieved by changing to a basis of orthogonal vectors composed of the eigenvectors of $\widetilde{A}$. Therefore, the problem lies in the diagonalization of $\widetilde{A}$ or A. It is considerably simplified owing to the translational symmetry of the original function U, which has already been shown to produce the dependence of $\widetilde{A}$ on $\mathbf{n} - \mathbf{n}'$. We have mentioned already that a translation of an eigenvector of the matrix $\widetilde{A}$ is simply equivalent to the multiplication of the vector by $\exp(i\mathbf{k}\mathbf{R}_m)$

$$T_{\mathbf{m}}\{\tilde{u}_{\mathbf{n}s\alpha}\} = \{\tilde{u}_{\mathbf{n}+\mathbf{m}s\alpha}\} = \exp(i\mathbf{k}\mathbf{R}_{\mathbf{m}})\{\tilde{u}_{\mathbf{n}s\alpha}\}$$

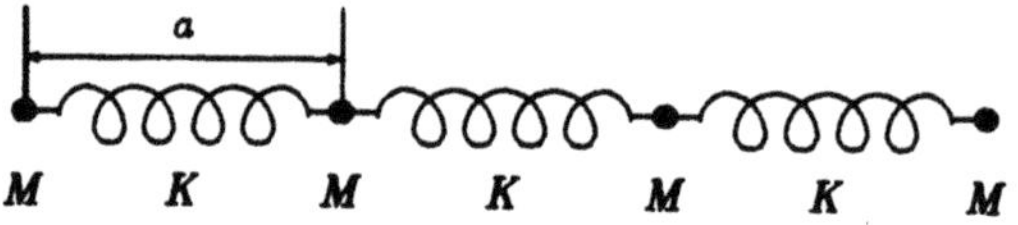

Figure 5.1. Monatomic string.

that is,

$$\tilde{u}_{\mathbf{n}+\mathbf{m}s\alpha} = \exp(i\mathbf{k}\mathbf{R}_{\mathbf{m}})\tilde{u}_{\mathbf{n}s\alpha} \qquad \tilde{u}_{\mathbf{m}s\alpha} = \tilde{u}_{0s\alpha}\exp(i\mathbf{k}\mathbf{R}_{\mathbf{m}})$$

The values assumed by the components of the eigenvector $\tilde{u}_{\mathbf{n}s\alpha}$ at $\mathbf{n} = 0$ determine all its other components.

Let us first consider the lattice dynamics of the simplest one-dimensional models, the Born string and the Karman strings. Let the potential energy be of the form

$$U = \frac{1}{2}\sum_n K(u_n - u_{n+1})^2$$

where the index n enumerates masses and their displacements and K is the force constant. The masses M of all atoms are assumed to be equal (Figure 5.1). Hamilton's function is

$$\mathscr{H} = \frac{1}{2M}\sum_n P_n^2 + \frac{K}{2}\sum_n (u_n - u_{n+1})^2$$

This is an analogue of the familiar formula (5.2.1). A specific feature of the model is its one-dimensionality and close range of interaction (interaction with the nearest atoms only).

The classical equations of motion are

$$M\ddot{u}_n = -\frac{\partial U}{\partial u_n} = K(u_{n-1} - u_n) + K(u_{n+1} - u_n)$$

We seek the solution to these equations in the form

$$u_n(t) = u_0 \exp(ikna - i\omega t) \tag{5.2.3}$$

where a is the spatial lattice period (Figure 5.1).

Substituting the solution in this form, we obtain, instead of a set of differential equations, a single algebraic equation

$$-M\omega^2 = K[(\exp(-ika) - 1) + (\exp(ika) - 1)]$$

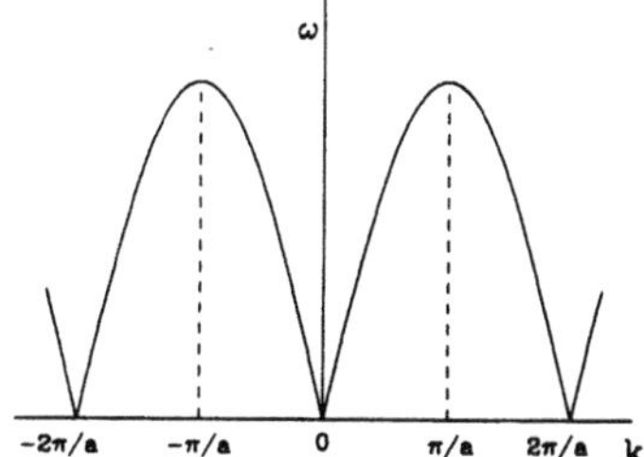

Figure 5.2. Dispersion law for monatomic string.

This is the equation of the dispersion law. Solving it for ω, we obtain

$$\omega_k = \sqrt{\frac{2K}{M}(1 - \cos ka)} = 2\sqrt{\frac{K}{M}}\left|\sin\frac{ka}{2}\right|$$

The dispersion law for a monatomic string is shown in Figure 5.2. Each value of k corresponds to a particular frequency ω_k, and ω is a periodic function of k, with a period $2\pi/a$. Hence, only one period can be considered in this dependence of ω on k. The physical meaning of this periodicity becomes clear if we consider a solution corresponding to $k' = k + 2\pi/a$:

$$u_n'(t) = u_0 \exp(\mathrm{i}(k + 2\pi/a)an - \mathrm{i}\omega t) = u_0 \exp(\mathrm{i}kna - \mathrm{i}\omega t) = u_n(t)$$

We see that the displacements of all atoms remain unchanged if k changes by $2\pi/a$. Hence, k must be found in the interval $(-\pi/a, \pi/a)$; in the three-dimensional case, it must be taken from the first Brillouin zone, in accordance with Chapter 1. Sometimes it is more convenient to work with a wider range of K, but the picture repeats itself outside of the first Brillouin zone. In the limit of long wavelengths, when $ka \ll 1$, that is, when $2\pi a/\lambda \ll 1$, we have

$$\omega_k = \sqrt{\frac{K}{M}}ak$$

Therefore, $S = \sqrt{\frac{K}{M}}a$ is the sound velocity in the crystal. This relation to the theory of continuous condensed medium is easily established. Indeed, in elasticity theory we have $S = \sqrt{E/\rho}$, where E is Young's modulus and ρ is the density of the medium. The ratio M/a can be interpreted as the density (linear) and Ka as the elasticity modulus. This last statement is implied by the expression for the elastic energy per unit volume,

$$\frac{U}{V} = \frac{1}{2}E\varepsilon^2$$

where ε is the relative deformation. If we write the expression for U of a monatomic string as

$$\frac{U}{a} = \sum_n \frac{1}{2}Ka\left(\frac{u_n - u_{n+1}}{a}\right)^2$$

then it becomes obvious that Ka can be interpreted as the elasticity modulus.

However, these waves coincide with the sound only in the long-wavelength limit. The minimal admissible wavelength is

$$\lambda_{\min} = \frac{2\pi}{k_{\max}} = 2a$$

and the dispersion law deviates significantly from acoustic law when λ becomes of the order of a. If λ is of the order of a, the group velocity of the wave

$$\frac{\partial \omega}{\partial k} = \sqrt{\frac{K}{M}}\, a \cos \frac{ka}{2}$$

is less than S and vanishes at $\lambda = \lambda_{\min}$. This last fact is implied by the Bragg reflection of waves from the periodic lattice at $\lambda = 2a$.

The result is a particular solution (5.2.3) of the initial dynamic equations. The general solution is the superposition of solutions of the type of (5.2.3). For a finite-length string, we also need to specify boundary conditions at the ends of the string.

The simplest boundary conditions for a string of N atoms are $u_0 = U_N = 0$ (fixed end points). Therefore, the first boundary condition is satisfied by a superposition of solutions with a wave vector $\pm k$, representing a standing wave:

$$u_n(t) = u_0 \sin(kna) \exp(-i\omega_k t)$$

The second boundary condition implies

$$kNa = \pi l$$

where l is an integer. Hence, $k_l = \pi l / Na$; the entire number of modes within the first Brillouin zone is

$$\frac{\pi}{a} \Big/ \frac{\pi}{Na} = N$$

The values $k = 0$ and $k = \pi/a$ must be rejected since they correspond to $u_n = 0$. There remain $(N-1)$ modes for $(N-1)$ degrees of freedom. This is what we should have expected; the fact that we find $(N-1)$ solutions for a system with $(N-1)$ degrees of freedom indicates that the set of solutions obtained is complete and that any solution can be written as a linear combination of those in the obtained set.

Different boundary conditions can be used; for instance, it is convenient to use periodic boundary conditions $u_0 = u_N$. They lead to solutions in the form of traveling waves and to N independent solutions for N degrees of freedom. This is also a complete system in the class of solutions which satisfy the condition of spatial periodicity.

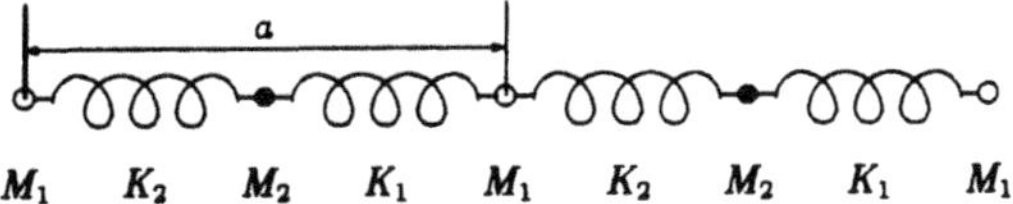

Figure 5.3. Diatomic string.

The results obtained for a diatomic string are qualitatively different (see Figure 5.3). This is the model of a crystal whose basis contains two atoms of unequal masses M_s $(s = 1, 2)$.

In this case, the equations of motion are

$$M_1\ddot{u}_n^{(1)} = K_1(u_{n-1}^{(2)} - u_n^{(1)}) + K_2(u_n^{(2)} - u_n^{(1)})$$
$$M_2\ddot{u}_n^{(2)} = K_2(u_n^{(1)} - u_n^{(2)}) + K_1(u_{n+1}^{(1)} - u_n^{(2)})$$

where $u_n^{(1)}$ and $u_n^{(2)}$ are the displacements of the atoms M_1 and M_2 from their respective equilibrium positions in the nth cell. We again seek a solution in the form

$$u_n^{(s)} = u_0^{(s)} \exp(ikna - i\omega_k t)$$

which gives a set of algebraic equations

$$\begin{aligned} (-M_1\omega_k^2 + K)u_0^{(1)} - [K_2 + K_1 \exp(-ika)]u_0^{(2)} &= 0 \\ -[K_2 + K_1 \exp(ika)]u_0^{(1)} + (-M_2\omega_k^2 + K)u_0^{(2)} &= 0 \end{aligned} \tag{5.2.4}$$

where $K = K_1 + K_2$. The condition under which this homogeneous set has a solution (its determinant must vanish) yields a quadratic equation for ω_k^2, whose solutions are

$$\omega_k^2 = \frac{K}{2M^*} \pm \sqrt{\left(\frac{K}{2M^*}\right)^2 - \frac{4K_1K_2}{M_1M_2}\sin^2\frac{ka}{2}}$$
$$M^{*-1} = M_1^{-1} + M_2^{-1}$$

The dispersion law is shown in Figure 5.4; it shows two branches. The lower branch passes through the point $k = 0$, $\omega = 0$ and is known as the acoustic branch; the second starts at the boundary frequency $\omega_0 = \sqrt{K/M^*}$ at $k = 0$. As before, ω is a periodic function of k, so that it is sufficient to consider it within the first Brillouin zone. The acoustic wave displays the same behavior as the dispersion law for the monatomic lattice and corresponds to sound waves for low k. The upper wave, known as the optical wave, corresponds to zero group velocity both at $k = 0$ and at the boundaries of the Brillouin zone.

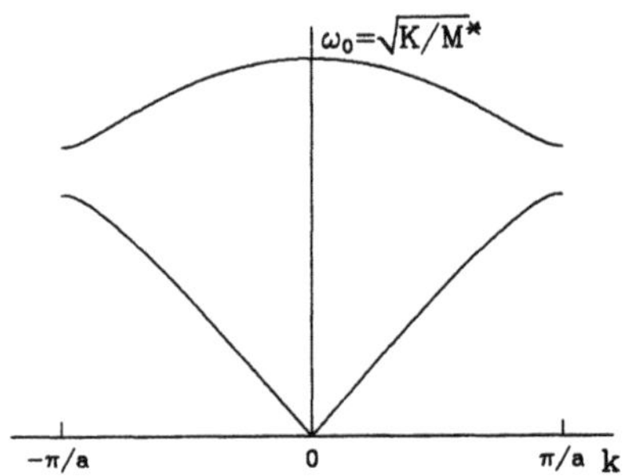

Figure 5.4. Dispersion law for diatomic string.

The term "optical" branch arose because, as we will see later, vibrations of this type efficiently interact with the optical (infrared) radiation.

If $ka \ll 1$, we have

$$\omega_k^2 \simeq \frac{K}{2M^*} \pm \frac{K}{2M^*}\left(1 - \frac{2K_1K_2M^{*2}}{M_1M_2K^2}\,k^2a^2\right)$$

If the minus sign is taken, a term is left in this expression which is proportional to ka^2a^2, which corresponds to acoustic waves; for the plus sign, $\omega^2 \simeq K/M^* = \omega_0^2$.

The number of degrees of freedom in a diatomic lattice composed of N unit cells is $2N$; correspondingly, the number of degrees of freedom is doubled, since there are two branches in the dispersion law, each branch again possessing n vibration modes.

We will now find the ratio of amplitudes $u_0^{(1)}/u_0^{(2)}$ for each of the waves as $\varkappa \to 0$. Equations (5.2.4) give

$$\frac{u_0^{(1)}}{u_0^{(2)}} = \frac{K_2 + K_1 \exp(-ika)}{K - M_1\omega_k^2}$$

On the acoustic wave, $\omega_k^2 \to 0$ and $u_0^{(1)}/u_0^{(2)} \to 1$; on the optical branch, this ratio tends to $-M_2/M_1$.

Therefore, in acoustic vibrations the atoms M_1 and M_2 move in the long-wavelength limit in phase and with identical amplitudes, while in optical vibrations they move in antiphase, so that the center of mass of the unit cell does not move.

For instance, in acoustic vibrations of a hydrogen crystal (which is a molecular crystal), a molecule moves as a whole while in optical vibrations its atoms oscillate relative to one another. In acoustic vibrations at $k \to 0$ molecules move as units and no appreciable forces are generated in unit cells, but in optical vibrations forces do arise within a unit cell. If $k \to 0$, optical vibrations are vibrations in the unit cells but all unit cells are in phase. In an ionic crystal, optical vibrations are relative oscillations of two sublattices (e.g., Na^+ and Cl^-). These vibrations generate an oscillating dipole moment, which explains the interaction of optical vibrations with the electromagnetic field.

So far we considered the solutions of the equations for atomic vibrations in a lattice, represented by traveling waves. Solutions of a different type are possible, given by waves which grow and decay in space.

These solutions correspond to the case in which k is not real any more; they are significant for bounded crystals (strings), because of the effect of boundaries, and for lattices with inhomogeneities.

A typical example of inhomogeneity is isotopic impurity. Such impurities are present in any crystal, they violate the homogeneity since the mass of an impurity atom differs from that of the host lattice atom.

Let the lattice site $n = 0$ correspond to an isotopic impurity atom of mass M', while the sites $n \neq 0$ are occupied by host lattice atoms of mass M. The equations of motion for $|n| \geq 1$ are the same as we had for the monatomic string,

$$M\ddot{u}_n = K(u_{n+1} + u_{n-1} - 2u_n) \tag{5.2.5}$$

while for $n = 0$ we have

$$M'\ddot{u}_0 = K(u_{-1} + u_{+1} - 2u_0) \tag{5.2.6}$$

since we assume that the electron state is not changed in the isotopic impurity, and neither are the energy U and the force constant K.

If we exclude the triple of atoms $n = 0$, $n = \pm 1$, the solutions of equations for the remaining atoms need no modification:

$$\begin{aligned} u_n &= \exp[ik(n+1)a]u_{-1} \qquad & n < -1 \\ u_n &= \exp[ik(n-1)a]u_{+1} \qquad & n > 1 \end{aligned}$$

The frequency ω corresponds to two vectors, $\pm k$, so that the general solution for the frequency ω is a superposition of solutions which differ in the sign of k. It can be rewritten as a wave partially reflected by the impurity atom:

$$\begin{aligned} u_n &= (\alpha \exp(ikna) + \beta \exp(-ikna)) \exp(-i\omega_k t) \qquad & n < 0 \\ u_n &= \gamma \exp(ikna - i\omega_k t) \qquad & n > 0 \end{aligned}$$

The three equations for $n = 0$, $n = \pm 1$ can be used to find the ratio of the three coefficients α, β and γ, that is, to find the coefficient of reflection of the wave by the impurity and the transmission coefficient.

In fact, this does not exhaust the possible solutions in the case of isotopic impurity. Solutions of the type

$$u_n = u_0 \exp(\mathrm{i}kna) \exp(-\mathrm{i}\omega t)$$

that have complex k can arise. Such solutions cannot exist in an infinite homogeneous lattice or string (they are not normalizable and tend to ∞ as $n \to \pm\infty$). They are admissible, however, in a nonideal lattice (e.g. with an impurity or a boundary).

Equations (5.2.5) give a dispersion law

$$\omega^2 = \frac{2K}{M}(1 - \cos ka)$$

Let $k = k' + \mathrm{i}k''$, which gives

$$\begin{aligned} \omega^2 &= \frac{2K}{M}(1 - \cos(k' + \mathrm{i}k'')a) \\ &= \frac{K}{M}(2 - \exp(-k''a)\exp(\mathrm{i}k'a) - \exp(k''a)\exp(-\mathrm{i}k'a)) \end{aligned}$$

For real frequencies ω

$$\mathrm{Im}\omega^2 = \frac{K}{M}(\exp(k''a) - \exp(-k''a)) \sin k'a = 0$$

that is, either $k'' = 0$ or $k' = \pm\pi l/a$ ($l = 0, \pm 1, \pm 2, \dots$). The case of real k has already been discussed. The second possibility gives

$$\omega^2 = \frac{2K}{M}(1 - (-1)^l \cosh k''a)$$

The condition $\omega^2 > 0$ yields $l = 1$, since all other odd l yield the same solutions.

Solutions of the following type are thus possible:

$$u_n = u_0 \exp(\mathrm{i}\pi n) \exp(\pm k''na) \exp(-\mathrm{i}\omega t)$$

If a crystal is bounded by the surface $n = 0$, so that $n \geq 0$, then the solutions possible for $k'' > 0$ are

$$u_n = u_0 \exp(\mathrm{i}\pi n) \exp(-k''na) \exp(-\mathrm{i}\omega t)$$

They damp out away from the surface, and their oscillation phase changes by π from one atom to another. The boundary conditions make it possible to find these solutions; the dispersion law for them is

$$\omega^2 = \frac{2K}{M}(1 - \cosh k''a) = 4\frac{K}{M}\cosh^2\frac{k''a}{2} \qquad (5.2.7)$$

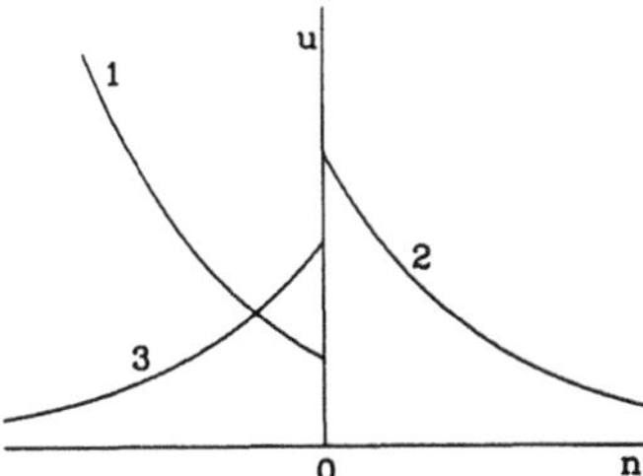

Figure 5.5. Oscillations in the neighborhood of an isotopic impurity at $\mathbf{n} = 0$: 1—"incident" wave, 2—"transmitted" wave and "reflected" wave.

This is the situation at the crystal surface. If we now return to the isotopic impurity, then wave-type solutions arise near the impurity atom in the general case of arbitrary k'', as they did for real k: the "incident" wave which decays exponentially as it approaches the impurity, the "transmitted" and the "reflected" waves which are decaying exponentially as we move away from the impurity (Figure 5.5).

If the impurity lies inside the crystal not too close to its boundary, the amplitude of its vibrations, and thus the amplitudes of the "transmitted" and the "reflected" waves are negligibly (exponentially) small; this case is physically of no interest. At a certain k'', however, the solution consists of only the "transmitted" and "reflected" waves, while the "incident" wave vanishes; this means that the first two generally exist independently of the incident wave. Vibrations also damp out on both sides of the impurity and are thus localized in its neighborhood, with the solution in the form

$$u_n = u_0 \exp(i\pi n) \exp(-k''|n|a)$$

Equation (5.2.6) for $n = 0$ gives

$$\omega^2 = \frac{2k}{M} [1 + \exp(-k''a)] \tag{5.2.8}$$

This, however, is not any more the dispersion law. Together with (5.2.7), equation (5.2.8) determines k'' and ω, that is, the degree of localization and the frequency. This is an eigenvalue problem. By subtracting equation (5.2.7) from (5.2.8), we obtain

$$2\Big(\frac{1}{M'} - \frac{1}{M}\Big) + \frac{2}{M'} \exp(-k''a) - \frac{1}{M} \exp(-k''a) - \frac{1}{M} \exp(k''a) = 0$$

or, introducing $y = e^{k''a}$, arrive at a quadratic equation

$$y^2 - \frac{2(M - M')}{M'} y - \frac{2M - M'}{M'} = 0$$

Its solution is

$$y = \frac{2M - M'}{M'} = \frac{M + \Delta M}{M - \Delta M}$$

where $\Delta M = M - M' > 0$ (because we require $y > 1$). This means that localized solutions exist only for the lighter isotope. We will take into account that $\Delta M \ll M$ holds in practically all cases, except the hydrogen atom. Hence,

$$k'' = \frac{1}{a} \ln \frac{M + \Delta M}{M - \Delta M} \simeq \frac{2\Delta M}{M}$$

and the frequency of localized vibrations,

$$\begin{aligned} \omega^2 = \frac{2K}{M'}(1 + y) = \frac{4KM}{M^2 - \Delta M^2} &= \omega_{\text{max}}^2 \frac{M^2}{M^2 - \Delta M^2} \\ \simeq \omega_{\text{max}}^2 \left(1 + \left(\frac{\Delta M}{M}\right)^2\right) \end{aligned}$$

is somewhat higher than the maximum frequency $\omega_{\text{max}} = 2\sqrt{K/M}$ of traveling waves at the boundary of the Brillouin zone. As a result, the frequency of local vibrations splits off the frequency band of traveling waves. This is physically quite clear: if the frequency belonged to the traveling wave frequency band, vibrations would excite traveling waves, the excitation would be transferred from atom to atom and no localized vibrations could exist. This also explains the condition of existence of localized vibrations ($M' < M$), since a frequency above all vibration frequencies of a homogeneous lattice can arise only under these conditions.

Practically not only M but also the other constant, K, are changed in the presence of an impurity (if this impurity is not isotopic). However, localized vibrations for monatomic lattices always correspond to frequencies above the maximum frequency for the homogeneous lattice. In lattices with a more complicated basis, localized vibrations can also appear in the spectrum gaps between the acoustic and optical branches.

5.3. Oscillations of Three-Dimensional Lattices. Quantization of Vibrations

Let us now generalize to three-dimensional lattices the results obtained for one-dimensional strings.

The form of Hamilton's function is given by (5.2.1). From this function we derive the dynamic equations

$$\begin{aligned} \dot{u}_{\mathbf{n}s\alpha} &= \frac{\partial \mathscr{H}}{\partial P_{\mathbf{n}s\alpha}} = \frac{P_{\mathbf{n}s\alpha}}{M_s} \\ \dot{P}_{\mathbf{n}s\alpha} &= -\frac{\partial \mathscr{H}}{\partial u_{\mathbf{n}s\alpha}} = -\sum_{\mathbf{n}'s'\alpha'} A_{\mathbf{n}-\mathbf{n}'}^{\substack{ss'\\ \alpha\alpha'}} u_{\mathbf{n}'s'\alpha'} \end{aligned}$$

Or, if we eliminate $P_{\mathbf{n}s\alpha}$,

$$M_s \ddot{u}_{\mathbf{n}s\alpha} + \sum_{\mathbf{n}'s'\alpha'} A_{\mathbf{n}-\mathbf{n}'}^{\alpha\alpha' \, ss'} u_{\mathbf{n}'s'\alpha'} = 0 \tag{5.3.1}$$

The solution to (5.3.1) is sought, in accordance with Section 5.1, in the form

$$u_{\mathbf{n}s\alpha} = u_{s\alpha} \exp(\mathrm{i}\mathbf{k}\mathbf{R}_{\mathbf{n}} - \mathrm{i}\omega_k t) \tag{5.3.2}$$

Substituting it into (5.3.1), we obtain

$$-M_s \omega_{\mathbf{k}}^2 u_{s\alpha} + \sum_{\mathbf{n}'s'\alpha'} A_{\mathbf{n}-\mathbf{n}'}^{\alpha\alpha' \, ss'} u_{s'\alpha'} \exp[-\mathrm{i}\mathbf{k}(\mathbf{R}_{\mathbf{n}} - \mathbf{R}_{\mathbf{n}'})] = 0$$

We will use the notation

$$A_{\alpha\alpha'}^{ss'}(\mathbf{k}) = \sum_{\mathbf{m}} A_{\mathbf{m}}^{\alpha\alpha' \, ss'} \exp(-\mathrm{i}\mathbf{k}\mathbf{R}_{\mathbf{m}})$$

which is a discrete analogue of the Fourier transformation. The result is

$$-M_s \omega_{\mathbf{k}}^2 u_{s\alpha} + \sum_{s'\alpha'} A_{\alpha\alpha'}^{ss'}(\mathbf{k}) u_{s'\alpha'} = 0 \tag{5.3.3}$$

This gives us equations for the vector $\mathbf{u}_{s\alpha}$ with the number of components $3r$, where r is the number of atoms in a unit cell ($s = 1, \ldots, r$).

The condition of solvability of (5.3.3) consists in setting the determinant of the $3r \times 3r$ matrix to zero:

$$\det \| M_s \omega_{\mathbf{k}}^2 \delta_{ss'} \delta_{\alpha\alpha'} - A_{\alpha\alpha'}^{ss'}(\mathbf{k}) \| = 0 \tag{5.3.4}$$

This is an equation of degree $3r$ with respect to $\omega_{\mathbf{k}}^2$. We denote the roots of equation (5.3.4) by $\omega_{\mathbf{k}j}$ where $j = 1, \ldots, 3r$ enumerates the roots of the equation and the corresponding vectors $\mathbf{u}_{s\alpha}$ which satisfy (5.3.3).

Qualitatively, the spectrum has the same form as for a diatomic string (Figure 5.4). There are altogether $3r$ spectrum branches, of which three are acoustic (Figure 5.6). The presence of acoustic waves, for which $\omega \to 0$ as $\mathbf{k} \to 0$, follows from general arguments.

If $u_{s\alpha} = c_\alpha$, where c_α are constants independent of the indices $\mathbf{n}$ and s, the energy $U(\mathbf{R} + \mathbf{u})$ is independent of c_α and equals $U(\mathbf{R})$ since such atomic displacements correspond to a translational displacement of the lattice as a whole. This means that the force acting on an atom is

$$F_{\mathbf{n}s\alpha} = \sum_{\mathbf{n}'s'\alpha'} A_{\mathbf{n}-\mathbf{n}'}^{\alpha\alpha' \, ss'} u_{\mathbf{n}'s'\alpha'} = \sum_{\mathbf{n}'s'\alpha'} A_{\mathbf{n}-\mathbf{n}'}^{\alpha\alpha' \, ss'} c_{\alpha'} = 0$$

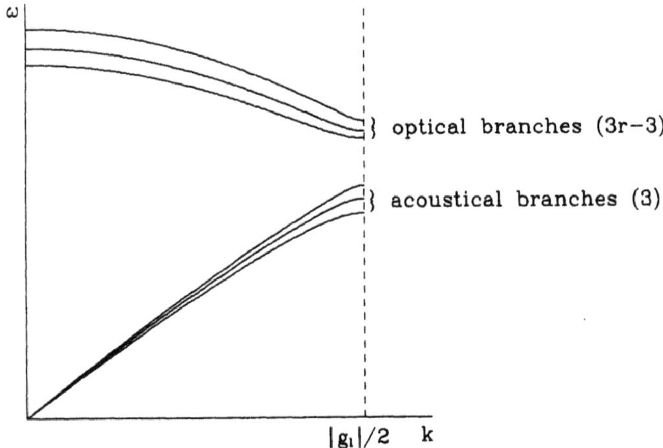

Figure 5.6. Dispersion law for vibrational modes in a crystal.

This implies, for instance, that

$$\sum_{\mathbf{n}' s' \alpha'} A_{\mathbf{n}-\mathbf{n}'}^{\substack{ss' \\ \alpha\alpha'}} = 0$$

that is,

$$\sum_{s'\alpha'} A_{\alpha\alpha'}^{ss'}(\mathbf{k} = 0) = 0$$

This last equality signifies that the sum of columns of the matrix $A_{\alpha\alpha'}^{ss'}(\mathbf{k} = 0)$ is zero, that is, that they are linearly dependent and therefore

$$\det \|A_{\alpha\alpha'}^{ss'}(\mathbf{k} = 0)\| = 0$$

Owing to (5.3.4), $\omega = 0$ at $\mathbf{k} = 0$ satisfies the dispersion equation, and the frequency $\omega = 0$ is thrice degenerate. The three acoustic waves of an isotropic solid correspond to longitudinal and twice-degenerate transverse waves.

The frequencies $\omega_{\mathbf{k}j}$ are functions of the wave vector $\mathbf{k}$. This dependence is periodical, with the reciprocal lattice period. As in the case of strings, the periodicity arises because

$$A_{\alpha\alpha'}^{ss'}(\mathbf{k} + \mathbf{g}_1) = \sum_{\mathbf{m}} A_{\mathbf{m}}^{\substack{ss' \\ \alpha\alpha'}} \exp(-i(\mathbf{k} + \mathbf{g}_1)\mathbf{R}_{\mathbf{m}})$$
$$= \sum_{\mathbf{m}} A_{\mathbf{m}}^{\substack{ss' \\ \alpha\alpha'}} \exp(-i\mathbf{k}\mathbf{R}_{\mathbf{m}}) = A_{\alpha\alpha'}^{ss'}(\mathbf{k})$$

since, by the definition of the reciprocal lattice vector $\mathbf{g}_1$,

$$\exp(-i\mathbf{g}_1\mathbf{R}_{\mathbf{m}}) = 1$$

Hence, $\omega_{\mathbf{k}+\mathbf{g}_l} = \omega_{\mathbf{k}}$ and the dispersion law can be considered only in the first Brillouin zone. Physically, this means that the plane waves $\exp(i\mathbf{kr})$ and $\exp(i(\mathbf{k}+\mathbf{g}_l)\mathbf{r})$ have identical values (are indistinguishable) if considered only at discrete points $\mathbf{r} = \mathbf{R_n}$. The values $\mathbf{k}$ and $\mathbf{k}+\mathbf{g}_l$ of the wave vector correspond to the same vector $u_{\mathbf{n}s\alpha}$ and the same function $u_{\mathbf{n}s\alpha}(t)$ of the type of (5.3.2).

In an anisotropic crystal, $\omega_{\mathbf{k}}$ is a function not only of the modulus of the wave vector but also of its direction. Hence, we can show on a planar diagram only the dependence of ω on $\mathbf{k}$ for some selected crystallographic direction, for instance, for a crystallographic axis. In this case we obtain the picture shown in Figure 5.6.

Each frequency $\omega_{\mathbf{k}j}$ corresponds to a polarization vector $e^j_{s\alpha}(\mathbf{k})$ which satisfies equations (5.3.3). This is a set of $3r$ numbers that give the ratio of displacements $u_{s\alpha}$ of atoms for a given type of vibrations, enumerated by the number j and the vector $\mathbf{k}$; j runs through $3r$ values and $\mathbf{k}$ varies almost continuously if the number N of unit cells in the crystal is large.

Equations (5.3.3) and (5.3.4) cannot be solved in the general case. However, it is possible to classify vibrations with frequencies $\omega_{\mathbf{k}j}$ according to symmetry. Each vibration, and, therefore, the frequency $\omega_{\mathbf{k}j}$ and the polarization vector $e^j_{s\alpha}$, corresponds to one irreducible representation of the symmetry group of the crystal. As a result, it is possible to determine the possible degrees of degeneracy of eigenfrequencies and establish the selection rules.

Let us evaluate the order of magnitude of the vibration frequencies $\omega_{\mathbf{k}j}$. The force constants are

$$A \sim \frac{\partial^2 U}{\partial u^2} \sim \frac{I_0}{a_0^2}$$

so that

$$\omega^2 \sim \frac{A}{M} \sim \frac{I_0}{a_0^2 M}$$

and the energy is

$$(\hbar\omega)^2 \sim \frac{\hbar^2}{2a_0^2 m}\,\frac{I_0 m}{M} = I_0^2\,\frac{m}{M}$$

Therefore, the energy is

$$\hbar\omega \sim \sqrt{\frac{m}{M}}\, I_0$$

that is, it is of the order of several hundredths of one electron volt. For the optical branches of the spectrum, the frequencies $\omega_{\mathbf{k}j}$ lie in the infrared region, while for the acoustic waves they reach this order of magnitude at $\mathbf{k}$ of the order of the wave vector at the boundaries of the Brillouin zone.

The matrices $A^{\alpha\alpha'}_{\mathbf{n}-\mathbf{n}'}{}^{ss'}$ are symmetric with respect to the operation $\mathbf{n}s\alpha \leftrightarrow \mathbf{n}'s'\alpha'$, so that their Fourier transforms $A^{ss'}_{\alpha\alpha'}(\mathbf{k})$ are Hermitian. If we use (5.2.2) and introduce modified vectors

$$\tilde{e}^j_{s\alpha} = e^j_{s\alpha}\sqrt{\frac{M_s}{M}}$$

where $M = \sum M_s$ is the mass of all atoms in a unit cell, the equation (5.3.3) changes to

$$\sum_{s'\alpha'} \frac{M}{\sqrt{M_s M_{s'}}} A^{ss'}_{\alpha\alpha'}(\mathbf{k})\tilde{e}^j_{s'\alpha'} = M\omega^2_{\mathbf{k}j}\tilde{e}^j_{s\alpha} \tag{5.3.5}$$

This is an eigenvalue problem for the Hermitian matrix

$$\tilde{A}^{ss'}_{\alpha\alpha'}(\mathbf{k}) = \frac{M}{\sqrt{M_s M_{s'}}} A^{ss'}_{\alpha\alpha'}(\mathbf{k})$$

The eigenvectors of this matrix are orthogonal and we can choose a normalization for them:

$$\sum_{s\alpha} \tilde{e}^{j'*}_{s\alpha}\tilde{e}^j_{s\alpha} = \delta_{j'j} \tag{5.3.6}$$

This means that the vectors $e^j_{s\alpha}$ are orthogonal with a weight:

$$\sum_{s\alpha} M_s e^{j'*}_{s\alpha} e^j_{s\alpha} = \delta_{j'j} M$$

For a finite crystal, we additionally need to use the boundary conditions at the surface. By analogy to what we had for one-dimensional strings, it is most convenient to use periodic boundary conditions on a parallelepiped consisting of $N = N_1N_2N_3$ unit cells; the edge lengths of the parallelepiped are N_1a_1, N_2a_2 and N_3a_3, respectively, with the edges pointing along the basis vectors of the Bravais lattice, $\mathbf{a}_1$, $\mathbf{a}_2$, and $\mathbf{a}_3$.

For such boundary conditions, we have a discrete set of wave vectors,

$$\mathbf{k} = \frac{\mathbf{g}_1}{N_1}m_1 + \frac{\mathbf{g}_2}{N_2}m_2 + \frac{\mathbf{g}_3}{N_3}m_3$$

where m_1, m_2, m_3 are integers and $\mathbf{g}_1$, $\mathbf{g}_2$, $\mathbf{g}_3$ are the basis vectors of the reciprocal lattice.

The density of modes in the wave vector space (in the reciprocal lattice space) is $(2\pi)^3/N\Omega$, where Ω is the unit cell volume, that is,

$$\Omega = (\mathbf{a}_1\mathbf{a}_2\mathbf{a}_3) = (2\pi)^{-3}(\mathbf{g}_1\mathbf{g}_2\mathbf{g}_3)^{-1}$$

On the other hand, the unit cell volume in the reciprocal lattice space (the volume of the first Brillouin zone) equals $(\mathbf{g}_1\mathbf{g}_2\mathbf{g}_3) = (2\pi)^3/\Omega$. Therefore, the total number of modes for one branch of the spectrum is N and the total number of all modes is $3rN$, in other words, it is exactly equal to the number of degrees of freedom

Periodic boundary conditions correspond to the total set of vectors

$$\tilde{\mathbf{e}}^{j}_{\mathbf{n}s\alpha}(\mathbf{k}) = \frac{1}{\sqrt{N}} \tilde{\mathbf{e}}^{j}_{s\alpha}(\mathbf{k}) \exp(\mathrm{i}\mathbf{k}\mathbf{R}_{\mathbf{n}})$$

that are orthogonal and normalized:

$$\sum_{\mathbf{n}s\alpha} \tilde{\mathbf{e}}^{j'*}_{\mathbf{n}s\alpha}(\mathbf{k}')\tilde{\mathbf{e}}^{j}_{\mathbf{n}s\alpha}(\mathbf{k}) = \delta_{j'j}\delta_{\mathbf{k}'\mathbf{k}} \tag{5.3.7}$$

The number of these vectors is $3Nr$; any vector $\tilde{u}_{\mathbf{n}s\alpha}$ can be expanded in these vectors. If we take into account that $\tilde{u}_{\mathbf{n}s\alpha} = (M_s/M)^{1/2}u_{\mathbf{n}s\alpha}$ is real, the expansion takes the form

$$\tilde{u}_{\mathbf{n}s\alpha} = \sum_{\mathbf{k}j}(u_{\mathbf{k}j}\tilde{\mathbf{e}}^{j}_{\mathbf{n}s\alpha}(\mathbf{k}) + u^{*}_{\mathbf{k}j}\tilde{\mathbf{e}}^{j*}_{\mathbf{n}s\alpha}(\mathbf{k})) \tag{5.3.8}$$

Correspondingly,

$$u_{\mathbf{n}s\alpha} = \sum_{\mathbf{k}j}(u_{\mathbf{k}j}\mathbf{e}^{j}_{\mathbf{n}s\alpha}(\mathbf{k}) + u^{*}_{\mathbf{k}j}\mathbf{e}^{j*}_{\mathbf{n}s\alpha}(\mathbf{k})) \tag{5.3.9}$$

where

$$\mathbf{e}^{j}_{\mathbf{n}s\alpha}(\mathbf{k}) = \frac{1}{\sqrt{N}} \mathbf{e}^{j}_{s\alpha}(\mathbf{k}) \exp(\mathrm{i}\mathbf{k}\mathbf{R}_{\mathbf{n}})$$

is the complete set of vectors which are orthogonal and normalized with a weight.

In expression (5.3.8), $u_{\mathbf{k}j} \sim \mathrm{e}^{-\mathrm{i}\omega_{\mathbf{k}j}t}$, $u^{*}_{\mathbf{k}j} \sim \mathrm{e}^{\mathrm{i}\omega_{\mathbf{k}j}t}$, and $\omega_{\mathbf{k}j} > 0$. Therefore, the summation in (5.3.8) is carried over all $\mathbf{k}$; the modes $\mathbf{k}$ and $-\mathbf{k}$ differ. However, equations (5.3.3) and (5.3.5) together with the definition of the matrix $A(\mathbf{k})$ imply that

$$\begin{aligned} A^{ss'}_{\alpha\alpha'*}(-\mathbf{k}) &= A^{ss'}_{\alpha\alpha'}(\mathbf{k}) \\ \mathbf{e}^{j}_{s\alpha}(-\mathbf{k}) &= \mathbf{e}^{j*}_{s\alpha}(\mathbf{k}) \\ \mathbf{e}^{j}_{\mathbf{n}s\alpha}(-\mathbf{k}) &= \mathbf{e}^{j*}_{\mathbf{n}s\alpha}(\mathbf{k}) \\ \omega^{2}_{-\mathbf{k}j} &= \omega^{2}_{\mathbf{k}j} \end{aligned} \tag{5.3.10}$$

Symmetry relations (5.3.10) follow from $A^{ss'}_{\mathbf{n}-\mathbf{n}'}{}^{\alpha\alpha'}$ being real and from the fact that the matrix $M^{-1}\tilde{A}^{*}(\mathbf{k})$ has eigenvectors $\tilde{\mathbf{e}}^{j*}_{s\alpha}(\mathbf{k})$ and eigenvalues

$\omega_{\mathbf{k}j}^{2*} = \omega_{\mathbf{k}j}^2$. $\omega_{\mathbf{k}j}^2$ are real because the matrix $\tilde{A}$ is Hermitian; it is nonnegative, $\omega_{\mathbf{k}j}^2 \geq 0$, because the potential energy is positive-definite, with the minimal value vanishing at the equilibrium position (the energy $U(\mathbf{R})$ at the equilibrium point can be ignored when the oscillational energy is considered). The matrix $\tilde{A}$ is one of a positive-definite quadratic form and thus has only nonnegative eigenvalues.

By virtue of dynamic equations, the momenta $P_{\mathbf{n}s\alpha}$ are

$$P_{\mathbf{n}s\alpha} = M_s \dot{u}_{\mathbf{n}s\alpha} = -\mathrm{i}M_s \sum_{\mathbf{k}j} \omega_{\mathbf{k}j}(u_{\mathbf{k}j}\mathbf{e}_{\mathbf{n}s\alpha}^{j}(k) - u_{\mathbf{k}j}^{*}\mathbf{e}_{\mathbf{n}s\alpha}^{j*}(\mathbf{k}))$$

that is,

$$\widetilde{P}_{\mathbf{n}s\alpha} = \sqrt{\frac{M}{M_s}} P_{\mathbf{n}s\alpha} = -\mathrm{i}M \sum_{\mathbf{k}j} \omega_{\mathbf{k}j}(u_{\mathbf{k}j}\tilde{\mathbf{e}}_{\mathbf{n}s\alpha}^{j}(\mathbf{k}) - u_{\mathbf{k}j}^{*}\tilde{\mathbf{e}}_{\mathbf{n}s\alpha}^{j*}(\mathbf{k})) \quad (5.3.11)$$

Equations (5.3.8) and (5.3.11) can be solved for the complex amplitudes $u_{\mathbf{k}j}$. Correspondingly, we multiply (5.3.8) and (5.3.11) by $\tilde{\mathbf{e}}_{\mathbf{n}s\alpha}^{j*}$ and sum up over $\mathbf{n}s\alpha$. Using (5.3.7) and (5.3.10), we obtain

$$\sum_{\mathbf{n}s\alpha} \tilde{u}_{\mathbf{n}s\alpha}\tilde{\mathbf{e}}_{\mathbf{n}s\alpha}^{j*}(\mathbf{k}) = u_{\mathbf{k}j} + u_{-\mathbf{k}j}^{*}$$

$$\sum_{\mathbf{n}s\alpha} \widetilde{P}_{\mathbf{n}s\alpha}\tilde{\mathbf{e}}_{\mathbf{n}s\alpha}^{j*}(\mathbf{k}) = -\mathrm{i}M\omega_{\mathbf{k}j}(u_{\mathbf{k}j} - u_{-\mathbf{k}j}^{*})$$

Therefore,

$$\begin{aligned} u_{\mathbf{k}j} &= \frac{1}{2}\sum_{\mathbf{n}s\alpha}\left(\tilde{u}_{\mathbf{n}s\alpha} + \frac{\mathrm{i}}{M\omega_{\mathbf{k}j}}\widetilde{P}_{\mathbf{n}s\alpha}\right)\tilde{\mathbf{e}}_{\mathbf{n}s\alpha}^{j*}(k) \\ u_{\mathbf{k}j}^{*} &= \frac{1}{2}\sum_{\mathbf{n}s\alpha}\left(\tilde{u}_{\mathbf{n}s\alpha} - \frac{\mathrm{i}}{M\omega_{\mathbf{k}j}}\widetilde{P}_{\mathbf{n}s\alpha}\right)\tilde{\mathbf{e}}_{\mathbf{n}s\alpha}^{j}(\mathbf{k}) \end{aligned} \quad (5.3.12)$$

Using the dynamic equations for $\tilde{u}_{\mathbf{n}s\alpha}$ and $\widetilde{P}_{\mathbf{n}s\alpha}$ (5.3.1) and $\widetilde{P}_{\mathbf{n}s\alpha} = M\dot{\tilde{u}}_{\mathbf{n}s\alpha}$, we find that $u_{\mathbf{k}j}$ satisfy the equations

$$\frac{\mathrm{d}u_{\mathbf{k}j}}{\mathrm{d}t} = -\mathrm{i}\omega_{\mathbf{k}j}u_{\mathbf{k}j}$$

$$\frac{\mathrm{d}u_{\mathbf{k}j}^{*}}{\mathrm{d}t} = \mathrm{i}\omega_{\mathbf{k}j}u_{\mathbf{k}j}^{*}$$

and that $\omega_{\mathbf{k}j} \geq 0$.

Substituting (5.3.8) and (5.3.11) into Hamilton's function (5.2.1), we arrive at an expression for energy in terms of the complex amplitudes $u_{\mathbf{k}j}$.

Using (5.3.5), (5.3.7) and (5.3.10), we obtain

$$\begin{aligned}\mathscr{H} &= \frac{1}{2M}\sum_{\mathbf{n}s\alpha}\widetilde{P}^2_{\mathbf{n}s\alpha} + \frac{1}{2}\sum_{\substack{\mathbf{n}s\alpha\\ \mathbf{n}'s'\alpha'}}\widetilde{A}^{ss'}_{\alpha\alpha'\,\mathbf{n}-\mathbf{n}'}\tilde{u}_{\mathbf{n}s\alpha}\tilde{u}_{\mathbf{n}'s'\alpha'}\\
&= \frac{M}{2}\sum_{\mathbf{k}j}\omega^2_{\mathbf{k}j}(2u_{\mathbf{k}j}u^*_{\mathbf{k}j} - u_{\mathbf{k}j}u_{-\mathbf{k}j} - u^*_{\mathbf{k}j}u^*_{-\mathbf{k}j})\\
&\quad + \frac{M}{2}\sum_{\mathbf{k}j}\omega^2_{\mathbf{k}j}(2u_{\mathbf{k}j}u^*_{\mathbf{k}j} + u_{\mathbf{k}j}u_{-\mathbf{k}j} + u^*_{\mathbf{k}j}u^*_{-\mathbf{k}j})\\
&= 2M\sum_{\mathbf{k}j}\omega^2_{\mathbf{k}j}u_{\mathbf{k}j}u^*_{\mathbf{k}j}\end{aligned}$$

When deriving the final expression

$$\mathscr{H} = 2M\sum_{\mathbf{k}j}\omega^2_{\mathbf{k}j}u_{\mathbf{k}j}u^*_{\mathbf{k}j} \tag{5.3.13}$$

we made use of the fact that $\tilde{e}^j_{\mathbf{n}s\alpha}(\mathbf{k})$ are eigenvectors of the matrix $\tilde{A}(\mathbf{k})$ and that

$$\sum_{\mathbf{n}}\exp(\mathrm{i}(\mathbf{k}+\mathbf{k}')\mathbf{R_n}) = N\delta_{-\mathbf{k},\mathbf{k}'}$$

In view of this we have, for example,

$$\begin{aligned}&\sum_{\substack{\mathbf{n}s\alpha\\ \mathbf{n}'s'\alpha'}}\widetilde{A}^{ss'}_{\alpha\alpha'\,\mathbf{n}-\mathbf{n}'}\tilde{e}^j_{\mathbf{n}s\alpha}(\mathbf{k})\tilde{e}^{j'}_{\mathbf{n}'s'\alpha'}(\mathbf{k})\\
&= \frac{1}{N}\sum\widetilde{A}^{ss'}_{\alpha\alpha'\,\mathbf{n}-\mathbf{n}'}\tilde{e}^j_{s\alpha}(\mathbf{k})\tilde{e}^{j'}_{s'\alpha'}(\mathbf{k}')\exp(\mathrm{i}\mathbf{k}\mathbf{R_n} + \mathrm{i}\mathbf{k}'\mathbf{R_{n'}})\\
&= \frac{1}{N}\sum_{\substack{\mathbf{n}s\alpha\\ \mathbf{n}'s'\alpha'}}\widetilde{A}^{ss'}_{\alpha\alpha'}(\mathbf{k}')\tilde{e}^j_{s\alpha}(\mathbf{k})\tilde{e}^{j'}_{s'\alpha'}(\mathbf{k}')\exp(\mathrm{i}(\mathbf{k}+\mathbf{k}')\mathbf{R_n})\\
&= \sum_{s\alpha,s'\alpha'}\widetilde{A}^{ss'}_{\alpha\alpha'}(\mathbf{k}')\tilde{e}^{j'}_{s'\alpha'}(\mathbf{k}')\tilde{e}^{j*}_{s\alpha}(\mathbf{k}')\delta_{-\mathbf{k},\mathbf{k}'}\\
&= M\omega^2_{\mathbf{k}j}\sum_{s\alpha}\tilde{e}^{j'}_{s\alpha}(\mathbf{k}')\tilde{e}^{j*}_{s\alpha}(\mathbf{k}')\delta_{-\mathbf{k},\mathbf{k}'} = M\omega^2_{\mathbf{k}j}\delta_{jj'}\delta_{-\mathbf{k},\mathbf{k}'}\end{aligned}$$

Now it is not difficult to turn to a quantum description. It suffices to replace $\tilde{u}_{\mathbf{n}s\alpha}$ and $\widetilde{P}_{\mathbf{n}s\alpha}$ by the Hermitian operators which satisfy the commutation relations

$$\begin{aligned}&[\tilde{u}_{\mathbf{n}s\alpha}, \tilde{u}_{\mathbf{n}'s'\alpha'}] = 0\\
&[\widetilde{P}_{\mathbf{n}s\alpha}, \widetilde{P}_{\mathbf{n}'s'\alpha'}] = 0\\
&[\widetilde{P}_{\mathbf{n}s\alpha}, \tilde{u}_{\mathbf{n}'s'\alpha'}] = -\mathrm{i}\hbar\delta_{\mathbf{nn}'}\delta_{ss'}\delta_{\alpha\alpha'}\end{aligned} \tag{5.3.14}$$

By virtue of (5.3.12), the amplitudes $u_{\mathbf{k}j}$ and $u^*_{\mathbf{k}j}$ then become operators $u_{\mathbf{k}j}$ and $u^+_{\mathbf{k}j}$ which satisfy the commutation relations

$$[u_{\mathbf{k}j}, u^+_{\mathbf{k}'j'}] = \frac{\hbar}{2M\omega_{\mathbf{k}j}}\,\delta_{\mathbf{k}\mathbf{k}'}\delta_{jj'}$$
$$[u_{\mathbf{k}j}, u_{\mathbf{k}'j'}] = 0$$

which are readily verified.

Instead of $u_{\mathbf{k}j}$ and $u^+_{\mathbf{k}j}$, it is more convenient to introduce the operators

$$a_{\mathbf{k}j} = \sqrt{\frac{2M\omega_{\mathbf{k}j}}{\hbar}}\,u_{\mathbf{k}j}$$
$$a^+_{\mathbf{k}j} = \sqrt{\frac{2M\omega_{\mathbf{k}j}}{\hbar}}\,u^+_{\mathbf{k}j} \qquad (5.3.15)$$

These operators satisfy the commutation relations for the quasiparticle creation and annihilation operators:

$$[\mathbf{a}_{\mathbf{k}j}, \mathbf{a}^+_{\mathbf{k}'j'}] = \delta_{\mathbf{k}\mathbf{k}'}\delta_{jj'}$$
$$[\mathbf{a}_{\mathbf{k}j}, \mathbf{a}_{\mathbf{k}'j'}] = 0 \qquad (5.3.16)$$

Formulas (5.3.8) and (5.3.11) now make it possible to again express the operators $\tilde{u}_{\mathbf{n}s\alpha}$ and $\widetilde{P}_{\mathbf{n}s\alpha}$ via the operators $u_{\mathbf{k}j}$ and $u^+_{\mathbf{k}j}$, and hence, the operators $a_{\mathbf{k}j}$ and $a^+_{\mathbf{k}j}$.

When $\tilde{u}_{\mathbf{n}s\alpha}$ and $\widetilde{P}_{\mathbf{n}s\alpha}$ are replaced with the corresponding operators, Hamilton's function transforms into the Hamiltonian operator, or simply the Hamiltonian. In the resulting Hamiltonian, the operators $\tilde{u}_{\mathbf{n}s\alpha}$ and $\widetilde{P}_{\mathbf{n}s\alpha}$ can be expressed in terms of $u_{\mathbf{k}j}$ and $u^+_{\mathbf{k}j}$. All the calculations correspond to those in the derivation of formula (5.3.13), but if we take into account that $u_{\mathbf{k}j}$ and $u^+_{\mathbf{k}j}$ are not commuting, we find

$$\mathscr{H} = M\sum_{\mathbf{k}j}\omega^2_{\mathbf{k}j}(u_{\mathbf{k}j}u^+_{\mathbf{k}j} + u^+_{\mathbf{k}j}u_{\mathbf{k}j})$$

Using (5.3.15) and (5.3.16), we finally obtain

$$\mathscr{H} = \frac{1}{2}\sum_{\mathbf{k}j}\hbar\omega_{\mathbf{k}j}(\mathbf{a}_{\mathbf{k}j}\mathbf{a}^+_{\mathbf{k}j} + \mathbf{a}^+_{\mathbf{k}j}\mathbf{a}_{\mathbf{k}j}) = \sum_{\mathbf{k}j}\hbar\omega_{\mathbf{k}j}\left(n_{\mathbf{k}j} + \frac{1}{2}\right) \qquad (5.3.17)$$

Here $n_{\mathbf{k}j} = a^+_{\mathbf{k}j}a_{\mathbf{k}j}$ is the operator of the number of phonons for the mode $\mathbf{k}j$. This Hermitian operator has integral and nonnegative eigenvalues. The cause of nonnegativity is the fact that the mean value of the operator $\mathbf{n}_{\mathbf{k}j}$ for any state $|\psi\rangle$ is nonnegative since

$$\langle\psi|\mathbf{n}_{\mathbf{k}j}|\psi\rangle = \langle\psi|a^{+}_{\mathbf{k}j}a_{\mathbf{k}j}|\psi\rangle = \langle\psi'|\psi'\rangle \geq 0$$

where $|\psi\rangle = a_{\mathbf{k}j}|\psi\rangle$.

Commutation relations (5.3.16) yield

$$[n_{\mathbf{k}j}, \mathbf{a}_{\mathbf{k}j}] = [a^{+}_{\mathbf{k}j}a_{\mathbf{k}j}, a_{\mathbf{k}j}] = a^{+}_{\mathbf{k}j}[a_{\mathbf{k}j}, a_{\mathbf{k}j}] + [a^{+}_{\mathbf{k}j}, a_{\mathbf{k}j}]a_{\mathbf{k}j} = -a_{\mathbf{k}j}$$
$$[n_{\mathbf{k}j}, a^{+}_{\mathbf{k}j}] = a^{+}_{\mathbf{k}j}$$

Therefore, if $|m_{\mathbf{k}j}\rangle$ is an eigenvector of the operator $n_{\mathbf{k}j}$ which corresponds to the eigenvalue $|m_{\mathbf{k}j}\rangle$, then

$$n_{\mathbf{k}j}a_{\mathbf{k}j}|m_{\mathbf{k}j}\rangle - a_{\mathbf{k}j}n_{\mathbf{k}j}|m_{\mathbf{k}j}\rangle = -a_{\mathbf{k}j}|m_{\mathbf{k}j}\rangle$$

that is,

$$n_{\mathbf{k}j}a_{\mathbf{k}j}|m_{\mathbf{k}j}\rangle = (m_{\mathbf{k}j} - 1)a_{\mathbf{k}j}|m_{\mathbf{k}j}\rangle$$

This means that $a_{\mathbf{k}j}|m_{\mathbf{k}j}\rangle$ is an eigenvector of the operator $n_{\mathbf{k}j}$ whose eigenvalue is $(m_{\mathbf{k}j} - 1)$. By analogy, we find that $a^{+}_{\mathbf{k}j}|m_{\mathbf{k}j}\rangle$ is an eigenvector of $n_{\mathbf{k}j}$ whose eigenvalue is $(m_{\mathbf{k}j} + 1)$.

It follows from the nonnegativity of the eigenvalues of the operator $n_{\mathbf{k}j}$ and from the arguments above that there exists a vector $|0_{\mathbf{k}j}\rangle \neq 0$, such that $a_{\mathbf{k}j}|0_{\mathbf{k}j}\rangle = 0$ and $n_{\mathbf{k}j}|0_{\mathbf{k}j}\rangle = 0$, while the vectors $(a^{+}_{\mathbf{k}j})^m|0_{\mathbf{k}j}\rangle$ are eigenvectors of $n_{\mathbf{k}j}$ with nonnegative eigenvalues. Consequently, $m_{\mathbf{k}j}$ is a nonnegative integer ($m_{\mathbf{k}j} = 0, 1, 2, \ldots$).

The energy of lattice vibrations can assume the values

$$E = \sum_{\mathbf{k}j} \hbar\omega_{\mathbf{k}j}\left(m_{\mathbf{k}j} + \frac{1}{2}\right)$$

and can vary only stepwise, by quanta $\hbar\omega_{\mathbf{k}j}$.

These energy quanta correspond to phonons, and $m_{\mathbf{k}j}$ is the number of phonons in the mode $\mathbf{k}j$. The operators $a^{+}_{\mathbf{k}j}$ and $a_{\mathbf{k}j}$ are the operators of creation and annihilation of phonons; this is clear from their effect on states with a specific value of $n_{\mathbf{k}j}$. From commutation relations, we can find the norm of the vectors $(a^{+}_{\mathbf{k}j})^m|0_{\mathbf{k}j}\rangle = 0$ and normalized vectors $|m_{\mathbf{k}j}\rangle$ with a specific number of phonons:

$$|m_{\mathbf{k}j}\rangle = \frac{1}{\sqrt{m_{\mathbf{k}j}!}}(a^{+}_{\mathbf{k}j})^{m_{\mathbf{k}j}}|0_{\mathbf{k}j}\rangle$$

Hence,

$$a_{\mathbf{k}j}|m_{\mathbf{k}j}\rangle = \frac{\sqrt{(m_{\mathbf{k}j}+1)!}}{\sqrt{m_{\mathbf{k}j}!}}|m_{\mathbf{k}j}+1\rangle = \sqrt{m_{\mathbf{k}j}+1}\,|m_{\mathbf{k}j}+1\rangle \qquad (5.3.18)$$

Using the commutation properties, we obtain

$$a_{\mathbf{k}j}|m_{\mathbf{k}j}\rangle = \sqrt{m_{\mathbf{k}j}}\,|m_{\mathbf{k}j} - 1\rangle \tag{5.3.19}$$

Finally, we rewrite (5.3.9), introducing the operators $a_{\mathbf{k}j}$:

$$u_{\mathbf{n}s\alpha} = \frac{1}{\sqrt{2}} \sum_{\mathbf{k}j} \sqrt{\frac{\hbar}{M\omega_{\mathbf{k}j}}}\,(\mathbf{e}^{j}_{\mathbf{n}s\alpha}(\mathbf{k})a_{\mathbf{k}j} + \mathbf{e}^{j*}_{\mathbf{n}s\alpha}(\mathbf{k})a^{+}_{\mathbf{k}j}) \tag{5.3.20}$$

The oscillation quanta, that is, phonons, are very similar to photons, the quanta of electromagnetic field. The main difference is that the momentum of a phonon is actually zero. If we sum up the momenta of individual atoms over the entire lattice, then the sum $\sum P_{\mathbf{n}s\alpha}$ vanishes owing to the exponential factors in $P_{\mathbf{n}s\alpha}$. The momentum is always connected with mass transfer (in relativistic physics, with energy transfer), but since oscillations do not lead to matter transfer, the total momentum is zero.

However, a phonon in a mode with wave vector $\mathbf{k}$ is put in correspondence with a quasimomentum $\hbar\mathbf{k}$. It is found that quasimomentum is conserved in various electron–phonon, photon–phonon, and phonon-phonon interactions. This will be shown later in this chapter; for the moment, it should be noted that the true momentum of an electron is nonzero, so that the law of momentum conservation implies that this momentum changes when a phonon is emitted or absorbed. Where is the momentum transferred then, if phonons carry zero momentum? Each emission or absorption of a phonon results in momentum transfer to the lattice as a whole.

The law of dispersion (i.e., the dependence of $\omega_{\mathbf{k}j}$ on $\mathbf{k}$) is transformed in quantum terms into a dependence of the phonon energy $E = \hbar\omega_{\mathbf{k}j}$ on quasimomentum $\mathbf{p} = \hbar\mathbf{k}$. The velocity of a phonon is $\mathbf{v} = \partial E/\partial\mathbf{p} = \partial\omega_{\mathbf{k}j}/\partial\mathbf{k}$, which coincides with the classic result operating with the group velocity. Let us recall that E is a periodic function of quasimomentum $\mathbf{p}$ and the period is dictated by the reciprocal lattice of the crystal, so that this dependence need not be considered beyond the first Brillouin zone.

5.4. Anharmonism. Phonon Lifetime

As we have demonstrated in the previous section, lattice oscillations can be treated in quantum terms as an ensemble of quasiparticles, that is, phonons, which in the first approximation do not interact with one another (the Hamiltonian splits into a sum of Hamiltonians for distinct modes; for each mode, the energy is proportional to the number of phonons, with the accuracy of zero energy, $\hbar\omega_{\mathbf{k}}/2$). The behavior of an ensemble of phonons is that of the ideal phonon gas.

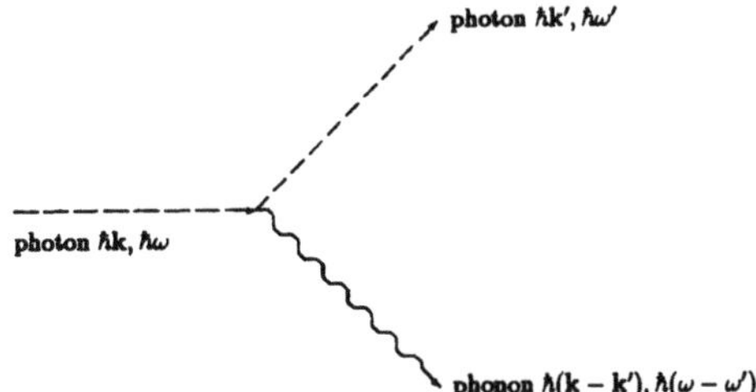

Figure 5.7. Diagram of photon–phonon scattering.

The interaction of phonons with quasiparticles of different nature results in scattering of quasiparticles, absorption and emission of phonons, with energy and quasimomentum conserved. Among such processes we find the scattering of optical radiation by optical phonons (Raman scattering of light in crystals) and by acoustic phonons (Mandelshtam–Brillouin scattering). This photon–phonon scattering is shown by the diagram of Figure 5.7. A reverse process in which a phonon is absorbed, as well as induced processes, are also possible.

Experiments on light scattering, based on energy and quasimomentum conservation, make it possible to derive the phonon dispersion law from measurements of energy and momentum of scattered photons. Unfortunately, since $\hbar\mathbf{k}$ of a photon is very low in the optical range ($k \ll 2\pi/a_0$), the phonon dispersion law can be found only for $\mathbf{k}$ much smaller than the size of the Brillouin zone. In order to extract information for higher $\mathbf{k}$, one has to switch to x-ray-range photons.

The phonon dispersion law can be found by measuring the change in frequency (inelastic scattering of x-rays) against the scattering angle. Such measurements are indeed carried out, but the main difficulty here is to determine small frequency shifts. A more convenient method is an analysis of neutron scattering, when neutron energy losses in the interactions with lattice atoms are measured as a function of the scattering angle.

The aspect that has to be taken into account here is that, as we have mentioned already for photon–photon interactions in Chapter 1, the quasimomentum conservation law is

$$\hbar\mathbf{q} + \hbar\mathbf{q}_l = \hbar\mathbf{q}' \pm \hbar\mathbf{k}$$

where $\hbar\mathbf{q}$ and $\hbar\mathbf{q}'$ are quasimomenta of the photon or neutron, $\hbar\mathbf{k}$ is the phonon quasimomentum and $\mathbf{g}_l$ is an arbitrary reciprocal lattice vector. If $\mathbf{g}_l \neq 0$, one speaks of umklapp processes.

In classical physics terms, the scattering of a photon by a phonon is the diffraction on a moving diffraction grating.

The scattering of phonons by crystal defects and its boundaries determines the heat conduction by the crystal (at least at low temperatures).

Indeed, a phonon gas can be treated in the framework of the kinetic theory of gases, so that its thermal conductivity is

$$\eta = \tfrac{1}{3}Cvl$$

where C is the specific heat of the phonon gas, v is the mean propagation velocity and l is the free path length. At low temperatures, the Debye theory predicts that the heat capacitance C of the phonon gas should increase proportionally to T^3, so that the thermal conductivity η of a solid also increases. Then η begins to decrease in response to the free path length l decreasing at high temperatures. The point is that the phonon free path length at high temperatures is determined by phonon decay processes and by their scattering due to the anharmonism of the lattice.

The anharmonism is the effect of higher, nonquadratic terms in the expansion of potential energy in powers of atomic displacements. If these terms are taken into account in (5.1.1), we obtain instead of (5.2.1)

$$\begin{aligned}\mathscr{H} &= \sum_{\mathbf{n}s\alpha} \frac{P^2_{\mathbf{n}s\alpha}}{2M_s} + \frac{1}{2} \sum_{\substack{\mathbf{n}s\alpha \\ \mathbf{n}'s'\alpha'}} A^{ss'}_{\alpha\alpha'\,\mathbf{n}-\mathbf{n}'} u_{\mathbf{n}s\alpha} u_{\mathbf{n}'s'\alpha'} \\ &+ \frac{1}{3!} \sum_{\substack{\mathbf{n}s\alpha \\ \mathbf{n}'s'\alpha' \\ \mathbf{n}''s''\alpha''}} B^{ss's''}_{\alpha\alpha'\alpha''\,\mathbf{n}-\mathbf{n}',\mathbf{n}'-\mathbf{n}''} u_{\mathbf{n}s\alpha} u_{\mathbf{n}'s'\alpha'} u_{\mathbf{n}''s''\alpha''} + \ldots \end{aligned} \tag{5.4.1}$$

Substituting into (5.4.1) expression (5.3.20) for the operators $u_{\mathbf{n}s\alpha}$ in terms of the phonon creation and annihilation operators, we find

$$\mathscr{H} = \mathscr{H}_0 + \mathscr{H}'$$

where $\mathscr{H}_0$ is the lattice Hamiltonian in the harmonic approximation (5.3.17) and $\mathscr{H}'$ is a perturbation:

$$\begin{aligned}\mathscr{H}' &= \frac{1}{6}(2N)^{-3/2} \sum B^{ss's''}_{\alpha\alpha'\alpha''\,\mathbf{n}-\mathbf{n}',\mathbf{n}'-\mathbf{n}''} \sum_{\substack{\mathbf{k}\mathbf{k}'\mathbf{k}'' \\ jj'j''}} \sqrt{\frac{\hbar^3}{M^3 \omega_{\mathbf{k}j} \omega_{\mathbf{k}'j'} \omega_{\mathbf{k}''j''}}} \\ &\times \mathbf{e}^{j}_{s\alpha}(\mathbf{k}) \mathbf{e}^{j'}_{s'\alpha'}(\mathbf{k}') \mathbf{e}^{j''}_{s''\alpha''}(\mathbf{k}'') \exp(\mathrm{i}\mathbf{k}\mathbf{R}_{\mathbf{n}} + \mathrm{i}\mathbf{k}'\mathbf{R}_{\mathbf{n}'} + \mathrm{i}\mathbf{k}''\mathbf{R}_{\mathbf{n}''}) \\ &\times a_{\mathbf{k}j} a_{\mathbf{k}'j'} a_{\mathbf{k}''j''} + 7 \text{ terms}\end{aligned}$$

The seven terms not elaborated above have the same structure as the term with the product of operators $a_{\mathbf{k}j} a_{\mathbf{k}'j'} a_{\mathbf{k}''j''}$, but one or several annihilation operators $a_{\mathbf{k}j}$ are replaced with phonon creation operators $a^+_{\mathbf{k}j}$.

Introducing $\mathbf{m}' = \mathbf{n} - \mathbf{n}'$ and $\mathbf{m}'' = \mathbf{n} - \mathbf{n}''$, and, correspondingly, $\mathbf{R}_{\mathbf{n}'} = \mathbf{R}_{\mathbf{n}} - \mathbf{R}_{\mathbf{m}'}$, $\mathbf{R}_{\mathbf{n}''} = \mathbf{R}_{\mathbf{n}} - \mathbf{R}_{\mathbf{m}''}$, and using

$$\sum_{\mathbf{n}} \exp[\mathrm{i}(\mathbf{k} - \mathbf{k}' - \mathbf{k}'')\mathbf{R}_{\mathbf{n}}] = N\delta_{\mathbf{k},\mathbf{k}'+\mathbf{k}''+\mathbf{g}_1} \tag{5.4.2}$$

we obtain

$$\mathscr{H}' = \frac{1}{6\sqrt{8N}} \sum_{\substack{\mathbf{k}'\mathbf{k}'' \\ jj'j''}} \sqrt{\frac{\hbar^3}{M\omega_{\mathbf{k}j}\omega_{\mathbf{k}'j'}}} B^{jj'j''*}(\mathbf{k}', \mathbf{k}'')$$
$$\times \left(a_{\mathbf{k}j}a_{\mathbf{k}'j'}a_{\mathbf{k}''j''} + a_{\mathbf{k}j}\overset{+}{a}_{\mathbf{k}'j'}\overset{+}{a}_{\mathbf{k}''j''}\right) + 6 \text{ terms} \tag{5.4.3}$$

The Fourier transforms B have been introduced in (5.4.3):

$$B^{jj'j''*}(\mathbf{k}', \mathbf{k}'') = \sum_{\substack{\mathbf{m}'\mathbf{m}'' \\ ss's'' \\ \alpha\alpha'\alpha''}} B^{\substack{ss's'' \\ \alpha\alpha'\alpha''}}_{\mathbf{m}'\mathbf{m}''} \exp(-i\mathbf{k}'\mathbf{R}_{\mathbf{m}'} - i\mathbf{k}''\mathbf{R}_{\mathbf{m}''})$$
$$\times \mathbf{e}^{j*}_{s\alpha}(\mathbf{k})\mathbf{e}^{j'}_{s'\alpha'}(\mathbf{k}')\mathbf{e}^{j''}_{s''\alpha''}(\mathbf{k}'') \tag{5.4.4}$$

By virtue of (5.4.2), we need to set $\mathbf{k} = \mathbf{k}' + \mathbf{k}''$ in (5.4.3) and (5.4.4), since here umklapp processes ($\mathbf{g}_l \neq 0$) are not taken into account. Umklapp processes are sometimes very important but can be ignored in the case of processes discussed below (no qualitative difference exists between them and normal processes with $\mathbf{g}_l = 0$).

The first term in (5.4.3) corresponds to the process of annihilation of three phonons. The conservation laws that must be satisfied are

$$\mathbf{k} = \mathbf{k}' + \mathbf{k}''$$
$$\hbar\omega_{\mathbf{k}j} + \hbar\omega_{\mathbf{k}'j'} + \hbar\omega_{\mathbf{k}''j''} = 0$$

The energy conservation law forbids this process because, as we have already remarked, $\omega_{\mathbf{k}j}$, $\omega_{\mathbf{k}'j'}$, $\omega_{\mathbf{k}''j''} \geq 0$.

The second term in (5.4.3), which contains $a_{\mathbf{k}j}\overset{+}{a}_{\mathbf{k}'j'}\overset{+}{a}_{\mathbf{k}''j''}$, results in a conservation law of the type

$$\hbar\omega_{\mathbf{k}j} = \hbar\omega_{\mathbf{k}'j'} + \hbar\omega_{\mathbf{k}''j''}$$

It corresponds to the process of annihilation of the phonon $\mathbf{k}j$ and creation of a pair of phonons $\mathbf{k}'j'$ and $\mathbf{k}''j''$. Allowed simultaneously are the processes corresponding to the other terms of (5.4.3), for example, those with a product of operators $\overset{+}{a}_{\mathbf{k}j}a_{\mathbf{k}'j'}a_{\mathbf{k}''j''}$, which represents the process of creation of a $\mathbf{k}j$ phonon and annihilation (merging) of $\mathbf{k}'j'$ and $\mathbf{k}''j''$ phonons.

The process of transformation (decay) of a $\mathbf{k}j$ phonon into a pair of $\mathbf{k}'j'$ and $\mathbf{k}''j''$ phonons is shown by the diagram of Figure 5.8.

Similarly, the fourth order term in (5.4.1) leads to expressions that contain products of operators of the type $a_{\mathbf{k}j}a_{\mathbf{k}'j'}\overset{+}{a}_{\mathbf{k}''j''}\overset{+}{a}_{\mathbf{k}'''j'''}$, which correspond to scattering of phonons by phonons. This process is shown in

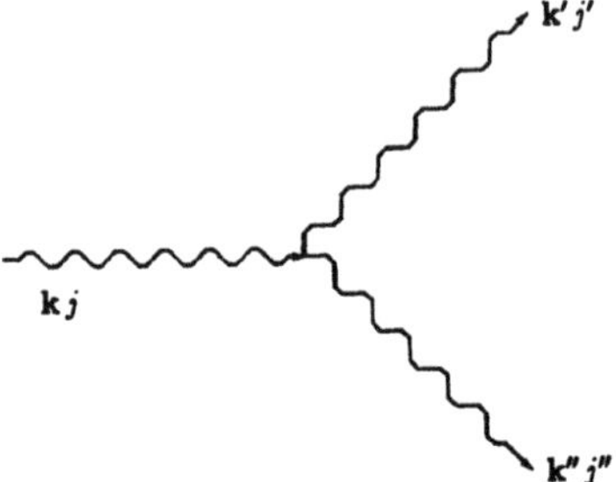

Figure 5.8. Diagram of decay of a phonon with quasimomentum $\hbar\mathbf{k}$ to a pair of phonons with quasimomenta $\hbar\mathbf{k}'$ and $\hbar\mathbf{k}''$.

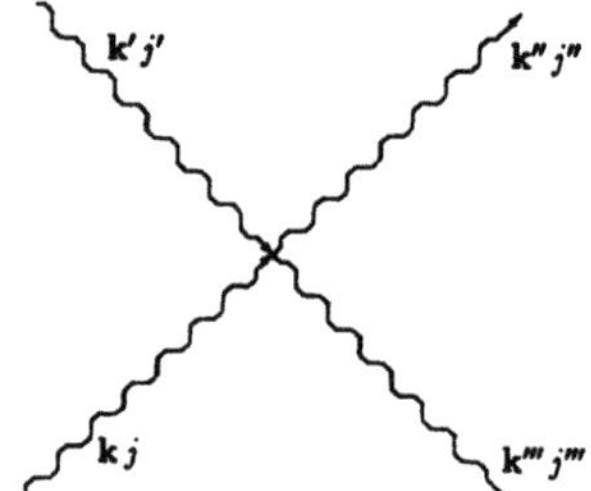

Figure 5.9. Diagram of phonon–phonon scattering.

Figure 5.9. At the same time, the fourth-order term in (5.4.1) describes processes like phonon decay into three phonons etc.

Let us evaluate the phonon lifetime, using perturbation operator $\mathscr{H}'$ (5.4.3). We need to know the lifetime not only to evaluate the thermal conductivity of the crystal but also to estimate the absorption linewidth in the interaction of infrared radiation with lattice vibrations, and also the linewidth of Raman scattering of light in crystals. We will calculate the lifetime of a long-wavelength optical phonon with $\mathbf{k} \simeq 0$, responsible for absorption and scattering of light.

The quantity inverse to the lifetime is the probability of decay in unit time. According to the quantum-mechanical perturbation theory, this probability is

$$\begin{aligned}\nu_j(\mathbf{k}) &= \frac{1}{\tau_j(\mathbf{k})} = \sum_{\mathbf{k}'j'j''} w_{\mathbf{k}'j',\mathbf{k}''j'',\mathbf{k}j} \\ &= \sum_{\mathbf{k}'j'j''} \frac{2\pi}{\hbar} |\mathscr{H}'_{\mathbf{k}'j',\mathbf{k}''j'',\mathbf{k}j}|^2 \delta(\hbar\omega_{\mathbf{k}j} - \hbar\omega_{\mathbf{k}'j'} - \hbar\omega_{\mathbf{k}''j''})\end{aligned}$$

Here $w_{\mathbf{k}'j',\mathbf{k}''j'',\mathbf{k}j}$ is the probability of decay of a $\mathbf{k}j$ phonon to a pair of $\mathbf{k}'j'$ and $\mathbf{k}''j''$ phonons, and $\mathscr{H}'_{\mathbf{k}'j',\mathbf{k}''j'',\mathbf{k}j}$ is the corresponding matrix element of the perturbation operator. The decay probability is found by summing up w over all possible final states.

The matrix element of the operator $\mathcal{H}'$, written in the form (5.4.3), is easily calculated since the matrix elements of the phonon creation and annihilation operators are known and described by formulas (5.3.18) and (5.3.19).

For the initial state in which the number of phonons in the state $\mathbf{k}j$ is $m_{\mathbf{k}j}$ and the numbers of phonons in the states $\mathbf{k}'j'$ and $\mathbf{k}''j''$ are $m_{\mathbf{k}'j'}$ and $m_{\mathbf{k}''j''}$, respectively, we have

$$(a_{\mathbf{k}j}a^+_{\mathbf{k}'j'}a^+_{\mathbf{k}''j''})_{\mathbf{k}'j',\mathbf{k}''j'',\mathbf{k}j} = \sqrt{m_{\mathbf{k}j}(m_{\mathbf{k}'j'}+1)(m_{\mathbf{k}''j''}+1)}$$

Therefore,

$$\nu_j(\mathbf{k}) = \frac{2\pi}{\hbar}\frac{1}{36\cdot 8N}\sum_{\mathbf{k}j'j''}\sqrt{\frac{\hbar^3}{M^3\omega_{\mathbf{k}j}\omega_{\mathbf{k}'j'}\omega_{\mathbf{k}''j''}}}|B^{jj'j''}(\mathbf{k}',\mathbf{k}'')|^2$$
$$\times\, m_{\mathbf{k}j}(m_{\mathbf{k}'j'}+1)(m_{\mathbf{k}''j''}+1)\delta(\hbar\omega_{\mathbf{k}j}-\hbar\omega_{\mathbf{k}'j'}-\hbar\omega_{\mathbf{k}''j''}) \qquad (5.4.5)$$

Since we calculate the decay probability per phonon in the $\mathbf{k}j$ mode in the initial state, (5.4.5) must be divided by $m_{\mathbf{k}j}$, that is, we need to set $m_{\mathbf{k}j} = 1$. If there are no phonons in the initial state in the modes $\mathbf{k}'j'$, $\mathbf{k}''j''$, then $m_{\mathbf{k}'j'} = m_{\mathbf{k}''j''} = 0$ and the factors $(1+m_{\mathbf{k}'j'})$ and $(1+m_{\mathbf{k}''j''})$ in (5.4.5) turn to unities.

Let us evaluate the probability of decay of an optical phonon with $\mathbf{k} \simeq 0$. In this case $\mathbf{k}'+\mathbf{k}'' = 0$, that is, $\mathbf{k}' = -\mathbf{k}''$. The frequency of phonons with $\mathbf{k}'$ and $\mathbf{k}''$ will both be equal to one-half of that of the $\mathbf{k}$ phonon. If the initial phonon is optical, this condition is usually satisfied by acoustic phonons; we will thus evaluate the probability of decay of an optical phonon into two acoustic phonons.

The summation over $\mathbf{k}'$ in (5.4.5) can be replaced with integration,

$$\sum_{\mathbf{k}'} \to \int d^3k'\,\frac{N\Omega}{(2\pi)^3} \qquad (5.4.6)$$

where $N\Omega = V$ is the volume of a crystal consisting of N unit cells, each of volume Ω. Rule (5.4.6) follows from the fact that, according to Section 5.3, the wave vector $\mathbf{k}'$ assumes the values

$$\mathbf{k}' = \frac{\mathbf{g}_1}{N_1}m_1' + \frac{\mathbf{g}_2}{N_2}m_2' + \frac{\mathbf{g}_3}{N_3}m_3'$$

where m_1', m_2', m_3' are integers, so that the mode density in the $\mathbf{k}'$ space is $(2\pi)^3/N\Omega$.

For the sake of evaluation, we assume that the first Brillouin zone is a sphere of radius $k_{\max}$ in the $\mathbf{k}$ space (in the reciprocal lattice space); we choose the dispersion law for acoustic phonons in the form $\omega_{\mathbf{k}'} = s|\mathbf{k}'|$, where s is the velocity of sound. This dispersion law holds strictly for small $\mathbf{k}'$ but can also be used at high $\mathbf{k}'$ for order-of-magnitude estimates. Instead of (5.4.5) we then obtain

$$\nu_j(0) = \frac{2\pi}{\hbar} \frac{4\pi\Omega}{36 \cdot 8(2\pi)^3} \frac{4\hbar^3}{M^3\omega_0^3} \int_0^{k_{\max}} k'^2 |B(\mathbf{k}', \mathbf{k}'')|^2 \times \delta(\hbar\omega_0 - 2\hbar s k')\, dk'$$

where ω_0 is the optical phonon frequency at $\mathbf{k} = 0$. We evaluate the coefficient $|B(\mathbf{k}', \mathbf{k}'')|$ as

$$|B(\mathbf{k}', \mathbf{k}'')| \sim \left| \frac{\partial^3 U}{\partial u \partial u \partial u} \right| \sim \frac{I_0}{a_0^3} \qquad I_0 \sim \frac{\hbar^2}{m a_0^2}$$

Omitting now multipliers of the order of unity, we obtain

$$\nu_j(0) \sim \frac{\hbar I_0^2}{M^3 \omega_0 a_0^3 s^3}$$

By the order of magnitude,

$$s \sim \frac{\omega_0}{k_{\max}} \sim \frac{\omega_0 a_0}{2\pi}$$

and in accordance with the estimates obtained above, the optical phonon frequency is

$$\omega_0 \sim \frac{1}{\hbar} \sqrt{\frac{m}{M}} I_0$$

Using the above estimates, we arrive at

$$\nu_j \sim \frac{1}{\hbar} \frac{m}{M} I_0$$

that is,

$$\hbar \nu_j \sim \frac{m}{M} I_0 \sim \sqrt{\frac{m}{M}} \hbar \omega_0$$

We have thus shown that the optical phonon linewidth ν is of the order of $\sqrt{m/M} \sim 0.01$ of the optical phonon frequency.

Laser spectroscopy experiments confirm this estimate: the optical phonon lifetime is of the order of 10^{-11} s, and the vibration frequency for the optical phonon is of the order of $\omega_0 \sim 10^{-13}$ s^{-1}.

This estimate also holds for short-wavelength longitudinal acoustic phonons. Longitudinal acoustic phonons in an isotropic or almost isotropic crystal are phonons of that acoustic branch which for small $\mathbf{k}$ is described by the dispersion law of longitudinal (or almost longitudinal) elastic waves in the crystal. The derivative $d\omega/dk$ for this wave at $k = 0$ equals the velocity of longitudinal elastic waves. Short-wavelength longitudinal acoustic phonons are phonons with k of the order of $k_{\max} \sim \pi/a_0$. If the anisotropy of the elastic properties of a crystal does not allow us to treat it as nearly isotropic, then the acoustic wave with the maximum propagation velocity is regarded as the longitudinal acoustic wave.

The decay of long-wavelength acoustic phonons is very unlikely since the number of modes into which they can decay is very small and since their lifetime is dictated by scattering and mixing with other phonons. At sufficiently low temperatures, however, the probability of such processes is low and consequently the lifetime of acoustic phonons is large.

Furthermore, even short-wavelength transverse phonons also have large lifetimes since in view of energy and quasimomentum conservation, they have nowhere to decay: their velocity is lower than that of longitudinal acoustic phonons. As follows from general formula (5.4.5), the phonon lifetime is a function of temperature, since at $\mathrm{k}T \gg \hbar\omega_{\mathbf{k}'j'}$ the number

$$\bar{m}_{\mathbf{k}'j'} = \frac{1}{\exp(\hbar\omega_{\mathbf{k}'j'}/\mathrm{k}T) - 1} \simeq \frac{\mathrm{k}T}{\hbar\omega_{\mathbf{k}'j'}}$$

depends on temperature.

Therefore, $\nu_j \sim (1 + m_{\mathbf{k}'j'})^2 \sim T^2$, that is, the linewidth increases at high temperatures as T^2. If fourth-order anharmonism plays the main role, then $\nu \sim T^3$.

Anharmonism results not only in a finite linewidth but also in phonon frequency shift. This shift can be estimated using (second-order) perturbation theory:

$$\Delta\hbar\omega_{\mathbf{k}j} = \sum \frac{|\mathcal{H}'_{\mathbf{k}'j',\mathbf{k}''j'',\mathbf{k}j}|^2}{\hbar\omega_{\mathbf{k}j} - \hbar\omega_{\mathbf{k}'j'} - \hbar\omega_{\mathbf{k}''j''}} \tag{5.4.7}$$

We can estimate this frequency correction using the estimates of the matrix elements of $\mathcal{H}'$ obtained earlier. This gives

$$\Delta\hbar\omega_{\mathbf{k}j} \sim \sqrt{\frac{m}{M}}\,\hbar\omega_{\mathbf{k}j}\left(\frac{\mathrm{k}T}{\hbar\omega_{\mathbf{k}j}}\right)^2$$

All calculations are made by analogy to the above calculations of linewidth since the imaginary part of (5.4.7) is precisely the expression for linewidth ν_j, provided the correct contour around the poles is chosen. The estimate of $\Delta\omega_{\mathbf{k}'j'}$ evaluated its absolute value. Its sign may vary; experimentally, frequencies may decrease or increase with temperature. Together with the temperature dependence of the average number $m_{\mathbf{k}'j'}$ of photons in the mode, the anharmonism results in a dependence of phonon frequencies on temperature, although the unperturbed phonon frequencies of the original Hamiltonian, fixed by the elastic constants matrix A and atomic masses, are temperature-independent.

If $\omega_{\mathbf{k}j}$ is small, this frequency may drop to zero at some temperature. As temperature changes further, frequency squared becomes negative, $\omega^2_{\mathbf{k}j} < 0$, which indicates instability (that is, the energy in this mode has a local maximum, not a local minimum) at which the lattice starts to reconstruct. The displacement $e^j_{s\alpha}$ grows spontaneously until a new equilibrium position is reached.

This structural transformation occurs, for example, in a segnetoelectric phase transition. If the original mode is dipole-active, a spontaneous dipole moment occurs. Such modes, whose frequency tends to zero as temperature is changed, are known as "soft modes". They determine the segnetoelectric phase transition and some other, so-called structural phase transitions.

5.5. Interaction of Lattice Vibrations with Electromagnetic Field

Let us now consider the interaction of lattice vibrations with electromagnetic fields. We will carry out the analysis first in the framework of the pointlike ions model, that is, we assume ions to be pointlike charges $Z_s e$, located at the equilibrium positions $\mathbf{R}_{\mathbf{n}s}$. Despite the obvious limitations of this model, it gives correct representation of the qualitative features of the electromagnetic response of ionic crystals in the infrared spectral range.

Taking into account the electric field $\mathcal{E}(\mathbf{r}, t)$, the equations of motion of the ionic lattice are

$$M_s \ddot{u}_{\mathbf{n}s\alpha} + \sum_{\mathbf{n}'s'\alpha'} A'^{ss'\,\alpha\alpha'}_{\mathbf{n}-\mathbf{n}'} u_{\mathbf{n}'s'\alpha'} = Z_s e \mathcal{E}^{(0)}_{\alpha}(\mathbf{R}_{\mathbf{n}s}, t) \tag{5.5.1}$$

The force constants matrix A' is generally not identical to the matrix A which determines the free vibrations of the same lattice. The reason for this difference is closely connected with the question of which field $\mathcal{E}$ enters the righthand sides of equations (5.5.1): the field of external charges, or the true field at the locations of ions, or the field averaged over a unit cell. The point is that the true field $\mathcal{E}(\mathbf{r})$ differs from the field of external charges, $\mathcal{E}^{(e)}(\mathbf{r})$ by the contribution of the intrinsic fields produced when ions are displaced (the fields of ions in their equilibrium positions are static and have nothing to do with the response to time-dependent fields).

The displacement of an sth ion by a vector $u_{\mathbf{n}s\alpha}$ from the equilibrium position is equivalent to adding a dipole $Z_s e u_{\mathbf{n}s\alpha}$ to the initial static charge $Z_s e$. The corresponding contribution to the field, summed up over all ions, has the form

$$\delta\mathcal{E}_\alpha(\mathbf{r},t) = -\sum_{\mathbf{n}s\beta} \frac{Z_s e}{|\mathbf{r}-\mathbf{R}_{\mathbf{n}s}|^3} \left\{ \delta_{\alpha\beta} - \frac{3(r_\alpha - R_{\mathbf{n}s\alpha})(r_\beta - R_{\mathbf{n}s\beta})}{|\mathbf{r}-\mathbf{R}_{\mathbf{n}s}|^2} \right\} u_{\mathbf{n}s\alpha} \tag{5.5.2}$$

The field $\delta\mathcal{E}(\mathbf{r},t)$ is a linear function of displacements $u_{\mathbf{n}s\alpha}$ and always accompanies lattice vibrations regardless of whether these are free or induced vibrations. When this field is substituted into the equations of motion of the type of (5.5.1), the righthand sides take the form $\sum_{\mathbf{n}'s'\alpha'} C^{ss'}_{\alpha\alpha'\,\mathbf{n}-\mathbf{n}'} u_{\mathbf{n}'s'\alpha'}$, where

$$\begin{aligned} C^{ss'}_{\alpha\alpha'\,\mathbf{n}-\mathbf{n}'} &= -\frac{Z_s Z_{s'} e^2}{|\mathbf{R}_{\mathbf{n}s} - \mathbf{R}_{\mathbf{n}'s'}|^3} \\ &\quad \times \left\{ \delta_{\alpha\alpha'} - \frac{3(\mathbf{R}_{\mathbf{n}s} - \mathbf{R}_{\mathbf{n}'s'})_\alpha (\mathbf{R}_{\mathbf{n}s} - \mathbf{R}_{\mathbf{n}'s'})_{\alpha'}}{|\mathbf{R}_{\mathbf{n}s} - \mathbf{R}_{\mathbf{n}'s'}|^2} \right\} \\ &= -Z_s Z_{s'} e^2 \frac{\partial^2}{\partial R_{\mathbf{n}s\alpha} \partial R_{\mathbf{n}'s'\alpha'}} \frac{1}{|\mathbf{R}_{\mathbf{n}s} - \mathbf{R}_{\mathbf{n}'s'}|} \end{aligned}$$

The structure of this expression is identical to that of the term which describes quasielastic forces in the lefthand side of (5.5.1). It is easy to show that $C^{ss'}_{\alpha\alpha'\,\mathbf{n}-\mathbf{n}'}$ is the contribution of the dipole–dipole interaction, which arises when ions are displaced from their equilibrium positions, to the complete force constants matrix $A^{ss'}_{\alpha\alpha'\,\mathbf{n}-\mathbf{n}'}$. Other contributions to this matrix are reflecting forces arising from exchange-type interactions, van der Waals forces, and so forth.

The dipole–dipole interaction in (5.5.1)-type equations of motion can thus be taken into account in two different ways: either as a contribution to quasielastic forces, so that the complete matrix A of force constants enters the lefthand side of the equation while the righthand side has only the field of external charges, or as a contribution to the field, so that the righthand side contains the total (true) field and the lefthand side contains the matrix $A'' = A + C$, which does not include the contribution of the dipole–dipole interaction to the quasielastic forces. There is also a third approach that we are going to use in this section. Namely, we split the field $\delta\mathcal{E}(\mathbf{r},t)$ into two parts, one with slow spatial variation (averaged over a unit cell), the other being a small-scale part which describes the structure of the field in a unit cell, after which we take the former into account in the righthand side of (5.5.1) and the latter in the quasielastic forces in the lefthand side of (5.5.1).

For a formal implementation of this program, we expand (5.5.2) in a Fourier series in accordance with the general theory of electromagnetic response in crystals:

$$\delta\mathcal{E}_\alpha(\mathbf{r},t) = \sum_{\mathbf{l}} \frac{1}{\sqrt{N}} \sum_{\mathbf{q}s\alpha'} \frac{4\pi Z_s e}{\Omega} \frac{(\mathbf{q}+\mathbf{g_l})_\alpha(\mathbf{q}+\mathbf{g_l})_{\alpha'}}{|\mathbf{q}+\mathbf{g_l}|^2} \times \exp[\mathrm{i}(\mathbf{q}+\mathbf{g_l})(\mathbf{r}-\boldsymbol{\rho}_s)] u_{\mathbf{q}s\alpha'}(t) \tag{5.5.3}$$

Here $\mathbf{g_l}$ are the reciprocal lattice vectors, Ω is the volume of the unit cell of the lattice, and $\boldsymbol{\rho}_s$ is the relative position of the sth ion in the unit cell: $\mathbf{R}_{\mathbf{n}s} = \mathbf{R_n} + \boldsymbol{\rho}_s$. In going from (5.5.2) to (5.5.3), the displacements are written in the form

$$u_{\mathbf{n}s\alpha} = \frac{1}{\sqrt{N}} \sum_{\mathbf{q}} u_{\mathbf{q}s\alpha}(t) \exp(\mathrm{i}\mathbf{q}\mathbf{R_n})$$

where the values of the vector $\mathbf{q}$ are limited to the first Brillouin zone and the field of the dipole $\mathbf{d}$ in the Fourier representation is

$$-\frac{\partial^2}{\partial r_\alpha \partial r_\beta} \frac{d_\beta}{|\mathbf{r}|} = \frac{1}{V} \sum_{\mathbf{k}} \frac{4\pi k_\alpha k_\beta}{k^2} \exp(\mathrm{i}\mathbf{k}\mathbf{r}) d_\beta$$

where $V = N\Omega$ is the total volume. The range of variation of $\mathbf{k}$ is not restricted to the first Brillouin zone. Correspondingly, (5.5.3) contains $\mathbf{k} = \mathbf{q} + \mathbf{g_l}$.

The splitting of the field, mentioned above, into a slowly varying part and a small-scale part now simply means that we need to single out in the sum over $\mathbf{l}$ in (5.5.3) the main spatial harmonic ($\mathbf{l} = 0$) $\delta\mathcal{E}^{(0)}$. All other terms ($\mathbf{l} \neq 0$) are included into quasielastic forces in the lefthand side of (5.5.1). The force constants matrix A' which then arises differs from the total matrix A in the equation of motion for free vibrations, namely, there is no contribution of the long-range part of the dipole–dipole interaction. In the Fourier representation,

$$A'^{ss'}_{\mathbf{q}\,\alpha\alpha'} = A^{ss'}_{\mathbf{q}\,\alpha\alpha'} - \frac{4\pi Z_s Z_{s'} e^2}{\Omega} \frac{q_\alpha q_{\alpha'}}{q^2} \exp[\mathrm{i}\mathbf{q}(\boldsymbol{\rho}_s - \boldsymbol{\rho}_{s'})]$$

The slowly varying part $\delta\mathcal{E}^{(0)}$ merges in the righthand side of (5.5.1) with the external field $\mathcal{E}^{(e)}$ and gives the slow (averaged) part of the total field, $\mathcal{E}^{(0)} = \mathcal{E}^{(e)} + \delta\mathcal{E}^{(0)}$. Despite apparent artificiality of this procedure, it is in complete accord with the approach in standard crystal optics. From equations (5.5.1), we find not the total susceptibility matrix $\chi^{\mathbf{ll}'}_{\alpha\beta}(\mathbf{q},\omega)$ but a response to the macroscopic field $\overline{\chi}_{\alpha\beta}(\mathbf{q},\omega)$ averaged over the unit cell. We will consider in Section 5.7 the other two possible approaches under which the righthand sides of the equations of motion include the total microscopic field $\mathcal{E}(\mathbf{r},t)$ or, alternatively, only the external field $\mathcal{E}^{(e)}(\mathbf{r},t)$, and also the connection of these approaches to the calculation of the total susceptibility matrices $\chi^{\mathbf{ll}'}_{\alpha\beta}(\mathbf{q},\omega)$ and of the response to the external field $\chi^{\mathbf{ll}'(e)}_{\alpha\beta}$. Here we will only mention that all these approaches are essentially equivalent, since they differ only in introducing the field or its parts into different terms of the equations of motion.

In order to solve equations (5.5.1), we introduce the eigenfrequencies $\omega'_{\mathbf{q}j}$ and the polarization vectors $e'^j_{s\alpha}(\mathbf{q})$ of the corresponding homogeneous equations ($\mathcal{E}^{(0)}=0$). It must be emphasized that they do not in general coincide with the above-defined eigenfrequencies $\omega_{\mathbf{k}}$ and polarization vectors $e^j_{s\alpha}(\mathbf{k})$ of free lattice vibrations, since the matrices A and A' are not identical. The procedures of finding them and their general properties are, however, identical to those described for free oscillations. The only important difference is that $\omega'_{\mathbf{q}j}$ are not necessarily real ($\omega'^2_{\mathbf{q}j}>0$), since they are not true eigenfrequencies of the ideal lattice and thus do not determine their instability.

Rewriting now the field in the form

$$\mathcal{E}^{(0)}_\alpha(\mathbf{r},t)=\sum_{\mathbf{q}}\mathcal{E}_{\mathbf{q}\alpha}\exp(\mathrm{i}\mathbf{q}\mathbf{r})$$

we expand equations (5.5.1), as usual, in powers of eigensolutions of the homogeneous system, multiply them by $N^{1/2}e^{j*}_{s\alpha}(\mathbf{q})e^{-\mathrm{i}\mathbf{q}\mathbf{R}_\mathbf{n}}$ and sum up over $\mathbf{n}$, s and α $\left(v_{\mathbf{q}j}=\sum_{\mathbf{n}s\alpha}\tilde{u}_{\mathbf{n}s\alpha}\tilde{e}'^{j*}_{\mathbf{n}s\alpha}\right)$:

$$-M\ddot{v}_{\mathbf{q}j}+M\omega'^2_{\mathbf{q}j}v_{\mathbf{q}j}=\sqrt{N}\,Q'^*_{j\alpha}(\mathbf{q})\mathcal{E}_{\mathbf{q}\alpha}(t)$$

We have introduced here the "effective charge" of the jth mode of vibrations

$$Q'_{j\alpha}(\mathbf{q})=\sum_s eZ_s\exp(-\mathrm{i}\mathbf{q}\boldsymbol{\rho}_s)\mathrm{e}'^j_{s\alpha}(\mathbf{q}) \tag{5.5.4}$$

The dimension and the order of magnitude of this quantity are indeed those of the elementary charge. Nevertheless, the term "effective charge" borrowed from the theory of molecules should not accepted in its literal meaning, since $Q'_{j\alpha}$ is a vector, that is, the value assumed by the "charge" depends on the direction of the field. It would be more correct to say that in the long-wavelength limit, when $\mathbf{q}\boldsymbol{\rho}_s\to 0$, the product $N^{-1/3}Q'_{j\alpha}(\mathbf{q})v_{\mathbf{q}j}$ is a variable dipole moment of the unit cell in its vibrations in the $\mathbf{q}$, j mode.

The induced current density for the field $\mathcal{E}_\mathbf{q}(t)=\mathcal{E}_\mathbf{q}e^{\mathrm{i}\omega t}$ can now be written in the form

$$j_\alpha(\mathbf{r},t)=\sum_{\mathbf{n}s}Z_se\dot{u}_{\mathbf{n}s\alpha}\delta(\mathbf{r}-\mathbf{R}_{\mathbf{n}s})=\sum_{\mathbf{n}sj}Z_se(-\mathrm{i}\omega)\mathrm{e}'^j_{s\alpha}(\mathbf{q})$$
$$\times\left[Q'^*_{j\beta}(\mathbf{q})\mathcal{E}_{\mathbf{q}\beta}/M\left(\omega'^2_{\mathbf{q}j}-\omega^2\right)\right]\exp(\mathrm{i}\mathbf{q}\mathbf{R}_\mathbf{n})\delta(\mathbf{r}-\mathbf{R}_{\mathbf{n}s})\exp(-\mathrm{i}\omega t)$$

and its Fourier component, as

$$j_{\mathbf{q}\alpha}(t) = \frac{1}{V}\int j_\alpha(\mathbf{r},t)\exp(-\mathrm{i}\mathbf{q}\mathbf{r})\,\mathrm{d}^3r$$
$$= -\mathrm{i}\omega\sum_j \frac{Q'_{j\alpha}(\mathbf{q})Q'^*_{j\beta}(\mathbf{q})}{\Omega M\,(\omega'^2_{\mathbf{q}j}-\omega^2)}\exp(-\mathrm{i}\omega t)$$

This immediately gives us the final expression for the (averaged) susceptibility of the ionic lattice:

$$\bar{\chi}_{\alpha\beta}(\mathbf{q},\omega) = \frac{1}{\Omega}\sum_j \frac{Q'_{j\alpha}(\mathbf{q})Q'^*_{j\beta}(\mathbf{q})}{M\,(\omega'^2_{\mathbf{q}j}-\omega^2)} \tag{5.5.5}$$

Even though the spatial harmonics of field and current were not involved explicitly in the derivation of this expression, their role has been taken into account completely when the quantities $\omega'_{\mathbf{q}j}$ and $Q'_{j\alpha}$ were calculated. In this connection, we emphasize again that in contradiction to frequently encountered statements, the resonance frequencies of averaged susceptibility $\omega'_{\mathbf{q}j}$ *do not* coincide with the true lattice eigenfrequencies $\omega_{\mathbf{q}j}$. Sections 5.6 and 5.7 will supply mode details on the interaction of these frequencies.

Let us evaluate the susceptibility of the ion lattice at low frequencies ($\omega \lesssim \omega'_{\mathbf{q}j}$). Using the estimates $Q'_{j\alpha} \sim e$ and $\hbar^2\omega'^2_{\mathbf{q}j} \sim I_0^2 m/M$, we obtain

$$\bar{\chi} \sim \frac{e^2}{\Omega M\omega_j^2} \sim \frac{e^2\hbar^2}{a_0^3 m I_0^2} \sim \frac{e^2}{a_0}\frac{\hbar^2}{m a_0^2}I_0^{-2} \sim \frac{I_0^2}{I_0^2} = 1$$

Therefore, both the susceptibility of an ionic lattice at low frequencies and the electron susceptibility of condensed media (we evaluated it earlier) are of the order of unity, in agreement with the results of measurements.

To calculate the nonlinear (quadratic) susceptibility, we need, as implied by (5.4.1), to add to the equations of motion the term

$$\frac{1}{2}\sum B^{ss's''}_{\alpha\alpha'\alpha''\,\mathbf{n}-\mathbf{n}',\mathbf{n}-\mathbf{n}''}u_{\mathbf{n}'s'\alpha'}u_{\mathbf{n}''s''\alpha''}$$

Now the equations of motion are

$$M_s\ddot{u}_{\mathbf{n}s\alpha} = -\sum_{\mathbf{n}'s'\alpha'} A'^{ss'}_{\alpha\alpha'\,\mathbf{n}-\mathbf{n}'}u_{\mathbf{n}'s'\alpha'} - \frac{1}{2}\sum B^{ss's''}_{\alpha\alpha'\alpha''\,\mathbf{n}-\mathbf{n}',\mathbf{n}-\mathbf{n}''}u_{\mathbf{n}'s'\alpha'}u_{\mathbf{n}''s''\alpha''}$$
$$+ Z_s e\mathcal{E}^{(0)}_\alpha(\mathbf{R}_{\mathbf{n}s},t)$$

If, by virtue of (5.3.9), we substitute into this last expression

$$u_{\mathbf{n}s\alpha} = \sum_{\mathbf{q}j} v_{\mathbf{q}j}e'^{j}_{\mathbf{n}s\alpha}(\mathbf{q})$$

where $v_{\mathbf{q}j} = u_{\mathbf{q}j} + u^*_{-\mathbf{q}j}$, multiply the equation by $e'^{j*}_{\mathbf{n}s\alpha}$ and sum up over $\mathbf{n}s\alpha$, we arrive at

$$M\ddot{v}_{\mathbf{q}j} + M\omega'^2_{\mathbf{q}j}v_{\mathbf{q}j} = \frac{1}{2}\sqrt{N}\,\exp(-\mathrm{i}\omega t)\sum_\alpha E_{\mathbf{q}\alpha}Q'^*_{j\alpha}(\mathbf{q})$$
$$- N^{1/2}\sum_{j'j''} v_{\mathbf{q}'j'}v_{\mathbf{q}''j''}B'^{jj'j''}(\mathbf{q}',\mathbf{q}'') \tag{5.5.6}$$

In these equations, $B'(\mathbf{q}',\mathbf{q}'')$ is already defined by (5.4.4), with e^j being replaced by e'^j. It is assumed that $\mathbf{q} = \mathbf{q}' + \mathbf{q}''$ and $\omega = \omega' + \omega''$, and that the field is of the form

$$\boldsymbol{\mathcal{E}}^{(0)}_\alpha(\mathbf{r},t) = \frac{1}{2}(E_{\mathbf{q}\alpha}\exp(\mathrm{i}\mathbf{q}\mathbf{r} - \mathrm{i}\omega t) + E_{\mathbf{q}'\alpha'}\exp(\mathrm{i}\mathbf{q}'\mathbf{r} - \mathrm{i}\omega' t)$$
$$+ E_{\mathbf{q}''\alpha}\exp(\mathrm{i}\mathbf{q}''\mathbf{r} - \mathrm{i}\omega'' t) + \mathrm{C.C.})$$

Equations similar to (5.5.6) also hold for the amplitudes $v_{\mathbf{q}'j'}$ and $v_{\mathbf{q}''j''}$. If we solve these equations neglecting the quadratic term, we find

$$v_{\mathbf{q}'j'} = \frac{\sqrt{N}}{2M(\omega'^2_{\mathbf{q}'j'} - \omega'^2)}\sum_{\alpha'} E_{\mathbf{q}'\alpha'}Q'^*_{j'\alpha'}(\mathbf{q}')\exp(-\mathrm{i}\omega' t)$$

and a similar expression for $v_{\mathbf{q}''j''}$. Substituting these expressions into (5.5.6), we obtain from it a correction to $v_{\mathbf{q}j}$, quadratic in field. By calculating the corresponding quadratic correction to the Fourier component of current (in quite the same way as in deriving the formula for the linear susceptibility), we arrive at the quadratic susceptibility of the ionic lattice:

$$\chi^{(2)}_{\alpha\beta\gamma}(\mathbf{q} = \mathbf{q}' + \mathbf{q}'',\, \omega = \omega' + \omega'') = -\frac{1}{2\Omega}\sum_{jj'j''} B'^{jj'j''}(\mathbf{q}',\mathbf{q}'')$$
$$\times\, \frac{Q'_{j\alpha}(\mathbf{q})Q'^*_{j'\beta}(\mathbf{q}')Q'^*_{j''\gamma}(\mathbf{q}'')}{M^3(\omega'^2_{\mathbf{q}j} - \omega^2)(\omega'^2_{\mathbf{q}'j'} - \omega'^2)(\omega'^2_{\mathbf{q}''j''} - \omega''^2)} \tag{5.5.7}$$

We can now evaluate $\chi^{(2)}$. As usual, we assume that $B \sim I_0/a_0^3$, $Q \sim e$, $\hbar^2\omega_j^2 \sim I_0^2 m/M$, so that

$$\chi^{(2)} \sim \frac{1}{a_0^3}\,\frac{I_0}{a_0^3}\,\frac{e^3M^3\hbar^6}{M^3I_0^6m^3} = \left(\frac{\hbar^2}{ma_0^2}\right)^3\frac{e^3}{I_0^5}$$
$$\sim \frac{e^3}{I_0^2} \sim \frac{a_0^2}{e}\left(\frac{e^2}{a_0}\right)^2\frac{1}{I_0^2} \sim \frac{a_0^2}{e} = \boldsymbol{\mathcal{E}}_{\mathrm{at}}^{-1}$$

If we compare this estimate of the ionic quadratic susceptibility with the estimate, made in Chapter 2, of the electron quadratic susceptibility, it becomes clear that in one crystal, they are of the same order of magnitude. In fact, this is true at frequencies that are of the order of or less than $\omega_{\mathbf{q}j'}$ (infrared range). The ionic contribution becomes negligible in the visible spectral range.

As we see from (5.5.5) and (5.5.7), susceptibilities increase close to resonances when frequencies approach vibrational frequencies. The imaginary part of (5.5.5) determines absorption (after a standard replacement $\omega \to \omega + i\delta$, which creates $\delta(\omega - \omega'_{\mathbf{q}j})$ in the imaginary part of χ). It would be more correct to use a replacement $\omega \to \omega + i\nu_j(\mathbf{q})$ in (5.5.5), where the finite value of $\nu_j(\mathbf{q})$ due to the anharmonism and to the finite phonon lifetime was calculated in Section 5.4 for optical phonons. It is the optical modes that may have $Q \neq 0$ in the long-wavelength limit and can interact with the electromagnetic field. Taking into account $\nu_j(\mathbf{q})$, we find that the absorption line, within which $\mathrm{Im}\chi \neq 0$, has finite width and Lorentzian shape. We have mentioned already that $\nu_j \ll \omega'_{\mathbf{q}j}$ and that the ratio $\nu_j/\omega_{\mathbf{q}j} \sim \sqrt{m/M} \sim 0.01$.

We will demonstrate in conclusion that the results of this section are applicable in a considerably wider region than the pointlike ions model. In the spirit of the adiabatic approximation, displacements of nuclei from equilibrium positions induce a redistribution of the electron density. If displacements are small, the induced charge density is a linear function of displacements (the first term in the Taylor series expansion):

$$\delta\rho(\mathbf{r},t) = \sum_{\mathbf{n}s\alpha} \Phi^s_\alpha(\mathbf{r} - \mathbf{R}_\mathbf{n}) u_{\mathbf{n}s\alpha}(t)$$

Charge redistribution is obviously accompanied with the generation of local currents which can be written, for the same reasons, in the form

$$\delta j_\alpha(\mathbf{r},t) = \sum_{\mathbf{n}s\beta} (\Lambda^s_{\alpha\beta}(\mathbf{r} - \mathbf{R}_\mathbf{n}) \dot{u}_{\mathbf{n}s\alpha}(t) + \mu^s_{\alpha\beta}(\mathbf{r} - \mathbf{R}_\mathbf{n}) u_{\mathbf{n}s\alpha}(t)) \qquad (5.5.8)$$

It is obvious that derivatives of displacements with respect to time must appear in the righthand side of (5.5.8). Indeed, $\delta\rho(\mathbf{r},t)$ and $\delta\mathbf{j}(\mathbf{r},t)$ must satisfy the continuity equation. The continuity equations also imply that

$$\frac{\partial}{\partial x_\alpha} \Lambda^s_{\alpha\beta}(\mathbf{r} - \mathbf{R}_\mathbf{n}) = -\Phi^s_\beta(\mathbf{r} - \mathbf{R}_\mathbf{n})$$

$$\frac{\partial}{\partial x_\alpha} \mu_{\alpha\beta}(\mathbf{r} - \mathbf{R}_\mathbf{n}) = 0$$

We also see that the second term in the righthand side of (5.5.8) describes quasistatic vortex currents, that is, a generation of, or a change in, magnetization in response to ionic displacements. This term is negligibly small in nonmagnetic materials. For the sake of simplification, we will limit our analysis here to this case; we neglect the second term in (5.5.8).

The equations of motion of the lattice, which will be derived strictly below, are

$$M_s \ddot{u}_{\mathbf{n}s\alpha} + \sum_{\mathbf{n}'s'\alpha'} A'^{ss'}_{\mathbf{n}-\mathbf{n}'}{}^{\alpha\alpha'} u_{\mathbf{n}'s'\alpha'} = \int \Lambda^s_{\beta\alpha}(\mathbf{r} - \mathbf{R_n}) \mathcal{E}_\beta(\mathbf{r}, t)\, \mathrm{d}^3 r \tag{5.5.9}$$

Here A' is the force constants matrix which contains only the contribution of small-scale Coulomb fields ($\mathbf{l} \neq 0$) while the main spatial harmonic of the field ($\mathbf{l} = 0$) is included in the righthand side of the equation together with the field of external charges. We thus find from (5.5.9) the response $\overline{\chi}$ to the long-wavelength field averaged over the unit cell.

Acting now exactly as in the case of the pointlike ions model, we readily derive from (5.5.9) and (5.5.8) that

$$\overline{\chi}_{\alpha\beta}(\mathbf{q}, \omega) = \frac{1}{\Omega} \sum_j \frac{\Lambda_{j\alpha}(\mathbf{q}) \Lambda^*_{j\beta}(\mathbf{q})}{M\left(\omega'^2_{\mathbf{q}j} - \omega^2\right)} \tag{5.5.10}$$

where

$$\Lambda_{j\alpha}(\mathbf{q}) = \sum_{s\beta} \int \mathrm{d}^3 r \exp(-\mathrm{i}\mathbf{q}\mathbf{r}) \Lambda^s_{\alpha\beta}(\mathbf{r}) \mathbf{e}'^{j}_{s\beta}(\mathbf{q})$$

Formula (5.5.10) exactly coincides with (5.5.5), differing from it in the redefined "effective charge" of the mode $\mathbf{q}$, j: instead of $Q'_{j\alpha}$ determined by formula (5.5.4), it includes $\Lambda_{j\alpha}(\mathbf{q})$ defined for (5.5.10). We see from here and from (5.5.8) that $\Lambda_{\alpha\beta}(\mathbf{r} - \mathbf{R_n})$ plays the role of the "effective charge" density, generated when nuclei are displaced (and obviously includes the charge of nuclei themselves).

Let us now give the derivation of the equation of motion (5.5.9). Lagrange's function of the crystalline lattice and the field in the adiabatic approximation of electron motion can be written in the form

$$L = \frac{1}{2} \sum_{\mathbf{n}s\alpha} M_s (\dot{u}_{\mathbf{n}s\alpha})^2 - \frac{1}{2} \sum_{\substack{\mathbf{n}s\alpha \\ \mathbf{n}'s'\alpha'}} u_{\mathbf{n}s\alpha} A'^{ss'}_{\mathbf{n}-\mathbf{n}'}{}^{\alpha\alpha'} u_{\mathbf{n}'s'\alpha'}$$
$$+ \frac{1}{c} \int A_\alpha(\mathbf{r}, t) j_\alpha(\mathbf{r}, t)\, \mathrm{d}^3 r - \int \varphi(\mathbf{r}, t) \rho(\mathbf{r}, t)\, \mathrm{d}^3 r + L_f$$

where

$$L_f = (\operatorname{curl} \mathbf{A})^2 - \left(\operatorname{grad} \varphi + \frac{1}{c} \frac{\partial \mathbf{A}}{\partial t}\right)^2$$

is Lagrange's function for the field. When we say "field", we mean only its long-wavelength component, since the purely coulombic small-scale fields of higher spatial harmonics are already included into the potential energy of the lattice (matrix A'). Let us now substitute into this formula the above expressions for $\delta j_\alpha(\mathbf{r},t)$ and $\delta\rho(\mathbf{r},t)$. The equations of motion are then obtained in a standard manner:

$$\frac{\mathrm{d}}{\mathrm{d}t}\left(\frac{\partial L}{\partial \dot{u}_{\mathbf{n}s\alpha}}\right) - \frac{\partial L}{\partial u_{\mathbf{n}s\alpha}} = 0$$

Elementary manipulations taking into account the relation between fields and potentials, $\mathcal{E} = -\operatorname{grad}\varphi - \frac{1}{c}\frac{\partial A}{\partial t}$, and the relation between the matrices $\Lambda^s_{\alpha\beta}$ and Φ^s_β, transform the equations of motion to the form (5.5.9).

Note that adding the total field to the force constants matrix, that is, using the total matrix A, would be considerably less convenient in the problem as formulated here. The point is that if the transverse field is taken into account (as we shall see in the next section, this field becomes essential for sufficiently long waves), the true (total) interaction between ions becomes retarding. This retarding contribution can be easily found from Lagrange's function given above if we derive from it ordinary equations for the field. The righthand sides of these equations will include $\delta\rho(\mathbf{r},t)$ and $\delta j_\alpha(\mathbf{r},t)$. The solution of these equations must then be substituted into expressions for the interaction of $\delta\rho$ and δj with the field. As a result, we would have in the Fourier representation the relation

$$\Lambda^{ss'}_{\alpha\alpha'}(\mathbf{q},\omega) = A'^{ss'}_{\alpha\alpha'}(\mathbf{q},\omega) - \frac{4\pi\omega^2}{\Omega c^2}\Lambda^{s*}_{\alpha\mu}(\mathbf{q})D_{0\mu\nu}(\mathbf{q},\omega)\Lambda^{s'}_{\nu\alpha'}(\mathbf{q})$$

The frequency characteristic of the transverse part of D_0 signifies the retarding nature of the interaction described by the matrix A.

There exists, however, a much simpler and visually clearer method of taking into account the long-wavelength electromagnetic field in the problem of crystal lattice vibrations, based on using the above-calculated averaged polarizability $\overline{\chi}$ together with Maxwell's equations. This method is presented in the next section.

5.6. Polaritons

Expression (5.5.5) for the susceptibility of ionic lattice makes it possible to treat electromagnetic fields in crystals. For simplicity, we will only consider a cubic crystal which has only one dipole-active optical vibration. Knowing the crystal symmetry and applying group theory, one can identify which vibrational modes in a crystal are dipole-active (i.e., for which $Q' \neq 0$). These are always optical modes since the cell dipole moment in long-wavelength acoustic modes is not changed by vibrations even if it were nonzero in the equilibrium configuration (segnetoelectric crystals):

indeed, long-wavelength acoustic vibrations mean that a unit cell displaces as a whole, without deformations. This is what prompted the term "optical modes," since these are the only ones capable of interacting with the electromagnetic field and manifesting themselves in absorption spectra.

Obviously, not all optical modes are dipole-active but if a mode is active in a cubic crystal, it is triply degenerate at $\mathbf{k} = 0$, which corresponds to three orientations of the induced dipole moment. For the single active mode in a cubic crystal, we have instead of (5.5.5) an isotropic expression

$$\chi_{\alpha\beta}(\mathbf{k}, \omega) = \frac{1}{\Omega} \frac{Q^2(\mathbf{k})}{M(\omega_{\mathbf{k}}'^2 - \omega^2)} \delta_{\alpha\beta} \tag{5.6.1}$$

This is an elementary expression for an ensemble of oscillators, in which only the frequency $\omega_{\mathbf{k}}'$ and the oscillator strengths depend on the wave vector $\mathbf{k}$. As we have remarked in Chapter 3, the condition

$$\varepsilon(\mathbf{k}, \omega) = 1 + 4\pi\chi(\mathbf{k}, \omega) = 0$$

dictates the law of dispersion of longitudinal waves.

Substituting it into (5.6.1), we arrive at the equation for the frequency of longitudinal electromagnetic waves in the crystal as a function of the wave vector:

$$\omega_{\mathbf{k}}^2 = \omega_{\mathbf{k}}'^2 + \frac{4\pi Q^2(\mathbf{k})}{M\Omega}$$

The law of dispersion for the transverse electromagnetic waves is obtained from the condition

$$c^2k^2 = \omega^2\varepsilon(k, \omega)$$

We will now take into account the electron contribution to susceptibility. As a result of this, the susceptibility at frequencies that satisfy the inequality

$$\omega_{\mathbf{k}}' \ll \omega \ll I_0/\hbar$$

is equal to the electron susceptibility χ_∞ and $\varepsilon = 1 + 4\pi\chi_\infty = \varepsilon_\infty$.

For the frequencies $\omega \ll I_0/\hbar$ (infrared range), where $k \sim \omega/c \ll g_1 \sim 2\pi a_0^{-1}$, we can neglect the dependence of $\omega_{\mathbf{k}}'$ and $Q(\mathbf{k})$ on $\mathbf{k}$, and can set $\omega_{\mathbf{k}}' = \omega_0$. Then we have

$$\varepsilon(\omega) = \varepsilon_\infty + \frac{4\pi}{\Omega} \frac{Q^2}{M(\omega_0^2 - \omega^2)} \tag{5.6.2}$$

The dielectric permittivity at high frequencies, $\varepsilon_\infty = 1 + 4\pi\chi_\infty$, can be calculated from experimental data, so that it is expedient to express the parameter Q^2/M in the last formula also in terms of an experimentally

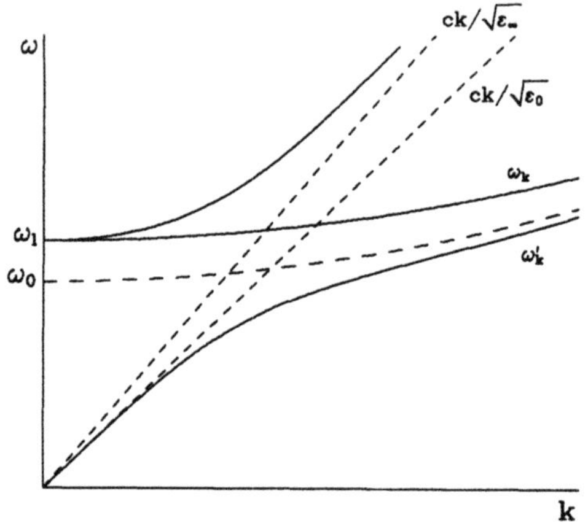

Figure 5.10. Law of dispersion for polaritons in crystals.

measured quantity. It will be convenient to use as such a quantity the low-frequency dielectric permittivity ε_0 ($\omega \to 0$). Then we have

$$\varepsilon_0 = \varepsilon_\infty + \frac{4\pi Q^2}{M\Omega\omega_0^2}$$

$$\varepsilon(\omega) = \varepsilon_\infty + \frac{(\varepsilon_0 - \varepsilon_\infty)\omega_0^2}{\omega_0^2 - \omega^2} = \frac{\varepsilon_0\omega_0^2 - \varepsilon_\infty\omega^2}{\omega_0^2 - \omega^2} \tag{5.6.3}$$

Substituting (5.6.3) into $c^2k^2 = \omega^2\varepsilon(\omega)$, we arrive at a biquadratic equation from which to find the law of dispersion, $\omega = \omega(\mathbf{k})$.

The frequency ω_l at which $\varepsilon(\omega) = 0$ carries the meaning of the frequency of longitudinal vibrations $\omega_{\mathbf{k}}$ at low $\mathbf{k}$. Therefore,

$$\omega_{\mathbf{k}}^2(\mathbf{k} \to 0) = \omega_0^2\,\frac{\varepsilon_0}{\varepsilon_\infty}$$

and since $\varepsilon_0 > \varepsilon_\infty$ (the lattice contributes to ε_0), we obtain $\omega_l = \omega_{\mathbf{k}}(\mathbf{k} \to 0) > \omega_0 = \omega'_{\mathbf{k}}(\mathbf{k} \to 0)$.

In the interval $\omega_0 < \omega < \omega_l$, the dielectric permittivity $\varepsilon(\omega) < 0$ and there are no propagating (bulk) modes in this frequency range.

Incident waves with these frequencies are totally reflected from the crystal. The refractive index $n = \sqrt{\varepsilon}$ becomes purely imaginary and the reflection coefficient is

$$R = \left|\frac{n-1}{n+1}\right|^2 = 1$$

Solving the quadratic equation for $\omega^2(\mathbf{k})$, we arrive at the law of dispersion for transverse waves:

$$\omega^2 = \frac{1}{2}(\omega_l^2 + \varepsilon_\infty^{-1}c^2k^2) \pm \frac{1}{2}\sqrt{(\omega_l^2 + \varepsilon_\infty^{-1}c^2k^2)^2 - 4\varepsilon_\infty^{-1}c^2k^2\omega_0^2}$$

Figure 5.10 schematically shows this solution. When the electromagnetic field interacts strongly with a dipole-active vibration in a crystal, one speaks of polariton waves (obviously referring to polarization), and of the corresponding elementary excitations as polaritons. We see from this figure that the dispersion law for polaritons has two (twice degenerate) branches for a single isolated dipole-active vibration in cubic crystal lattice. If $\mathbf{k}$ are small, the lower branch follows the law $\omega = ck/\sqrt{\varepsilon_0}$, which corresponds to photons in a medium with $\varepsilon = \varepsilon_0$. In the wave vector range where $\omega(\mathbf{k})$ becomes close to ω_0 (rather, to $\omega'(\mathbf{k})$), the curve $\omega(\mathbf{k})$ deviates from the law of dispersion for low-frequency photons and asymptotically approaches $\omega'(\mathbf{k})$, that is, the phonon frequency with the longitudinal long-wavelength field ignored. The actual upward (or downward) bend on the $\omega'(\mathbf{k})$ curve occurs at $k \gg \omega_0\sqrt{\varepsilon_0}/c$, so that, as we have assumed above, we can neglect the difference between $\omega'(\mathbf{k})$ and ω_0 at small k typical of phonons in crystals.

If $k \gg \omega_0\sqrt{\varepsilon_0}/c$, the contribution of the transverse electromagnetic field to the polariton wave is small and the dispersion law is determined by the frequency dependence of $\omega'(\mathbf{k})$ for the phonon subsystem.

The upper branch of the law of dispersion for polaritons begins at $\mathbf{k} = 0$, at the longitudinal wave frequency $\omega_l = \omega_{\mathbf{k}}$ ($\mathbf{k} \to 0$) and then deviates from the longitudinal wave frequency $\omega(\mathbf{k})$ toward the phonon dispersion law at high frequencies, $\omega = ck/\sqrt{\varepsilon_\infty}$. One should not forget again that on the scale of $\mathbf{k}$ typical of the optical range we have $\omega(\mathbf{k}) = \omega_l$.

If there is another dipole-active vibration at a frequency $\omega'(\mathbf{k})^{(2)}$ greater than ω_l, the upper branch undergoes a similar splitting into two branches at the intersection with the curve $\omega'(\mathbf{k})^{(2)}$, and so on.

For an individual dipole-active vibration in a cubic crystal, a thrice-degenerate vibration at $\mathbf{k} = 0$ generates three branches in the polariton spectrum shown in Figure 5.10: the longitudinal branch $\omega(\mathbf{k}) = \omega_l$ and the twice-degenerate transverse branch. In an anisotropic (noncubic) crystal, the behavior of polariton dispersion becomes considerably more complex and depends on the direction of $\mathbf{k}$ relative to the directions of the crystallographic axes. The propagation of longitudinal waves is a specific feature of cubic crystals; in anisotropic crystals, they arise only when $\mathbf{k}$ is oriented along certain axes.

In a uniaxial crystal, that is, in a crystal with higher than twofold symmetry axis, the frequency ω_0 which is thrice-degenerate in cubic crystals transforms into a nondegenerate frequency ω_1 with oscillations that are polarized along the symmetry axis of the crystal and a twice-degenerate frequency ω_2 polarized normally to the axis. The frequencies split even before the long-wavelength field has been taken into account. The pattern describing the propagation of a wave along the symmetry axis of the crystal is schematically plotted in Figure 5.11 for $\omega_1 > \omega_2$. The longitudinal wave is not connected with transverse waves. If the wave propagates along other

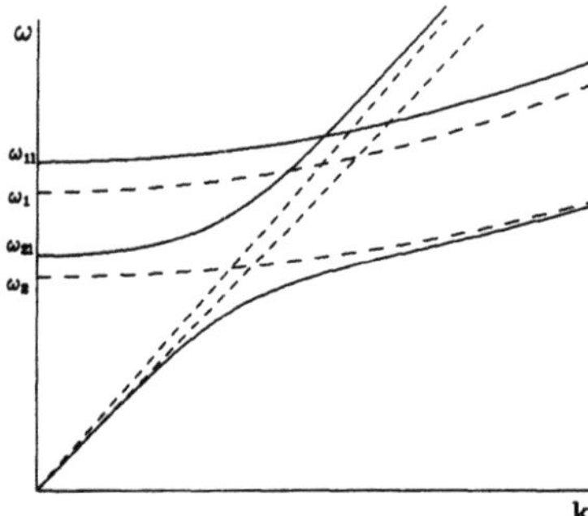

Figure 5.11. Law of dispersion for polaritons in a uniaxial crystal in the case of propagation along the symmetry axis.

directions, the picture becomes more complex; while the ordinary wave has singularities at frequencies about ω_2 (it is always polarized perpendicularly to the axis), the extraordinary wave interacts both with the frequency-ω_1 vibration and with the frequency-ω_2 vibration, and the pattern resembles the one obtained when two dipole-active modes are present.

As we see from figures 5.10 and 5.11, the spatial dispersion in a crystal with dipole-active optical modes, arising by virtue of (5.5.5) when the dependence of $\omega'_{\mathbf{k}}$ and $Q(\mathbf{k})$ on the wave vector is taken into account, may result in two or more values of $\mathbf{k}$ corresponding to one frequency ω. Several waves with different $\mathbf{k}$ can then propagate through the crystal. These additional waves, which arise when spatial dispersion is not ignored, create certain complications for the calculation of the reflection and transmission coefficients at an interface between two media. If the spatial dispersion is neglected, ordinary boundary conditions of electrodynamics are sufficient to calculate these coefficients in terms of the known values of ε of the two media (Fresnel's coefficients). When several waves arise in a medium with $\mathbf{k}$-dependent $\varepsilon(\mathbf{k}, \omega)$, these boundary conditions become insufficient. All attempts to identify general-case additional boundary conditions failed. The point is that in this situation, one has to take into account in the macroscopic problem not only the boundary conditions for electromagnetic fields but also the boundary conditions for mechanical displacements (free interface, fixed interface, etc.).

For crystals with a center of inversion, dipole-active vibrations are forbidden in the Raman spectrum by symmetry arguments and vice versa (alternative ban at low $\mathbf{k}$). For crystals without a center of inversion (such crystals are of special interest for nonlinear optics since they possess quadratic dipole polarizability), one and the same vibrational mode can manifest itself both in the Raman scattering and in the infrared band absorption. This dipole-active mode and the polaritons related to it can be observed and studied in the Raman scattering spectrum. The Raman scattering of light by polaritons offers an efficient method of determining absorption. This dipole-active mode and the polaritons related to it can be observed and studied in the Raman scattering spectrum. The Raman scattering of light by polaritons offers an efficient method of determining dispersion curves of polaritons in crystals.

Let us now return to the relationship between the true lattice oscillations and the poles of the response functions (poles of susceptibilities). It is readily demonstrated that the true normal lattice vibrations are exactly the waves identified in this section: the polaritons (including the longitudinal polaritons). Indeed, they are actually joint oscillations of lattice atoms and of the electromagnetic field they generate. In other words, these are lattice vibrations (lattice waves) taking into account all the fields they create—both small-scale and long-wavelength waves, both longitudinal and transverse. However, this comprehensive treatment was done in two stages. At the first stage, the system's response to the long-wavelength field is found taking into account only small-scale longitudinal fields. The expression for susceptibility then includes the fictitious frequencies $\omega'_{\mathbf{k}j}$ which would be the real lattice vibration frequencies if there were no long-wavelength field at all. At the second stage, the contribution of long-wavelength fields to wave propagation in the system under study is taken into account. This is achieved simply by solving Maxwell's equations for long-wavelength fields, since the susceptibility for these fields has already been found at the preceding stage.

Therefore, the true normal vibration frequencies are connected with the susceptibility poles only indirectly. We will presently show, however, that they exactly correspond to the poles of the function of response to the external field $\chi^{(\mathrm{e})}$. This can be qualitatively understood even in terms of the arguments of Section 5.5. Indeed, we have remarked there that if the contribution of long-wavelength fields was not singled out from the force constants, only the field of external charges, $\boldsymbol{\mathcal{E}}^{(\mathrm{e})}$, would appear in the righthand sides of the equations of motion. Having found the response in the problem thus formulated, we arrive at an expression for $\chi^{(\mathrm{e})}$ of the same type as (5.5.5) but with the denominators containing the eigenfrequencies of the total matrix A, that is, we find the true frequencies of lattice vibrations.

This reasoning, even though based on an impeccable idea, has one flaw. Strictly speaking, it is applicable only to longitudinal fields, since only the Coulomb interaction between ions was taken into account in the force constants matrix C. Therefore, the function $\chi^{(\mathrm{e})}$ thus calculated is the function of response only to the external longitudinal field and its poles coincide with the true frequencies of longitudinal waves. To turn it into the response function for external transverse fields as well, we need either to take into account the contribution of these fields to the interaction between the lattice atoms or, which is a simpler task, to resort to the general formulas of Sections 5.1.6 and 5.1.9, which make it possible to write the response to the external field in terms of the susceptibility. These formulas readily yield

$$\chi^{(e)00}_{\alpha\beta}(\mathbf{q},\omega) = \bar{\chi}_{\alpha\gamma}(\mathbf{q},\omega)\left(\left[1 - \frac{4\pi\omega^2}{c^2} D_0(\mathbf{q},\omega)\bar{\chi}(\mathbf{q},\omega)\right]^{-1}\right)_{\gamma\beta} \tag{5.6.4}$$

The frequencies $\omega'_{\mathbf{q}j}$ are not the poles of this expression since the poles $\overline{\chi}(\mathbf{q},\omega)$ in the first and second multipliers in the righthand side cancel out. Hence, the poles $\chi^{(e)00}_{\alpha\beta}(\mathbf{q},\omega)$ are determined only by the equation

$$\det\left\| D^{-1}_{0\alpha\beta}(\mathbf{q},\omega) - \frac{4\pi\omega^2}{c^2}\bar{\chi}_{\alpha\beta}(\mathbf{q},\omega)\right\| = 0 \tag{5.6.5}$$

which defines the poles of the second multiplier (the poles of the inverse matrix). However, this is the familiar equation of crystal optics used to determine the frequencies of normal waves in a medium whose susceptibility is $\overline{\chi}_{\alpha\beta}$. Thus, in the isotropic case with the spatial dispersion neglected, $\overline{\chi}_{\alpha\beta} = \chi(\omega)\delta_{\alpha\beta}$, it is reduced to the equations discussed above: $\varepsilon(\omega) = 0$ for longitudinal waves and $k^2 = \omega^2\varepsilon(\omega)/c^2$ for transverse waves.

We conclude that polaritons are the true normal lattice vibrations, with all the accompanying fields taken into account, and that their frequencies are given by the poles of the response to the external field. They may coincide with the frequencies $\omega'_{\mathbf{q}j}$ only in the exceptional case of vibrations that do not generate any long-wavelength fields (either longitudinal or transverse), that is, vibrations that do not manifest themselves in any way in optical absorption spectra. We need to emphasize again that as we see from the ratio of the first and second terms in the matrix $D_{0\alpha\beta}(q,\omega)$ (see formula (1.6.8)), the contribution of the transverse field to polariton frequencies is appreciable only for the longest waves $q^2 \lesssim \omega^2/c^2 \lesssim (10^3\ \mathrm{cm}^{-1})^2$.

5.7. Surface Polaritons

We have shown above that the dielectric permittivity in a crystal with a dipole-active vibration becomes negative in the frequency band $\omega_0 < \omega < \omega_l$. No bulk waves propagate in the frequency range but there exist special-type solutions of Maxwell's equations, which are nonzero only within the interface between two media: surface waves or surface polaritons. An interest to surface waves increased greatly in the recent period, in connection with the study of phenomena on the surfaces of solids.

First we consider longitudinal surface waves in a cubic crystal. We neglect the spatial dispersion in the infrared and optical bands, that is, we assume that $\varepsilon_{\alpha\beta} = \varepsilon(\omega)\delta_{\alpha\beta}$ and is independent of $\mathbf{k}$.

Consider a planar interface between two media, medium I and medium II. We denote the dielectric permittivities of the two media by ε_1 and ε_2 and the fields by $\boldsymbol{\mathcal{E}}_1$ and $\boldsymbol{\mathcal{E}}_2$, respectively. By definition, we have $\operatorname{curl} \boldsymbol{\mathcal{E}}_{1,2} = 0$. By virtue of Maxwell's equations in the case of no external charges or currents, $\nabla \boldsymbol{\mathcal{E}}_{1,2} = 0$. For a longitudinal field, $\boldsymbol{\mathcal{E}}_{1,2} = -\operatorname{grad} \varphi_{1,2}$, so that for a homogeneous medium we have

$$\nabla^2(\varepsilon_{1.2}\varphi_{1,2}) = 0$$

Together with the boundary conditions and the constraint of continuity of the potential φ at the interface, these equations determine the potentials and fields of surface waves.

We denote by z the axis perpendicular to the planar interface between the media I and II and directed from I into II, and by x and y, the axes in the plane of the interface. We seek the solution in the form

$$\varphi_1 = \varphi_0 \exp(ikx + \kappa_1 z) \exp(-i\omega t)$$
$$\varphi_2 = \varphi_0 \exp(ikx - \kappa_2 z) \exp(-i\omega t)$$

We assume here, without loss of generality, that $\mathbf{k}$ points along the axis x. The constants φ_0 and $\mathbf{k}$ in the functions φ_1 and φ_2 are identical, owing to the boundary condition. The damping constants κ_1 and κ_2 are positive, which ensure that for the direction of the z axis as chosen above, the field is localized in the interface. We find from the equations for the potentials φ_1 and φ_2 that

$$k^2 - \kappa_1^2 = 0 \qquad k^2 - \kappa_2^2 = 0$$

Hence, $\kappa_1 = \kappa_2 = |k|$, that is, the wave intensity falls off at identical rate on both sides of the interface.

The boundary condition

$$\varepsilon_1 \mathcal{E}_{1z} = \varepsilon_2 \mathcal{E}_{2z}$$

implies

$$\varepsilon_1(\omega) = -\varepsilon_2(\omega)$$

This is the existence condition for surface waves. While the condition of existence of longitudinal bulk waves was written as $\varepsilon(\mathbf{k}, \omega) = 0$, here we have $\varepsilon_1(\omega) + \varepsilon_2(\omega) = 0$. If one of these media is the vacuum, then the longitudinal bulk waves exist if $\varepsilon(\omega) = 0$ and the surface waves exist if $\varepsilon(\omega) = -1$, which in a cubic crystal is satisfied precisely in the gap between ω_0 and ω_1.

Note that the purely longitudinal field does not exactly satisfy the set of Maxwell's equations since the magnetic field of a longitudinal field is zero.

The solution in the form of a longitudinal field does approximately satisfy Maxwell's equations at sufficiently low frequencies, when $\omega \ll kc/\varepsilon$ or, which is the same, $k \gg \omega\varepsilon/c$. At high k, when $\kappa \simeq |k|$, the transverse field satisfying the condition $\nabla\boldsymbol{\mathcal{E}} = 0$ is at the same time almost longitudinal. Indeed, the wave vector with components k and $\mathrm{i}\kappa$ is orthogonal to itself:

$$k^2 + (\mathrm{i}\kappa)^2 \simeq k^2 - k^2 = 0$$

The longitudinal surface polariton is a convenient model for large wave vectors k.

Let us now consider transverse surface waves for which $\nabla\boldsymbol{\mathcal{E}}_{1,2} = 0$. Let the field be written as

$$\boldsymbol{\mathcal{E}}_{1,2} = \mathbf{E}_{1,2} \exp(\mathrm{i}kx \pm \kappa_{1,2}z) \exp(-\mathrm{i}\omega t)$$

For transverse waves, Maxwell's equations lead to the wave equations

$$\Delta\boldsymbol{\mathcal{E}}_{1,2} - \frac{\varepsilon_{1,2}}{c^2} \frac{\partial^2 \boldsymbol{\mathcal{E}}_{1,2}}{\partial t^2} = 0$$

which yield

$$\begin{aligned} k^2 - \kappa_1^2 &= \frac{\omega^2}{c^2} \varepsilon_1 \\ k^2 - \kappa_2^2 &= \frac{\omega^2}{c^2} \varepsilon_2 \end{aligned} \tag{5.7.1}$$

The condition of transverse waves, $\nabla\boldsymbol{\mathcal{E}}_{1,2} = 0$, gives

$$\begin{aligned} \mathrm{i}kE_{1x} + \kappa_1 E_{1z} &= 0 \\ \mathrm{i}kE_{2x} - \kappa_2 E_{2z} &= 0 \end{aligned} \tag{5.7.2}$$

The boundary conditions impose the continuity of the tangential components of the electric and magnetic fields at the interface. We know that by virtue of the field equations, the boundary conditions for the normal field components are automatically satisfied in the media I and II. This is implied by the fact that for time-dependent fields, Maxwell's equations containing $\nabla\boldsymbol{\mathcal{D}}$ and $\nabla\boldsymbol{\mathcal{H}}$ follow from the first two equations. Since

$$\operatorname{curl}\boldsymbol{\mathcal{E}}_{1,2} = -\frac{1}{c}\frac{\partial\boldsymbol{\mathcal{H}}}{\partial t}$$

these boundary conditions give

$$\begin{gathered} E_{1x} = E_{2x} \qquad E_{1y} = E_{2y} \qquad \kappa_1 E_{1y} = -\kappa_2 E_{2y} \\ \kappa_1 E_{1x} - \mathrm{i}kE_{1z} = -\kappa_2 E_{2y} - \mathrm{i}kE_{2z} \end{gathered}$$

The second and the third boundary conditions immediately imply $E_{1y} = E_{2y} = 0$, since κ_1, $\kappa_2 > 0$. Using the transversality condition (5.7.2), we derive from the first and the last boundary conditions

$$\kappa_1 - \frac{k^2}{\kappa_1} = -\kappa_2 + \frac{k^2}{\kappa_2} \tag{5.7.3}$$

which, together with (5.7.1), gives

$$\frac{\kappa_1}{\kappa_2} = -\frac{\varepsilon_1}{\varepsilon_2} \tag{5.7.4}$$

Hence, κ_1 and κ_2 for transverse waves generally differ.

Now we rewrite (5.7.3) as

$$\kappa_1 + \kappa_2 = k^2\left(\frac{1}{\kappa_1} + \frac{1}{\kappa_2}\right) = \frac{k^2(\kappa_1 + \kappa_2)}{\kappa_1\kappa_2}$$

whence

$$\kappa_1\kappa_2 = k^2$$

Together with (5.7.4), this allows us to write the expressions for κ_1 and κ_2 as

$$\kappa_1^2 = -k^2\frac{\varepsilon_1}{\varepsilon_2} \qquad \kappa_2^2 = -k^2\frac{\varepsilon_2}{\varepsilon_1}$$

Finally, we substitute these equalities into any equation of (5.7.1) and obtain

$$k^2 = \frac{\omega^2}{c^2}\frac{\varepsilon_1\varepsilon_2}{\varepsilon_1 + \varepsilon_2} = \frac{\omega^2}{c^2}\left(\frac{1}{\varepsilon_1(\omega)} + \frac{1}{\varepsilon_2(\omega)}\right)^{-1} \tag{5.7.5}$$

This is the law of dispersion for surface transverse waves. As we see from (5.7.4), surface transverse waves exist only in the region where $\varepsilon < 0$ for one of the media, since $\kappa_{1,2} > 0$.

Assume now that $\varepsilon_2 < 0$ and $\varepsilon_1 > 0$. For example, if medium I is the vacuum, so that $\varepsilon_1 = 1$, (5.7.5) implies the following condition of existence of surface transverse waves:

$$\varepsilon_2(\omega) < -\varepsilon_1(\omega)$$

The law of dispersion of surface waves is plotted in Figure 5.12. Surface waves exist only in the frequency gap from ω_0 to ω_l of bulk waves. The figure also shows schematically, for the sake of comparison, the law of dispersion for bulk polaritons. If $k \to \infty$, then $\varepsilon_1(\omega) + \varepsilon_2(\omega) \to 0$, that

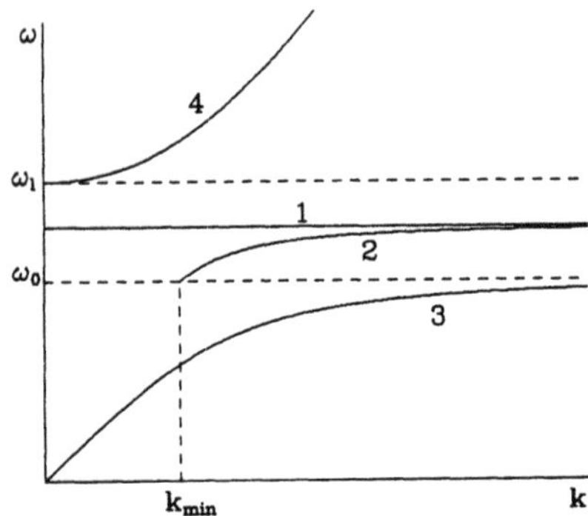

Figure 5.12. Law of dispersion for longitudinal surface polaritons (1) and transverse surface polaritons (2). Also plotted are the dispersion curves for bulk polaritons (3 and 4).

is, the branch of surface transverse polaritons tends to run close to that of surface longitudinal polaritons.

There exists a minimal $k = k_{\text{min}}$ for surface transverse polaritons, which corresponds to $\varepsilon_2(\omega) \to -\infty$, that is, $\omega \to \omega_0$ and

$$k_{\text{min}}^2 = \frac{\omega_0^2}{c^2} \varepsilon_1(\omega_0)$$

This value of k coincides with the wave vector of the bulk wave in medium I (e.g., the vacuum) for $\omega = \omega_0$.

5.8. Model of a Lattice with Pointlike Ions

Let us turn again to the pointlike ions lattice model. In Section 5.5, we analyzed the interaction of this lattice with long-wavelength field. The role of higher-order spatial field harmonics was taken into account implicitly, by determining the resonance frequencies $\omega'_{\mathbf{q}j}$ and the "effective charges" $Q'_{j\alpha}(\mathbf{q})$.

The short-wavelength fields can be regarded, as was shown in Section 5.1.9, as purely longitudinal and their interaction with lattice charges was already taken into account in the lattice force constants matrix A or A', but not in $A'' = A + C$ (see Section 5.5). In this section, we will find the total susceptibility matrices $\chi_{\alpha\beta}^{\mathbf{l}\mathbf{l}'}(\mathbf{q}, \omega)$ and the matrices $\chi_{\alpha\beta}^{(\mathrm{e})\mathbf{l}\mathbf{l}'}(\mathbf{q}, \omega)$ of response to the external field in the framework of the simple pointlike ions lattice model, and examine their relation to the force constant matrices A'' and A.

Consider a lattice whose lattice sites $\mathbf{R}_{\mathbf{n}s} = \mathbf{R}_{\mathbf{n}} + \boldsymbol{\rho}_s$ ($\mathbf{R}_{\mathbf{n}}$ being the radius vector of the central point of the unit cell) are occupied by pointlike ions with charges $Z_s e$. In addition to the direct Coulomb interaction between ions, there is also a different sort of interionic interaction, connected with van der Waals forces, the exchange interaction, and the repulsion of ions at short distances when the filled electron shells of the ions overlap. In contrast to the Coulomb interaction, these forces fall off rapidly with distance (as R^{-6} for the van der Waals interaction and exponentially for the exchange interaction, while the Coulomb interaction depends on distance as R^{-1}).

Assume first that the interionic Coulomb interaction is not specified in the force constants matrix A. By analogy to Section 5.5, the equations of motion for the harmonic field $\mathcal{E}^{(e)}(\mathbf{r},t) = \mathbf{E}_{\omega}^{(e)}e^{-i\omega t}$, we then have

$$\begin{aligned} -M_s\omega^2 u_{\mathbf{n}s\alpha} + \sum_{\mathbf{n}'s'\alpha'} A_{\mathbf{n}-\mathbf{n}'}^{\alpha\alpha'\,ss'} u_{\mathbf{n}'s'\alpha'} &= eZ_s E_{\omega\alpha}^{(e)}(\mathbf{R}_{\mathbf{n}s}) \\ &= eZ_s \int E_{\omega\alpha}^{(e)}(\mathbf{r}')\delta(\mathbf{r}' - \mathbf{R}_{\mathbf{n}s})\, d^3r' \end{aligned} \tag{5.8.1}$$

The charges eZ_s are not necessarily integral numbers of e. Even in such a typically ionic compound as NaCl, $Z_s \sim 0.8$, the charge transfer is even lower in other crystals.

The current density in the classical problem can be written as

$$j_{\omega\alpha}(\mathbf{r}) = \sum_{\mathbf{n}s\alpha} eZ_s(-i\omega)u_{\mathbf{n}s\alpha}\delta(\mathbf{r} - \mathbf{R}_{\mathbf{n}s}) \tag{5.8.2}$$

We perform the Fourier transformation of equations (5.8.1) introducing

$$u_{\mathbf{k}s\alpha} = \frac{1}{\sqrt{N}}\sum_{\mathbf{n}} \exp(-i\mathbf{k}\mathbf{R}_{\mathbf{n}})u_{\mathbf{n}s\alpha}$$

so that

$$u_{\mathbf{n}s\alpha} = \frac{1}{\sqrt{N}}\sum_{\mathbf{k}} \exp(i\mathbf{k}\mathbf{R}_{\mathbf{n}})u_{\mathbf{k}s\alpha} \tag{5.8.3}$$

Substituting this last equation into equation (5.8.1), multiplying it by $N^{-1/2}e^{i\mathbf{k}\mathbf{R}_{\mathbf{n}}}$ and summing up over $\mathbf{n}$, we obtain

$$\begin{aligned} -M\omega^2\tilde{u}_{\mathbf{k}s\alpha} + \sum_{s'\alpha'} \tilde{A}_{\alpha\alpha'}^{ss'}(\mathbf{k})u_{\mathbf{k}s'\alpha'} &= \frac{e\tilde{Z}_s}{\sqrt{N}} \\ \times \sum_{\mathbf{n}} \int d^3r' E_{\omega\alpha}^{(e)}(\mathbf{r}') \exp(-i\mathbf{k}\mathbf{R}_{\mathbf{n}})\delta(\mathbf{r}' - \mathbf{R}_{\mathbf{n}s}) \end{aligned} \tag{5.8.4}$$

where

$$\tilde{u}_{\mathbf{k}s\alpha} = \sqrt{\frac{M_s}{M}}u_{\mathbf{k}s\alpha} \qquad \tilde{Z}_s = \sqrt{\frac{M}{M_s}}Z_s$$

Equation (5.8.4) in the symbolic matrix notation is

$$(\widetilde{A}(\mathbf{k}) - M\omega^2)\tilde{u}_{\mathbf{k}} = \frac{1}{\Omega\sqrt{N}} \sum_{\mathbf{l}} \widetilde{Q}_{\beta}^{\mathbf{l}*} \int d^3r' \exp[-i(\mathbf{k}+\mathbf{g}_{\mathbf{l}})\mathbf{r}'] E_{\omega\beta}^{(e)}(\mathbf{r}') \quad (5.8.5)$$

where $\widetilde{A}(\mathbf{k})$ is a matrix of rank $3r$ and the rows and columns $\widetilde{Q}_{\alpha}^{\mathbf{l}}$ and $\widetilde{Q}_{\beta}^{\mathbf{l}*}$ are defined by the relations

$$\begin{aligned} (\widetilde{Q}_{\beta}^{\mathbf{l}}(\mathbf{k}))_{s\alpha} &= e\widetilde{Z}_s \exp[-i(\mathbf{k}+\mathbf{g}_{\mathbf{l}})\boldsymbol{\rho}_s]\delta_{\alpha\beta} \\ (\widetilde{Q}_{\beta}^{\mathbf{l}*}(\mathbf{k}))_{s\alpha} &= e\widetilde{Z}_s \exp[i(\mathbf{k}+\mathbf{g}_{\mathbf{l}})\boldsymbol{\rho}_s]\delta_{\alpha\beta} \end{aligned} \quad (5.8.6)$$

To write the equations in the form (5.8.5), we used the equality

$$\begin{aligned} \sum_{\mathbf{n}} \exp(-i\mathbf{k}\mathbf{R}_{\mathbf{n}})\delta(\mathbf{r}' - \mathbf{R}_{\mathbf{n}} - \boldsymbol{\rho}_s) &= \exp[-i\mathbf{k}(\mathbf{r}' - \boldsymbol{\rho}_s)] \\ \times \sum_{\mathbf{n}} \delta(\mathbf{r}' - \mathbf{R}_{\mathbf{n}} - \boldsymbol{\rho}_s) &= \frac{1}{\Omega} \sum_{\mathbf{l}} \exp[-i(\mathbf{k}+\mathbf{g}_{\mathbf{l}})(\mathbf{r}' - \boldsymbol{\rho}_s)] \end{aligned}$$

where $\sum_n \delta(\mathbf{r}' - \boldsymbol{\rho}_s - \mathbf{R}_{\mathbf{n}})$, a periodic function of $\mathbf{r}'$, is expanded in a Fourier series.

Likewise, the expression for current density (5.8.2) can be rewritten, after the substitution of (5.8.3) and (5.8.6) into it, as

$$j_{\omega\alpha}(\mathbf{r}) = -\frac{i\omega}{\sqrt{N}\Omega} \sum_{\mathbf{kl}} \tilde{u}_{\mathbf{k}} \widetilde{Q}_{\alpha}^{\mathbf{l}}(\mathbf{k}) \exp[i(\mathbf{k}+\mathbf{g}_{\mathbf{l}})\mathbf{r}]$$

Substituting here $\tilde{u}_{\mathbf{k}}$ from (5.8.5), we obtain

$$\begin{aligned} j_{\omega\alpha}(\mathbf{r}) = -\frac{i\omega}{N\Omega^2} \sum_{\mathbf{kll}'} \exp[i(\mathbf{k}+\mathbf{g}_{\mathbf{l}})\mathbf{r}] \int d^3r' \widetilde{Q}_{\alpha}^{l} \\ \times [\widetilde{A}(\mathbf{k}) - M\omega^2]^{-1} \widetilde{Q}_{\beta}^{\mathbf{l}'*} \exp[-i(\mathbf{k}+\mathbf{g}_{\mathbf{l}})\mathbf{r}'] E_{\omega\beta}^{(e)}(\mathbf{r}') \end{aligned}$$

If this formula is compared with the definition of susceptibility,

$$j_{\omega\alpha}(\mathbf{r}) = -i\omega \int d^3r' \chi_{\alpha\beta}^{(e)}(\mathbf{r}, \mathbf{r}'; \omega) E_{\omega\beta}^{(e)}(\mathbf{r}')$$

we obtain, in accordance with Section 5.1.8, the total matrix of response to the arbitrary external field

$$\chi_{\alpha\beta}^{(e)\mathbf{ll}'}(\mathbf{k}, \omega) = \frac{1}{\Omega} \widetilde{Q}_{\alpha}^{\mathbf{l}} [\widetilde{A}(\mathbf{k}) - M\omega^2]^{-1} \widetilde{Q}_{\beta}^{\mathbf{l}'*} \quad (5.8.7)$$

This gives us a compact expression (5.8.7) for $\chi^{(e)}$. Its poles are given by the equation $\det\|\tilde{A} - M\omega^2\| = 0$, which gives the natural frequencies of lattice vibrations. To derive a more explicit expression for $\chi^{(e)}$, we expand $\widetilde{Q}_\beta^{\mathbf{l}*}$ in the eigenvectors $\tilde{e}^j_{s\alpha}(\mathbf{k})$ of the matrix $\tilde{A}(\mathbf{k})$, by using the fact that this system of vectors is complete:

$$(\widetilde{Q}_\beta^{\mathbf{l}*}(\mathbf{k}))_{s\alpha} = \sum_j \tilde{e}^j_{s\alpha}(\mathbf{k}) Q^{\mathbf{l}*}_{j\beta}(\mathbf{k})$$

Hence,

$$\chi^{(e)\mathbf{ll}'}_{\alpha\beta}(\mathbf{k},\omega) = \frac{1}{\Omega}\sum_j \frac{Q^{\mathbf{l}}_{j\alpha}(\mathbf{k}) Q^{\mathbf{l}*}_{j\alpha}(\mathbf{k})}{M(\omega^2_{\mathbf{k}j} - \omega^2)} \tag{5.8.8}$$

This formula is similar to expression (5.5.5) but differs from it in the arrangement of poles ($\omega_{\mathbf{k}j} \neq \omega'_{\mathbf{k}j}$) and also in that it describes the relation between arbitrary, not only long-wavelength, spatial current and field harmonics.

Now it has become clear that the susceptibility matrix of response to the total field in a medium, $\chi^{\mathbf{ll}'}_{\alpha\beta}(\mathbf{k},\omega)$, is to be derived in a completely analogous manner from the equations of motion which differ from (5.8.1) only in that the total field — both the long-wavelength and the small-scale — are included in the righthand side, and that the matrix $\widetilde{A}'' = \widetilde{A} + \widetilde{C}$ defined in Section 5.5 (which contains no contribution from interatomic interactions through electric and electromagnetic fields) plays the role of force constants in the lefthand side of the equations. It is quite obvious that the result of this analysis is a function

$$\chi^{\mathbf{ll}'}_{\alpha\beta}(\mathbf{k},\omega) = \frac{1}{\Omega}\widetilde{Q}^{\mathbf{l}}_\alpha[\widetilde{A}''(\mathbf{k}) - M\omega^2]^{-1}\widetilde{Q}^{\mathbf{l}'*}_\beta \tag{5.8.9}$$

whose poles $\omega''_{\mathbf{k}j}$ are found from the equation $\det\|\tilde{A}'' - M\omega^2\| = 0$ which ignores the contribution of the dipole–dipole interaction between ions. Rather than reproducing here this derivation, it appears more instructive to analyze, using the formulas of the general theory outlined in Chapter 1, the relationship between formulas (5.8.7) and (5.8.9); as an illustration, we will derive (5.8.7) from (5.8.9).

We begin with the case of moderately long wavelengths, $q^2 \gtrsim \omega^2/c^2$, when the contribution of the transverse field is negligible even for the main spatial harmonic $\mathbf{l} = 0$.

In symbolic notation used in Section 5.1.8, we can rewrite (5.8.7) and (5.8.9) in a concise form:

$$\chi^{(e)} = \frac{1}{\Omega}\widetilde{Q}[\widetilde{A} - M\omega^2]^{-1}\widetilde{Q}^*$$

$$\chi = \frac{1}{\Omega}\widetilde{Q}[\widetilde{A}'' - M\omega^2]^{-1}\widetilde{Q}^*$$

The relation between χ and $\chi^{(e)}$ for a crystal is obtained by completely retracing the derivation of Section 5.1.6; it has the form

$$\chi^{(e)} = \chi \left[1 - \frac{4\pi\omega^2}{c^2} D_0 \chi\right]^{-1} \tag{5.8.10}$$

If $\mathbf{k}$ is not too small, D_0 must be replaced by its longitudinal component,

$$D_{0\parallel} = -\frac{c^2}{\omega^2}\Pi_\parallel$$

where $\Pi_\parallel$ is the operator of projection onto the direction of the vector $(\mathbf{k}+\mathbf{g}_l)$:

$$(\Pi_\parallel)_{\alpha\beta} = (\mathbf{k}+\mathbf{q}_l)_\alpha(\mathbf{k}+\mathbf{g}_l)_\beta\,|\mathbf{k}+\mathbf{g}_l|^{-2}$$

Now we expand the second multiplier in (5.8.10) in a series in powers of $4\pi\Pi_\parallel\chi$:

$$\chi^{(e)} = \chi\sum_{n=0}^{\infty}(-4\pi\Pi_\parallel\chi)^n$$

This expansion is the generalization of the ordinary expression for the sum of the geometrical progression, in which all multiplication operations are treated as multiplication of matrices. Substituting now the above expression for the susceptibility χ (5.8.9) and recollecting all multipliers in the numbers of the series for $\chi^{(e)}$, we rewrite this series as

$$\chi^{(e)} = \frac{1}{\Omega}\widetilde{Q}[\widetilde{A}'' - M\omega]^{-1}\sum_{n=0}^{\infty}\left(-\frac{4\pi}{\Omega}\widetilde{Q}^*\Pi_\parallel\widetilde{Q}[\widetilde{A}'' - M\omega^2]^{-1}\right)^n\widetilde{Q}^*$$

Finally, rearranging the series into the geometrical progression formula, we obtain

$$\chi^{(e)} = \frac{1}{\Omega}\widetilde{Q}[\widetilde{A}'' - \widetilde{C} - M\omega^2]^{-1}\widetilde{Q}^* \tag{5.8.11}$$

where $\widetilde{C} = -4\pi\Omega^{-1}\widetilde{Q}^*\Pi_\parallel\widetilde{Q}$ is the matrix of the dipole–dipole interaction introduced in Section 5.5. We have thus again derived formula (5.8.7) for $\chi^{(e)}$, since $\widetilde{A}'' - \widetilde{C} = \widetilde{A}$.

In the same manner, we can use the general formulas of Chapter 1 and derive expression (5.8.9) for χ from the expression for $\chi^{(e)}$. In a certain sense, therefore, these functions are equivalent: each one defines the other. However, they must be used differently.

The matrix χ that ignores the contribution of the field to interatomic interactions must be used in complete Maxwell's equations when the fields and currents in the sample are determined. However, the matrix $\chi^{(e)}$ which takes complete account of the field contribution already includes all this information. The function $\overline{\chi}$ (5.5.5) occupies a special position. It implicitly covers the contribution of the higher field harmonics and thus, jointly with Maxwell's equations, determines quite strictly the distribution of long-wavelength (spatially averaged) fields and currents; however, it does not make it possible to determine explicitly their small-scale structure.

These statements remain quite true when the transverse field is taken into account. For fields with wavelengths considerably greater than interatomic distances, that is, fields down to x-ray wavelengths, it becomes even easier to cover transverse fields in calculations. We have mentioned on a number of occasions that the transverse field plays practically no role in the formation of higher-order spatial harmonics. Taking this field into account is thus important only for the main spatial harmonic, $\chi^{(e)}$, of the response to the external field, since if the matrix χ (susceptibility) is calculated, contributions of all fields are ignored, and if $\overline{\chi}$ (spatially averaged susceptibility) is calculated, only the higher-order spatial harmonics are taken into account.

Hence, the simplest and most natural method of including the effects of the transverse field is to use formula (5.6.4). The latter expresses $\chi^{(e)00}_{\alpha\beta}(\mathbf{k},\omega)$ in terms of $\overline{\chi}_{\alpha\beta}(\mathbf{k},\omega)$, which is calculated neglecting both the transverse field effects and the matrix D_0 which contains both the longitudinal and the transverse terms. We have shown at the end of Section 5.6 that $\chi^{(e)00}_{\alpha\beta}(\mathbf{k},\omega)$ thus determined does indeed give the correct description of the propagation of transverse waves through a medium.

The reason for giving such a detailed analysis of the pointlike-ions model is that this model allows an analysis of the physical nature and the differences between various response functions in a most comprehensive and at the same time sufficiently simple way. The qualitative conclusions we have drawn actually cover any medium.

For characterizing a medium, $\overline{\chi}$ and χ have a greater physical meaning. Indeed, $\overline{\chi}$ and χ are determined only by the short-wavelength fields in the matter and therefore their nonlocality radius is of the order of a_0. This is clear from the expressions obtained in this section and in Sections 5.5 and 5.6. The matrix $\widetilde{A}'(\mathbf{k})$ and its eigenvectors $\tilde{e}'$ are independent of $\mathbf{k}$, up to $\mathbf{k}$ of the order of a_0^{-1}, because short-wavelength fields are mostly determined by displacements of ions within one unit cell as long as $k \ll g_1 \sim a_0^{-1}$.

This is not true, however, for $\chi^{(e)}$ and $\overline{\chi}^{(e)}$. Indeed, as a result of long-wavelength fields the nonlocality radius may grow to values of the order of macroscopic scales. Assume that we consider a cubic optically isotropic crystal for which $\overline{\chi}_{\alpha\beta} = \chi_0\delta_{\alpha\beta}$. As follows from the formulas of

Section 5.1.6,

$$\bar{\chi}^{(e)} = \bar{\chi}\left[1 - \frac{4\pi\omega^2}{c^2}\, D_0\bar{\chi}\right]^{-1}$$

$$= \chi_0\left[\left(1 - \frac{4\pi\chi_0}{k^2c^2\omega^{-2} - 1}\right)\Pi_\perp + (1 + 4\pi\chi_0)\Pi_\parallel\right]^{-1}$$

$$= \chi_0\left[\left(1 + \frac{4\pi\chi_0}{1 - k^2c^2\omega^{-2}}\right)^{-1}\Pi_\perp + (1 + 4\pi\chi_0)^{-1}\Pi_\parallel\right]$$

If $k \ll \omega/c$, then $\overline{\chi}^{(e)} = \chi_0/1 + 4\pi\chi_0 = \chi_0/\varepsilon$, but $\overline{\chi}^{(e)}$ now differs considerably from $\overline{\chi}^{(e)}(\mathbf{k} = 0)$ at $k \sim \omega/c$. The nonlocality radius for $\overline{\chi}^{(e)}$ is thus of the order of optical wavelength. For low frequencies, when the specimen size is less than c/ω, the nonlocality radius is determined by the specimen size and its geometric shape. The specimen polarization for elongated shapes is affected even by the orientation of the "bar" relative to the direction of the external field (the familiar depolarization factor).

At the same time, the theory calculates, as we have emphasized already, nothing different but $\chi_{\alpha\beta}^{(e)ll'}$ and $\overline{\chi}_{\alpha\beta}^{(e)} = \chi_{\alpha\beta}^{(e)00}$. Furthermore, it can be said that in the long run, experimental studies also analyze, as a rule, only $\chi^{(e)}$, since fields in optical and RF experiments are determined by external sources and are recorded by receivers which are external with respect to the medium. This was the reason for introducing $\chi^{(e)}$ at all.

In conclusion, we will consider another significant generalization of the pointlike ions model. So far we were considering the contribution to the susceptibility of the medium, caused by the displacement of ions from equilibrium positions. In fact, ions (and neutral atoms) also have the intrinsic (electron) polarizability $\varkappa(\omega)$, that is, the ability to generate a dipole moment proportional to the local value of the field on the ion: $d = \varkappa(\omega)\boldsymbol{\mathcal{E}}(\mathbf{R_n})$. If we again assume ions to be pointlike, then the electron polarization is the ensemble of the pointlike dipoles located at lattice sites $\mathbf{R}_{\mathbf{n}s}$. This model of electron polarizability has already been treated in Section 5.1.10 (the Lorentz model).

Assume now $\chi_{\rm i}$ to be the lattice susceptibility (5.8.9) and $\chi_{\rm el}$ to be the susceptibility of the electron subsystem, treated in Section 5.1.10. The total susceptibility is then $\chi = \chi_{\rm i} + \chi_{\rm el}$ and the mean susceptibility for long-wavelength field is

$$\bar{\chi} = \left\{(\chi_{\rm i} + \chi_{\rm el})\left[1 - \frac{4\pi\omega^2}{c^2}\,\widetilde{D}_0(\chi_{\rm i} + \chi_{\rm el})\right]^{-1}\right\}^{00}$$

For longitudinal fields of higher-order spatial harmonics we find

$$\widetilde{D}_0 = \widetilde{D}_{0\parallel} = -\frac{c^2}{\omega^2}\widetilde{\Pi}_\parallel$$

where $\widetilde{\Pi}_{\|}$ is the operator of projection onto the direction of the vector $\mathbf{k}+\mathbf{g_l}$ at $\mathbf{l} \neq 0$. Hence,

$$\bar{\chi} = \{(\chi_{\rm i} + \chi_{\rm el})[1 + 4\pi\widetilde{\Pi}_{\|}(\chi_{\rm i} + \chi_{\rm el})]^{-1}\}^{00} \tag{5.8.12}$$

We expand the matrix $[1+4\pi\widetilde{\Pi}_{\|}(\chi_{\rm i}+\chi_{\rm el})]^{-1}$ in powers of $4\pi\widetilde{\Pi}_{\|}\chi_{\rm i}$, in order to generate transformations of the same type as those that derived (5.8.11) from (5.8.9). Now the expansion becomes

$$\begin{aligned}
\left[1 + 4\pi\widetilde{\Pi}_{\|}(\chi_{\rm i} + \chi_{\rm el})\right]^{-1} &= \left[(1 + 4\pi\widetilde{\Pi}_{\|}\chi_{\rm el}) + 4\pi\widetilde{\Pi}_{\|}\chi_{\rm i}\right]^{-1} \\
&= \left[1 + (1 + 4\pi\widetilde{\Pi}_{\|}\chi_{\rm el})^{-1}4\pi\widetilde{\Pi}_{\|}\chi_{\rm i}\right]^{-1}(1 + 4\pi\widetilde{\Pi}_{\|}\chi_{\rm el})^{-1} \\
&= \Big[1 - (1 + 4\pi\widetilde{\Pi}_{\|}\chi_{\rm el})^{-1}4\pi\widetilde{\Pi}_{\|}\chi_{\rm i} + (1 + 4\pi\widetilde{\Pi}_{\|}\chi_{\rm el})^{-1} \\
&\times 4\pi\widetilde{\Pi}_{\|}\chi_{\rm i}(1 + 4\pi\widetilde{\Pi}_{\|}\chi_{\rm el})^{-1}4\pi\widetilde{\Pi}_{\|}\chi_{\rm i} - \ldots\Big](1 + 4\pi\widetilde{\Pi}_{\|}\chi_{\rm el})^{-1}
\end{aligned}$$

Substituting now $\chi_{\rm i}$ from (5.8.9) into the equation above and into (5.8.12) and then collecting the series back like we did to obtain (5.8.11), we find

$$\begin{aligned}
\bar{\chi} = \Big\{&\frac{1}{\Omega}(1 + 4\pi\chi_{\rm el}\widetilde{\Pi}_{\|})^{-1}\widetilde{Q}(\widetilde{A}'' + \delta\widetilde{A}' - M\omega^2)^{-1}\widetilde{Q}^* \\
&\times(1 + 4\pi\widetilde{\Pi}_{\|}\chi_{\rm el})^{-1} + \chi_{\rm el}(1 + 4\pi\widetilde{\Pi}_{\|}\chi_{\rm el})^{-1}\Big\}^{00}
\end{aligned} \tag{5.8.13}$$

where

$$\delta\widetilde{A}' = \frac{4\pi}{\Omega}\widetilde{Q}^*(1 + 4\pi\widetilde{\Pi}_{\|}\chi_{\rm el})^{-1}\widetilde{\Pi}_{\|}\widetilde{Q}$$

The second term in (5.8.13) is the averaged electron susceptibility, already derived in Section 5.1.10, which takes into account the difference between the actual and the mean fields. However, the presence of electron polarizability alters drastically the contribution of ion vibrations to the susceptibility. In accordance with the structure of the first term in (5.8.13), these changes can be interpreted as changes in the effective charges $\widetilde{Q} \to (1 + 4\pi\chi_{\rm el}\widetilde{\Pi}_{\|})^{-1}\widetilde{Q}$ and changes in the vibrational frequencies of ions, since $\widetilde{A}' \to \widetilde{A}' + \delta\widetilde{A}'$. It is not difficult to give a qualitative explanation of these changes.

The appearance of additional (electron) dipoles affects the local values of fields on the ions in comparison with the values in the absence of electron polarizability. This change in the acting field can be taken into account formally, assuming that the field remains the same but the ion charge is multiplied by a factor $(1+4\pi\chi_{\rm el}\widetilde{\Pi}_{\|})^{-1}$. The dipole interaction between ions is accordingly altered, since it now takes place in a medium with additional susceptibility χ_{el}, which changes the force constants matrix by a quantity $\delta\widetilde{A}'$.

Calculations similar to those given for the linear susceptibility and the use of Hamilton's function (5.4.1) with nonlinear corrections implied by the matrix B lead to an expression for the quadratic susceptibility:

$$\chi^{(2e)}(\mathbf{k}=\mathbf{k}'+\mathbf{k}'',\omega=\omega'+\omega'') = -\frac{1}{2\Omega}\widetilde{Q}(\mathbf{k})[\widetilde{A}(\mathbf{k})-M\omega^2]^{-1} \times \widetilde{B}(\mathbf{k}',\mathbf{k}'')[\widetilde{A}(\mathbf{k}')-M\omega'^2]^{-1}[\widetilde{A}(\mathbf{k}'')-M\omega''^2]^{-1}\widetilde{Q}^*(\mathbf{k}')\widetilde{Q}^*(\mathbf{k}'')$$

where $\widetilde{B}(\mathbf{k}',\mathbf{k}'')$ is the Fourier transform of the matrix

$$\widetilde{B}^{ss's''}_{\substack{\alpha\alpha'\alpha''\\ \mathbf{n}-\mathbf{n}',\mathbf{n}-\mathbf{n}''}} = B^{ss's''}_{\substack{\alpha\alpha'\alpha''\\ \mathbf{n}-\mathbf{n}',\mathbf{n}-\mathbf{n}''}}\left(\frac{M_sM_{s'}M_{s''}}{M^3}\right)^{-1/2}$$

Using the relation of $\chi^{(2e)}$ to the true susceptibility $\chi^{(2)}$ and the relation between $\overline{\chi}^{(2)}$ and $\chi^{(2)}$, we can find

$$\bar{\chi}^{(2)}(\mathbf{k}=\mathbf{k}'+\mathbf{k}'',\omega=\omega'+\omega'') = -\frac{1}{2\Omega}\widetilde{Q}^0(\mathbf{k})[\widetilde{A}'(\mathbf{k})-M\omega^2]^{-1} \times \widetilde{B}(\mathbf{k}',\mathbf{k}'')[\widetilde{A}'(\mathbf{k}')-M\omega'^2]^{-1}[\widetilde{A}'(\mathbf{k}'')-M\omega''^2]^{-1}\widetilde{Q}^{0*}(\mathbf{k}')\widetilde{Q}^{0*}(\mathbf{k}'')$$

In its matrix form, this expression corresponds to (5.5.7). Note that the expression for $\chi^{(2)}$ includes the matrix $\widetilde{A}'(\mathbf{k})$ and not the total matrix $\widetilde{A}(\mathbf{k})$ that we find in the expression for $\chi^{(2e)}$. The reader will recall that $\widetilde{A}'(\mathbf{k})$ neglects the long-wavelength field.

Another remark is in order to conclude the discussion of the electromagnetic response of ion lattices. The consideration above was carried out in classical terms. The same results can be obtained quantum-mechanically from formula (2.6.1) since the matrix elements for harmonic oscillators (normal modes of lattice oscillations) are well known. This is not necessary, however, since the harmonic oscillator is a unique system in the sense that its response to an external force is practically identical, after averaging over the wave function, to the result of classical analysis. Consequently, quantum effects manifest themselves appreciably when damping due to phonon decays is taken into account. Phenomenologically, they can be included into the formulas above by introducing imaginary increments like $2i\omega\gamma_{\mathbf{k}j}$ into all resonance denominators ($\gamma_{\mathbf{k}j}$ is the inverse phonon lifetime in the $\mathbf{k}j$ mode).

The quantum approach will be quite indispensable in the next section, where we turn to the electromagnetic response of electrons in a crystal.

5.9. Electrons in Crystals. Bloch Theorem

The type of bonding plays a decisive role in shaping the electronic properties of crystals. If the electron exchange is intense, electrons get delocalized. Contrariwise, electrons in molecular crystals are localized and the spectra differ only slightly from those of the constituent molecules.

One of the important achievements of quantum theory is the explanation of the fact that free path lengths of electrons in metals and semiconductors may greatly exceed interatomic distances. The potentials of electron–ion interaction being high, electrons constantly collide with ions and it can be expected that the free path length are of the order of distances between atoms. In fact, if the crystal is ideally periodic, an electron can travel without scattering; a finite free path length of macroscopic value is caused by deviations from periodicity: impurities, defects and distortions caused by vibrations (phonons).

Strictly speaking, the motion of electrons must be treated in terms of the wave function of a large number of electrons even in the adiabatic approximation (with fixed positions of nuclei). However, a one-electron approximation is often used, in which the motion of a single electron is considered in a self-consistent crystal field created by the lattice nuclei and all other electrons. It can be justified more rigorously by using the equation of motion for the density matrix of a single particle; this matrix can always be introduced. In the one-electron approximation, the Schrödinger equation for the wave function of an electron in the self-consistent field is

$$-\frac{\hbar}{2m}\nabla^2\psi(\mathbf{r}) + W(\mathbf{r})\psi(\mathbf{r}) = E\psi(\mathbf{r}) \tag{5.9.1}$$

where $W(\mathbf{r})$ is a periodic self-consistent potential and E is the energy. The potential $W(\mathbf{r})$ satisfies the periodicity condition $W(\mathbf{r}+\mathbf{R_n}) = W(\mathbf{r})$.

A (5.9.1)-type equation for the wave function cannot be derived in a rigorous manner but a corresponding equation for the density matrix can.

The group symmetry of the electron's Hamiltonian in a crystal, which results in (5.9.1), is the group of translations $T_\mathbf{n}$. The solutions of equation (5.9.1) realize irreducible representations of the group of translations $T_\mathbf{n}\psi(\mathbf{r}) = \exp(\mathbf{ikR_n})\psi$, as we have mentioned in Section 5.1. This means that the function $\psi\exp(-\mathbf{ikr})$ is invariant under all translations:

$$\begin{aligned} T_\mathbf{n}\psi\exp(-\mathbf{ikr}) &= \exp(\mathbf{ikR_n})\exp(-\mathbf{ikR_n})\psi\exp(-\mathbf{ikr}) \\ &= \psi\exp(\mathbf{ikr}) = u \end{aligned}$$

Hence, the function ψ, which is a solution of equation (5.9.1), must have the form

$$\psi = u\exp(\mathbf{ikr})$$

where $u(\mathbf{r} + \mathbf{R_n}) = u(\mathbf{r})$ is a periodic function.

Usually not the wave vector $\mathbf{k}$ but the quasimomentum $\mathbf{p} = \hbar\mathbf{k}$ is introduced for the electron. We will show now that equation (5.9.1) indeed has solutions having the form

$$\psi_{\mathbf{p}j}(\mathbf{r}) = \frac{1}{\sqrt{N}} \exp(\frac{i}{\hbar}\mathbf{pr})u_{\mathbf{p}j}(\mathbf{r}) \tag{5.9.2}$$

where the index j enumerates the solutions which correspond to a given $\mathbf{p}$. The factor $N^{-1/2}$ is introduced into (5.9.2) to transform the normalization condition to

$$\int_{\Omega} u^*_{\mathbf{p}j}(\mathbf{r})u_{\mathbf{p}j}(\mathbf{r})\, \mathrm{d}^3r = 1$$

where integration is carried over the volume of one unit cell and the crystal consists, as when phonons are considered, of N unit cells.

Substituting (5.9.2) into (5.9.1), we obtain the equation for the periodic function $u_{\mathbf{p}j}(\mathbf{r})$:

$$-\frac{\hbar^2}{2m}\nabla^2 u_{\mathbf{p}j} - \frac{i\hbar}{m}\mathbf{p}\nabla u_{\mathbf{p}j} + \left(W(\mathbf{r}) + \frac{p^2}{2m} - E_{\mathbf{p}j}\right)u_{\mathbf{p}j} = 0 \tag{5.9.3}$$

At first glance, this equation is more complex than (5.9.1) but this is an equation for the function $u_{\mathbf{p}j}$ given on a finite volume Ω of a unit cell and satisfying the periodic boundary conditions at its boundaries. This is an eigenvalue problem for a self-conjugate operator $L_{\mathbf{p}}$:

$$L_{\mathbf{p}}u_{\mathbf{p}j} = E_{\mathbf{p}j}u_{\mathbf{p}j}$$
$$L_{\mathbf{p}} = -\frac{\hbar^2}{2m}\nabla^2 - \frac{i\hbar}{m}\mathbf{p}\nabla + \frac{\mathbf{p}^2}{2m} + W(\mathbf{r})$$

The self-conjugation of the operator $L_{\mathbf{p}}$ means that the equality

$$\int_{\Omega} v^* L_{\mathbf{p}} u\, \mathrm{d}^3r = \int_{\Omega} (L_{\mathbf{p}}v)^* u\, \mathrm{d}^3r \tag{5.9.4}$$

holds for arbitrary periodic functions u and v for fixed $\mathbf{p}$. Since the functions u and v are periodic, the integration in (5.9.4) can be carried over an arbitrarily selected unit cell, for instance, over the Wigner–Seitz unit cell.

The self-conjugation of the operator $L_{\mathbf{p}}$ (5.9.4) is readily verified, using its definition, by integrating by parts. A self-conjugate operator, acting in a finite volume Ω, has a discrete spectrum $E_{\mathbf{p}j}$, its eigenfunctions $u_{\mathbf{p}j}$ are orthogonal and enumerated by the index j (for fixed $\mathbf{p}$). By contrast, the spectrum of (5.9.1) is definitely continuous or quasicontinuous, since

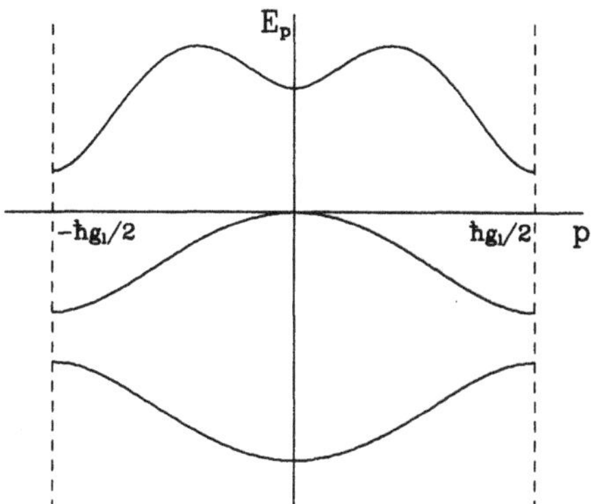

Figure 5.13. Law of dispersion for electrons in a crystal.

$\mathbf{p}$ for large N varies almost continuously. Functions of the type of (5.9.2) are known as the Bloch functions. The corresponding energy values $E_{\mathbf{p}j}$ for each j form bands (allowed bands), and the dependence of $E_{\mathbf{p}j}$ on $\mathbf{p}$ gives the law of dispersion for electrons in the jth band of the crystal. Qualitatively, the law of dispersion is shown in Figure 5.13.

We now assume that the Bloch functions are normalized by the conditions:

$$\int_\Omega u^*_{\mathbf{p}j} u_{\mathbf{p}j}\, \mathrm{d}^3r = \delta_{j'j} \qquad \int_V \psi^*_{\mathbf{p}'j'} \psi_{\mathbf{p}j}\, \mathrm{d}^3r = \delta_{j'j}\delta_{\mathbf{p}'\mathbf{p}} \tag{5.9.5}$$

where $V = N\Omega$.

The first condition in (5.9.5) stems from the self-conjugation of the operator $L_{\mathbf{p}}$, as analyzed above, and the second is implied by the equalities

$$\int_V \psi^*_{\mathbf{p}'j'} \psi_{\mathbf{p}j}\, \mathrm{d}^3r = \frac{1}{N} \sum_{\mathbf{n}} \exp\left(-\frac{\mathrm{i}}{\hbar}(\mathbf{p}' - \mathbf{p})\mathbf{R}_{\mathbf{n}}\right) \int_\Omega \exp\left(-\frac{\mathrm{i}}{\hbar}(\mathbf{p}' - \mathbf{p})\mathbf{r}\right)$$

$$\times\, u^*_{\mathbf{p}'j'} u_{\mathbf{p}j}\, \mathrm{d}^3r = \int_\Omega \exp\left(-\frac{\mathrm{i}}{\hbar}(\mathbf{p}' - \mathbf{p})\mathbf{r}\right) u^*_{\mathbf{p}'j'} u_{\mathbf{p}j} \delta_{\mathbf{p}'\mathbf{p}}\, \mathrm{d}^3r = \delta_{jj'}\delta_{\mathbf{p}'\mathbf{p}}$$

The energy $E_{\mathbf{p}j}$ is an even function of $\mathbf{p}$, periodical on the reciprocal lattice. Indeed, if we replace $\mathbf{p}$ in (5.9.3) by $-\mathbf{p}$, we obtain an equation which is a complex conjugate to (5.9.3); it is satisfied by the function $u_{-\mathbf{p}j}(\mathbf{r}) = u^*_{\mathbf{p}j}(\mathbf{r})$, and $E_{-\mathbf{p}j} = E_{\mathbf{p}j}$.

If we perform in (5.9.3) the replacement $\mathbf{p} \to \mathbf{p}' = \mathbf{p} + \hbar g_l$, where g_l is an arbitrary reciprocal lattice vector, (5.9.3) is satisfied by the function

$$u_{\mathbf{p}'} = u_{\mathbf{p}} \exp(-\mathrm{i}\mathbf{g}_l\mathbf{r})$$

This means that

$$\psi_{\mathbf{p}'} = \exp\left(-\frac{\mathrm{i}}{\hbar}(\mathbf{p} + \hbar\mathbf{g}_l)\mathbf{r}\right) u'_{\mathbf{p}} = \exp\left(\frac{\mathrm{i}}{\hbar}\mathbf{p}\mathbf{r}\right) u_{\mathbf{p}} = \psi_{\mathbf{p}}$$

that is, changing $\mathbf{p}$ by a vector $\hbar g_l$ does not affect either the wave function or its eigenvalue $E_{\mathbf{p}j}$. Therefore, $E_{\mathbf{p}'} = E_{\mathbf{p}+\hbar\mathbf{g}_l} = E_{\mathbf{p}}$, that is, $E_{\mathbf{p}j}$ is a periodic function; hence, it is sufficient to consider it only in the first Brillouin zone. The situation is the same as for photons (Chapter 1) and phonons in crystals.

The dependence of $E_{\mathbf{p}j}$ on $\mathbf{p}$ determines the properties of electrons in crystals, for example, the mean velocity of electrons. Indeed, the mean velocity for the Bloch function (5.9.2) is given by

$$\mathbf{v}_{\mathbf{p}j} = \int\limits_V \psi^*_{\mathbf{p}j}(\mathbf{r}) \left(-\frac{i\hbar}{m}\nabla\right) \psi_{\mathbf{p}j}(\mathbf{r})\, d^3r$$

$$= \frac{\mathbf{p}}{m} + \int\limits_\Omega u^*_{\mathbf{p}j} \left(-\frac{i\hbar}{m}\nabla\right) u_{\mathbf{p}j}\, d^3r \tag{5.9.6}$$

In order to calculate the resulting integral, we differentiate equation (5.9.3) with respect to $\mathbf{p}$. This gives

$$L_{\mathbf{p}} \frac{\partial u_{\mathbf{p}j}}{\partial \mathbf{p}} = E_{\mathbf{p}j} \frac{\partial u_{\mathbf{p}j}}{\partial \mathbf{p}} + \frac{i\hbar}{m} \nabla u_{\mathbf{p}j}(\mathbf{r}) - \left(\frac{\mathbf{p}}{m} - \nabla_{\mathbf{p}} E_{\mathbf{p}j}\right) u_{\mathbf{p}j}$$

Making use of self-conjugation (5.9.4) of the operator $L_{\mathbf{p}}$, we obtain

$$E_{\mathbf{p}j} \int\limits_\Omega u^*_{\mathbf{p}j} \frac{\partial u_{\mathbf{p}j}}{\partial \mathbf{p}}\, d^3r = \int\limits_\Omega (L_{\mathbf{p}} u_{\mathbf{p}j'})^* \frac{\partial u_{\mathbf{p}j'}}{\partial \mathbf{p}}\, d^3r$$

$$= E_{\mathbf{p}j'} \int\limits_\Omega u^*_{\mathbf{p}j} \frac{\partial u_{\mathbf{p}j'}}{\partial \mathbf{p}}\, d^3r + \frac{i\hbar}{m} \int\limits_\Omega u^*_{\mathbf{p}j} \nabla u_{\mathbf{p}j'}\, d^3r - \left(\frac{\mathbf{p}}{m} - \nabla_{\mathbf{p}} E_{\mathbf{p}j}\right) \delta_{jj'}$$

Therefore,

$$-\frac{i\hbar}{m} \int\limits_\Omega u^*_{\mathbf{p}j} \nabla u_{\mathbf{p}j'}\, d^3r$$

$$= (E_{\mathbf{p}j'} - E_{\mathbf{p}j}) \int\limits_\Omega u^*_{\mathbf{p}j} \frac{\partial u_{\mathbf{p}j'}}{\partial \mathbf{p}}\, d^3r + \left(\nabla_{\mathbf{p}} E_{\mathbf{p}j} - \frac{\mathbf{p}}{m}\right) \delta_{jj'} \tag{5.9.7}$$

Setting $j = j'$ in (5.9.7), we find from (5.9.6) that

$$\mathbf{v}_{\mathbf{p}j} = \nabla_{\mathbf{p}} E_{\mathbf{p}j} \tag{5.9.8}$$

This is the Bloch theorem, stating that the mean velocity for states of the type of (5.9.2) equals the derivative of energy with respect to quasimomentum, or the derivative of frequency with respect to the wave vector, that is, the group velocity:

$$\mathbf{v}_{\mathrm{gr}} = \frac{\partial \omega}{\partial k} = \frac{\partial(\hbar\omega)}{\partial(\hbar k)} = \frac{\partial E}{\partial \mathbf{p}}$$

It is important that $v_{\mathbf{p}j} \neq 0$. In other words, the state $\psi_{\mathbf{p}j}$ corresponds to a directed, not damped with time, motion of an electron even in a periodic potential $W(\mathbf{r})$. Strictly periodic lattice does not produce scattering of electrons: its effect is, ultimately, a change in the dispersion law.

The equality $E_{-\mathbf{p}j} = E_{\mathbf{p}j}$ implies that $\mathbf{v}_{-\mathbf{p}j} = -\mathbf{v}_{\mathbf{p}j}$ — the velocity is an odd function of quasimomentum.

5.10. Two Simplest Models for Calculation of Bloch Functions

Before considering the motion of electrons in a crystal in electromagnetic fields, we will look at two simple models which clarify the structure of the electron wave functions in a crystal and make it possible to calculate the Bloch functions and the law of dispersion in two limiting cases.

Perturbation theory proved to be an efficient method for solving numerous problems in quantum mechanics. This makes it very natural to use the so-called approximation of nearly free (or weakly bound) electrons, in which the periodic potential $W(\mathbf{r})$ is assumed to be small and is treated as perturbation.

If $W(\mathbf{r}) = 0$, (5.9.1) yields the Schrödinger equation for a free electron, which has a solution with a definite value of momentum $\mathbf{p}$:

$$\psi_{\mathbf{p}}^{(0)}(\mathbf{r}) = \frac{1}{\sqrt{V}} \exp\left(\frac{\mathrm{i}}{\hbar}\mathbf{pr}\right)$$

For this state, the energy E is

$$E_{\mathbf{p}}^{(0)} = \frac{p^2}{2m}$$

The periodic potential $W(\mathbf{r})$ can be expanded in a Fourier series

$$W(\mathbf{r}) = \sum_{\mathbf{l}} W_{\mathbf{l}} \exp(\mathrm{i}\mathbf{g}_{\mathbf{l}}\mathbf{r})$$

so that the first-order perturbative correction to energy is

$$E_{\mathbf{p}}^{(1)} = \int_V \psi_{\mathbf{p}}^{(0)*} W(\mathbf{r}) \psi_{\mathbf{p}}^{(0)} \, \mathrm{d}^3 r = W_0$$

The first-order correction to the wave function is

$$\begin{aligned}
\psi_{\mathbf{p}}^{(1)}(\mathbf{r}) &= \sum_{\mathbf{l}\neq 0} \frac{W_{\mathbf{l}}}{E_{\mathbf{p}}^{(0)} - E_{\mathbf{p}+\hbar\mathbf{g}_{\mathbf{l}}}^{(0)}} \psi_{\mathbf{p}+\hbar\mathbf{g}_{\mathbf{l}}}^{(0)} \\
&= \exp\left(\frac{\mathrm{i}}{\hbar}\mathbf{pr}\right) \frac{1}{\sqrt{V}} \sum_{\mathbf{l}\neq 0} \frac{W_{\mathbf{l}}}{E_{\mathbf{p}}^{(0)} - E_{\mathbf{p}+\hbar\mathbf{g}_{\mathbf{l}}}^{(0)}} \exp(\mathrm{i}\mathbf{g}_{\mathbf{l}}\mathbf{r}) \\
&= \frac{1}{\sqrt{N}} \exp\left(\frac{\mathrm{i}}{\hbar}\mathbf{pr}\right) u^{(1)}
\end{aligned}$$

where $u^{(1)}$ is a periodic function.

The condition of applicability of perturbation theory, $|\psi^{(1)}| \ll |\psi^0|$, is satisfied for all $\mathbf{p}$ except those in a narrow neighborhood of the resonances to be discussed below, provided

$$|W_1| \ll E^{(0)}_{\mathbf{p}+\hbar\mathbf{g}_1}$$

that is, if $|W_1| \ll |\hbar g_1|^{-2}/2m \sim I_0$. This is the condition of smallness, which must be imposed on the potential $W(\mathbf{r})$ in order to be able to use perturbation theory and the approximation of nearly free electrons.

First-approximation wave function in perturbation theory, $\psi_{\mathbf{p}} = \psi^{(0)}_{\mathbf{p}} + \psi^{(1)}_{\mathbf{p}}$, has the form of a Bloch function:

$$\psi_{\mathbf{p}} = \frac{1}{\sqrt{N}} \exp\left(\frac{i}{\hbar}\mathbf{pr}\right)\left(\frac{1}{\sqrt{\Omega}} + u^{(1)}\right)$$

The second-order correction to energy is

$$E^{(2)}_{\mathbf{p}} = \sum_{1\neq 0} \frac{|W_1|^2}{E^{(0)}_{\mathbf{p}} - E^{(0)}_{\mathbf{p}+\hbar\mathbf{g}_1}}$$

and it becomes infinite when $E^{(0)}_{\mathbf{p}} = E^{(0)}_{\mathbf{p}+\hbar\mathbf{g}_1}$, that is, when $p^2 = (\mathbf{p}+\hbar\mathbf{g}_1)^2$. This last equality is the Bragg reflection condition for the electron wave incident on a periodic perturbation. As we move closer to satisfying the Bragg conditions, perturbation theory becomes invalid for the nondegenerate state, because $E^{(2)}_{\mathbf{p}}$ and $\psi^{(1)}_{\mathbf{p}}$ cease to be small compared with $E^{(1)}_{\mathbf{p}}$ and $\psi^{(0)}_{\mathbf{p}}$.

In this case, we need to use perturbation theory for degenerate or almost degenerate nonperturbed states. Even in the zero approximation, the wave function must be taken in the form

$$\psi = c_1\psi^{(0)}_{\mathbf{p}} + c_2\psi^{(0)}_{\mathbf{p}+\hbar\mathbf{g}_1}$$

For c_1 and c_2, we obtain in an ordinary manner a set of linear homogeneous equations and the condition of existence of a nonzero solution yields the secular equation. The roots of the secular equation for energy (a quadratic equation) are

$$E_{\mathbf{p}} = \frac{1}{2}(E^{(0)}_{\mathbf{p}} + E^{(0)}_{\mathbf{p}+\hbar\mathbf{g}_1}) \pm \sqrt{\frac{1}{4}(E^{(0)}_{\mathbf{p}} - E^{(0)}_{\mathbf{p}+\hbar\mathbf{g}_1})^2 + |W_1|^2}$$

Figure 5.14 plots $E_{\mathbf{p}}$ as a function of $\mathbf{p}$. The condition $E^{(0)}_{\mathbf{p}} = E^{(0)}_{\mathbf{p}+\hbar\mathbf{g}_1}$ is met first at the boundary of the first Brillouin zone. Close to a Brillouin

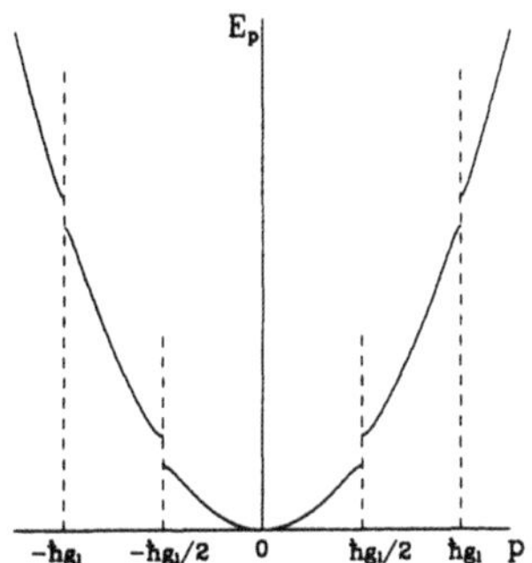

Figure 5.14. Law of dispersion in the approximation of nearly free electrons.

zone boundary, the law of dispersion for $E_{\mathbf{p}}$ deviates from the law for free electrons $E_{\mathbf{p}}^{(0)}$, and this occurs each time when $p^2 \simeq (\mathbf{p} + \hbar \mathbf{g}_l)^2$. For the direction along the vector $\mathbf{g}_l$ of the reciprocal lattice, this takes place first when $\mathbf{p} = \hbar \mathbf{g}_l/2$, then at $\mathbf{p} = \hbar \mathbf{g}_l$, and so forth.

If, keeping in mind the periodic dependence of $E_{\mathbf{p}}$ on $\mathbf{p}$, we transfer the curves of Figure 5.14 to the first Brillouin zone ($|\mathbf{p}| \leq \hbar g/2$), we arrive at the dispersion law shown in Figure 5.15. It shows a picture which is qualitatively similar to Figure 5.13; however, the nearly free electron approximation resulted in wide allowed bands but narrow forbidden bands (energy gaps, in which energy values do not correspond to any electron states). Forbidden gaps are $2|W_l|$ wide, which is small in comparison with the width of the allowed band, $(\hbar \mathbf{g}_l)^2/2m \sim I_0$. Electron waves are reflected at the band boundaries; in this case wave functions are standing waves and the velocity $\mathbf{v}_{\mathbf{p}} = \partial E_{\mathbf{p}}/\partial \mathbf{p}$ vanishes. Instead of momentum, we need to operate with the quasimomentum (defined up to $\hbar \mathbf{g}_l$), even if $W(\mathbf{r})$ is small.

The dispersion curve $E(\mathbf{p})$ has very considerable curvature in the neighborhood of the points of Bragg reflection, as we immediately see from figures 5.14 and 5.15. The inverse to the curvature of the $E(\mathbf{p})$ curve at its extremum points is traditionally called the *effective mass* of the electron. It will be possible to see below that in states that lie close to an extremum, electrons do behave in external fields as particles with mass $m_{\text{eff}} = (\partial^2 E/\partial p^2)^{-1}$. The formula shown above readily yields that in the nearly free electrons model close to a Bragg reflection, the mass ratio is

$$\frac{m_{\text{eff}}}{m} = \pm \frac{|W_l|}{g_l^2/4m \pm |W_l|} \simeq \pm \frac{|W_l|}{g_l^2/4m} \ll 1$$

where the plus sign refers to states above the corresponding forbidden band and the minus sign, below it. This result is in full agreement with numerous experimental data which confirm that very small effective electron masses ($m_{\text{eff}}/m \sim 10^{-1} - 10^{-2}$) are quite frequent in metals and semiconductors

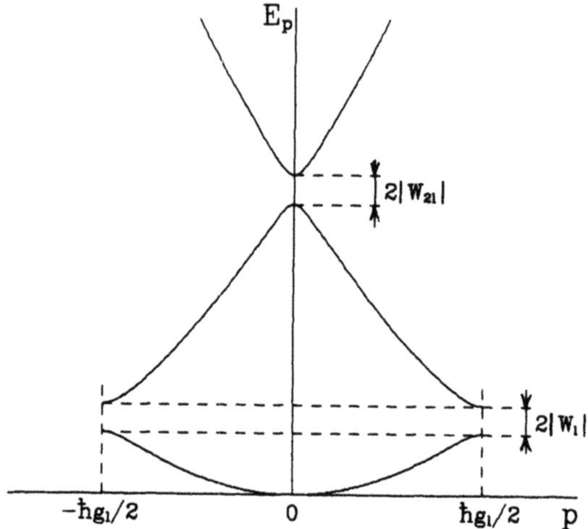

Figure 5.15. Same as in Figure 5.14 but the dispersion law has been transferred to the first Brillouin zone.

and that they always accompany narrow forbidden bands. Furthermore, $m_{eff}/m \sim E_g/I_0$, where E_g is the band gap width.

Another limiting case is the approximation of tight binding of electrons. In this case the potential $W(\mathbf{r})$ is written in the form

$$W(\mathbf{r}) = \sum_{\mathbf{n}} V(\mathbf{r} - \mathbf{R_n})$$

where $V(\mathbf{r})$ is the potential of the atoms in one unit cell. Obviously, the $W(\mathbf{r})$ potential is periodic. It is assumed that $V(\mathbf{r})$ is so strong that an electron stays mostly within one unit cell. Let there exist a solution of the Schrödinger equation for a single unit cell (or for a single atom in a one-atom lattice):

$$-\frac{\hbar}{2m}\nabla^2\varphi_j + V(\mathbf{r})\varphi_j(\mathbf{r}) = E_j\varphi_j(\mathbf{r})$$

If we completely neglect the effect of the potential of the adjacent unit cells, then $\varphi_j(\mathbf{r})$ is the solution of the Schrödinger equation (5.9.1) for an electron in a crystal but the corresponding energy level E_j is N-fold degenerate since $\varphi_j(\mathbf{r} - \mathbf{R_n})$ also satisfies (5.9.1) in this approximation with the same value of E_j. We are thus coming back to the perturbation theory problem for a (multiply) degenerate level. This is an argument in favor of choosing

$$\psi(\mathbf{r}) = \frac{1}{\sqrt{N}}\sum_{\mathbf{n}'} c_{\mathbf{n}'}\varphi_j(\mathbf{r} - \mathbf{R_{n'}}) \tag{5.10.1}$$

with arbitrary coefficients $c_{\mathbf{n}'}$ as a zeroth-approximation wave function. Using perturbation theory for an N-fold degenerate level E_j yields a set of N linear homogeneous equations for $c_{\mathbf{n}}$, whose solutions have the form $c_{\mathbf{n}} = c_0 \exp(i\hbar^{-1}\mathbf{p}\mathbf{R}_{\mathbf{n}})$; this could be expected on the basis of translational-symmetry arguments.

Substituting (5.10.1) into equation (5.9.1), multiplying it by $\varphi_j^*(\mathbf{r}-\mathbf{R}_{\mathbf{n}})$ and integrating, we obtain

$$(E_j - E)\sum_{\mathbf{n}'} A^j_{\mathbf{n}\mathbf{n}'} c_{\mathbf{n}'} + \sum_{\mathbf{n}'} B^j_{\mathbf{n}\mathbf{n}'} c_{\mathbf{n}'} = 0 \tag{5.10.2}$$

where

$$A^j_{\mathbf{n}\mathbf{n}'} = A^j_{\mathbf{n}-\mathbf{n}'} = \int \varphi_j^*(\mathbf{r} - \mathbf{R}_{\mathbf{n}})\varphi_j(\mathbf{r} - \mathbf{R}_{\mathbf{n}'})\,\mathrm{d}^3 r$$

$$B^j_{\mathbf{n}\mathbf{n}'} = B^j_{\mathbf{n}-\mathbf{n}'} = \int \varphi_j^*(\mathbf{r} - \mathbf{R}_{\mathbf{n}})(W(\mathbf{r}) - V(\mathbf{r} - \mathbf{R}_{\mathbf{n}'}))\varphi_j(\mathbf{r} - \mathbf{R}_{\mathbf{n}'})\,\mathrm{d}^3 r$$

For normalized functions φ_j, the integral $A_{\mathbf{n}\mathbf{n}} = 1$ and $A_{\mathbf{n}\mathbf{n}'}$ can be regarded as a small quantity at $\mathbf{n} \neq \mathbf{n}'$. Equations (5.10.2) indeed have solutions of the form $c_{\mathbf{n}} = c_0 \exp(\frac{i}{\hbar}\mathbf{p}\mathbf{R}_{\mathbf{n}})$, since by analogy to the equations for lattice vibrations, they are all converted by this substitution into one equation

$$(E_j - E)A_{\mathbf{p}j} + B_{\mathbf{p}j} = 0 \tag{5.10.3}$$

in which $A_{\mathbf{p}j}$ and $B_{\mathbf{p}j}$ are the Fourier transforms of $A^j_{\mathbf{n}-\mathbf{n}'}$ and $B^j_{\mathbf{n}-\mathbf{n}'}$:

$$A_{\mathbf{p}j} = \sum_{\mathbf{m}} A^j_{\mathbf{m}} \exp\left(-\frac{i}{\hbar}\mathbf{p}\mathbf{R}_{\mathbf{m}}\right)$$

$$B_{\mathbf{p}j} = \sum_{\mathbf{m}} B^j_{\mathbf{m}} \exp\left(-\frac{i}{\hbar}\mathbf{p}\mathbf{R}_{\mathbf{m}}\right)$$

From (5.10.3) we obtain

$$\begin{aligned} E_{\mathbf{p}j} &= E_j + \frac{B_{\mathbf{p}j}}{A_{\mathbf{p}j}} = E_j + \frac{\sum\limits_{\mathbf{m}} B^j_{\mathbf{m}} \exp\left(-\frac{i}{\hbar}\mathbf{p}\mathbf{R}_{\mathbf{m}}\right)}{1 + \sum\limits_{\mathbf{m}\neq 0} A^j_{\mathbf{m}} \exp\left(-\frac{i}{\hbar}\mathbf{p}\mathbf{R}_{\mathbf{m}}\right)} \\ &\simeq E_j + \sum_{\mathbf{m}} B^j_{\mathbf{m}} \exp\left(-\frac{i}{\hbar}\mathbf{p}\mathbf{R}_{\mathbf{m}}\right) = E_j + \sum_{\mathbf{m}} B^j_{\mathbf{m}} \cos\left(\frac{\mathbf{p}\mathbf{R}_{\mathbf{m}}}{\hbar}\right) \end{aligned} \tag{5.10.4}$$

This last equality is derived under the assumption that real functions $\varphi_j(\mathbf{r})$ were chosen, so that $A_{\mathbf{m}}$ and $B_{\mathbf{m}}$ are also real; furthermore, $\varphi_j(-\mathbf{r}) = \varphi_j(\mathbf{r})$, whence $B^j_{\mathbf{m}} = B^{j^*}_{-\mathbf{m}}$.

As we see from (5.10.4), the atomic level E_j, which is N-fold degenerate in the zeroth approximation, spreads into a band of a width of the order of $|B| \ll I_0$. This last inequality is the condition of validity of the tight-binding approximation for electrons, when the perturbation $W(\mathbf{r}) - V(\mathbf{r} - \mathbf{R}_{\mathbf{n}'})$ is small in comparison with $E_j \sim I_0$ for an electron in the state $\varphi_j(\mathbf{r} - \mathbf{R}_{\mathbf{n}'})$, that is, an electron localized in a unit cell with a number $\mathbf{n}'$.

Expression (5.10.4) in its explicit form shows that $E_{\mathbf{p}j}$ is a periodic function of $\mathbf{p}$ on the reciprocal lattice, and that if $c_{\mathbf{n}} = c_0 \exp(\frac{i}{\hbar}\mathbf{p}\mathbf{R}_{\mathbf{n}})$, expression (5.10.1) takes the form of the Bloch function:

$$\psi_{\mathbf{p}j}(\mathbf{r}) = \frac{c_0}{\sqrt{N}} \sum_{\mathbf{n}} \exp\left(\frac{i}{\hbar}\mathbf{p}\mathbf{R}_{\mathbf{n}}\right) \varphi_j(\mathbf{r} - \mathbf{R}_{\mathbf{n}})$$
$$= \exp\left(\frac{i}{\hbar}\mathbf{p}\mathbf{r}\right) \sum_{\mathbf{n}} \exp\left(-\frac{i}{\hbar}\mathbf{p}(\mathbf{r} - \mathbf{R}_{\mathbf{n}})\right) \varphi_j(\mathbf{r} - \mathbf{R}_{\mathbf{n}})$$

If the system allows using the tight-binding approximation, allowed band widths are much smaller than I_0 and hence, much smaller than the forbidden gap widths, which roughly equal $E_j - E_{j'} \sim I_0$. The situation is a reverse of that found in the nearly-free-electrons approximation, in which allowed band widths are of the order of I_0 while forbidden gap widths are much less than I_0. There can be no doubt that both these models are the limiting cases, remote from the real situation; it is found, however, that the nearly-free-electrons approximation works quite well for simple metals, where gap widths are of the order of $1\,\mathrm{eV} \sim 0.1 I_0$. The tight-binding approximation works rather well for molecular crystals and fairly frequently for semiconductors and dielectrics, which was rather unexpected. It is definitely adequate for describing electrons in internal atomic shells.

Formula (5.10.4) also implies that if the conditions are favorable for the tight-binding approximation, the electron effective masses are quite large and are, by the order of magnitude, inversely proportional to the width of the corresponding allowed band:

$$\frac{m_{\text{eff}}}{m} \sim \frac{\hbar}{m a_0^2 |B|} \sim \frac{I_0}{|B|} \gg 1$$

The tight-binding approximation for electrons also allows one to interpret qualitatively the physical meaning of two factors which form a Bloch wave function: the exponential factor $\exp(i\hbar^{-1}\mathbf{p}\mathbf{r})$ determines the change in the wave function as we go from a unit cell to the adjacent one, that is, it describes the propagation of an electron through a crystal by hopping from one unit cell to another, and the periodic factor $u_{pj}(\mathbf{r})$ (it is identical for all cells) which describes the internal motion of an electron in a unit cell; in the tight-binding approximation, it almost coincides with the corresponding intra-atomic and intramolecular wave functions.

5.11. Motion of Electrons in Crystals Placed in External Fields

Let a homogeneous external electric field $\mathcal{E}$ be applied to a crystal. The equation of motion for an electron is obtained from (5.9.1) by transforming it into a nonstationary equation and replacing $W(\mathbf{r})$ by $W(\mathbf{r}) - e\mathcal{E}\mathbf{r}$, since $-e\mathcal{E}\mathbf{r}$ is the energy of an electron in the external field. We obtain

$$\mathrm{i}\hbar \frac{\partial \psi}{\partial t} = -\frac{\hbar}{2m} \nabla^2 \psi + (W(\mathbf{r}) - e\mathcal{E}\mathbf{r})\psi \tag{5.11.1}$$

Assume now that the field $\mathcal{E}$ is much weaker than the intracrystalline field $e^{-1}\nabla W(\mathbf{r})$, whose strength is of the order of the intra-atomic field, that is, $\mathcal{E}_{\mathrm{at}} \sim 10^9$ V/cm. Assuming $-e\mathcal{E}\mathbf{r}$ to be a perturbation, we expand the function ψ in the eigenfunctions $\psi_{\mathbf{p}j}$ of equation (5.9.1); by virtue of (5.9.4), they form a complete orthonormal set of eigenfunctions.

Substituting

$$\psi = \sum_{\mathbf{p}'j'} c_{\mathbf{p}'j'}(t)\psi_{\mathbf{p}'j'}(\mathbf{r})$$

into equation (5.11.1), multiplying it by $\psi^*_{\mathbf{p}j}$, integrating over volume V, and taking into account (5.9.5), we obtain

$$\mathrm{i}\hbar \frac{\partial c_{\mathbf{p}j}}{\partial t} = E_{\mathbf{p}j} c_{\mathbf{p}j} - \sum_{\mathbf{p}'j'} c_{\mathbf{p}'j'} \int \psi^*_{\mathbf{p}j} e\mathcal{E}\mathbf{r}\psi_{\mathbf{p}'j'}\, \mathrm{d}^3 r \tag{5.11.2}$$

Now we substitute the explicit form of the Bloch functions (5.9.2) into the last term of (5.11.2) and perform some transformations:

$$\begin{aligned}
&\frac{1}{N}\sum_{\mathbf{p}'j'} c_{\mathbf{p}'j'} \int u^*_{\mathbf{p}j} e\mathcal{E}\mathbf{r} u_{\mathbf{p}'j'} \exp\left(-\frac{\mathrm{i}}{\hbar}(\mathbf{p}-\mathbf{p}')\mathbf{r}\right) \mathrm{d}^3 r \\
&= \mathrm{i}\hbar e\mathcal{E} \frac{\partial}{\partial \mathbf{p}} \sum_{\mathbf{p}'j'} c_{\mathbf{p}'j'} \frac{1}{N} \int u^*_{\mathbf{p}j} \exp\left(-\frac{\mathrm{i}}{\hbar}\mathbf{p}\mathbf{r}\right) u_{\mathbf{p}'j'} \exp\left(-\frac{\mathrm{i}}{\hbar}\mathbf{p}'\mathbf{r}\right) \mathrm{d}^3 r \\
&= \mathrm{i}\hbar e\mathcal{E} \sum_{\mathbf{p}'j'} c_{\mathbf{p}'j'} \frac{1}{N} \int \frac{\partial u^*_{\mathbf{p}j}}{\partial \mathbf{p}} u_{\mathbf{p}'j'} \exp\left(-\frac{\mathrm{i}}{\hbar}(\mathbf{p}-\mathbf{p}')\mathbf{r}\right) \mathrm{d}^3 r
\end{aligned}$$

Note that $\partial u^*_{\mathbf{p}j}/\partial \mathbf{p}$ is a periodic function of $\mathbf{p}$ and therefore

$$\begin{aligned}
&\frac{1}{N} \int \frac{\partial u^*_{\mathbf{p}j}}{\partial \mathbf{p}} u_{\mathbf{p}'j'} \exp\left(-\frac{\mathrm{i}}{\hbar}(\mathbf{p}-\mathbf{p}')\mathbf{r}\right) \mathrm{d}^3 r \\
&= \int_{\Omega} \frac{\partial u^*_{\mathbf{p}j}}{\partial \mathbf{p}} u_{\mathbf{p}'j'} \exp\left(\frac{\mathrm{i}}{\hbar}(\mathbf{p}-\mathbf{p}')\mathbf{r}\right) \mathrm{d}^3 r \frac{1}{N} \sum_{\mathbf{n}} \exp\left(-\frac{\mathrm{i}}{\hbar}(\mathbf{p}-\mathbf{p}')\mathbf{R}_{\mathbf{n}}\right) \\
&= \delta_{\mathbf{p}\mathbf{p}'} \int_{\Omega} \frac{\partial u^*_{\mathbf{p}j}}{\partial \mathbf{p}} u_{\mathbf{p}j'}\, \mathrm{d}^3 r = -\delta_{\mathbf{p}\mathbf{p}'} \int u_{\mathbf{p}j} \frac{\partial u^*_{\mathbf{p}j'}}{\partial \mathbf{p}}\, \mathrm{d}^3 r
\end{aligned}$$

The last equality follows from

$$\frac{\partial}{\partial \mathbf{p}} \int u^*_{\mathbf{p}j} u_{\mathbf{p}j'} \, \mathrm{d}^3 r = 0$$

In view of this result, equation (5.11.2) becomes

$$\mathrm{i}\hbar \frac{\partial c_{\mathbf{p}j'}}{\partial \mathbf{t}} = E_{\mathbf{p}j} c_{\mathbf{p}j} - \mathrm{i}\hbar e \boldsymbol{\mathcal{E}} \frac{\partial c_{\mathbf{p}j}}{\partial \mathbf{p}} - e\boldsymbol{\mathcal{E}} \sum_{j'} c_{\mathbf{p}j} \mathbf{R}_{jj'}(\mathbf{p}) \tag{5.11.3}$$

where

$$\mathbf{R}_{jj'} = \frac{\mathrm{i}\hbar}{2} \int\limits_{\Omega} \left(u^*_{\mathbf{p}j} \frac{\partial u_{\mathbf{p}j'}}{\partial \mathbf{p}} - \frac{\partial u^*_{\mathbf{p}j}}{\partial \mathbf{p}} u_{\mathbf{p}j'} \right) \mathrm{d}^3 r$$

The last term in (5.11.3) at $j \neq j'$ corresponds to interband transitions. This term can be ignored if the field is weak: $\boldsymbol{\mathcal{E}} \ll \boldsymbol{\mathcal{E}}_{\mathbf{at}} \sim E_{\mathbf{p}j}/ea_0$. Indeed, $|\partial u/\partial p| \sim ua_0/\pi\hbar$, since $u_{\mathbf{p}j}$ as a function of $\mathbf{p}$ varies within the Brillouin zone. Hence, the last term in (5.11.3) is of the order of $e\boldsymbol{\mathcal{E}}a_0$, that is, of the increment of the potential of the field over a distance of the order of a_0. This quantity is small in comparison with $E_{\mathbf{p}j}$ or $W(\mathbf{r})$ and we can drop the last term in (5.11.3).

All this reasoning holds for a constant or low-frequency field. Interband transitions begin to play an important role only when $\hbar\omega$ grows to a value of the order of the difference between the energies of states in the two bands ($\hbar\omega \sim I_0$). This is an immediate implication of energy conservation.

Once the last term in (5.11.3) is omitted, we obtain an equation which relates only the $c_{\mathbf{p}j}$ that belong to one band (the same j). Equations (5.11.3) split and give a first-order differential equation that can be integrated by the characteristic functions method. The general solution of this equation has the form

$$c_{\mathbf{p}j}(t) = c_j(\mathbf{p} - e\boldsymbol{\mathcal{E}}t) \exp\left(-\frac{\mathrm{i}}{\hbar} \int\limits_0^t E_{\mathbf{p}'j} \, \mathrm{d}t' \right) \tag{5.11.4}$$

where c_j is an arbitrary function defined by the initial condition for $c_{\mathbf{p}j}$ at $t = 0$, and $\mathbf{p}' = \mathbf{p} - e\boldsymbol{\mathcal{E}}t + e\boldsymbol{\mathcal{E}}t'$. If $\boldsymbol{\mathcal{E}}$ is a function of time, $\boldsymbol{\mathcal{E}}t$ must be replaced by $\int_0^t \boldsymbol{\mathcal{E}}(t') \, \mathrm{d}t'$.

In particular, there exist solutions

$$c_{\mathbf{p}j}(t) \sim \delta(\mathbf{p} - \mathbf{p}_0 - e\boldsymbol{\mathcal{E}}t)$$

which signify, as does the general solution (5.11.4), that $\mathbf{p}$ varies as dictated by the equation

$$\frac{\mathrm{d}\mathbf{p}}{\mathrm{d}t} = e\boldsymbol{\mathcal{E}}$$

This reasoning can be generalized to the case of a magnetic field $\mathcal{H}$ added to the electric field. If the fields $\mathcal{E}$ and $\mathcal{H}$ vary slowly in space and are weak and low-frequency in the sense defined above, we obtain

$$\frac{d\mathbf{p}}{dt} = e\mathcal{E} + \frac{e}{c}[\mathbf{v}\mathcal{H}] \tag{5.11.5}$$

At a glance, equation(5.11.5) is indistinguishable from the equation of motion of a free electron. The actual motion of an electron through a crystal in electric and magnetic fields is very different from the motion of a free electron, since by virtue of the Bloch theorem (5.9.7), the velocity $\mathbf{v}$ is a complicated periodic function of quasimomentum $\mathbf{p}$.

We conclude that the law of dispersion $E_{\mathbf{p}j}$ determines not only the kinematics of the electron (its velocity) but also, as follows from (5.11.5), its dynamics. Owing to the periodic dependence of $E_{\mathbf{p}j}$ and $\mathbf{v}_{\mathbf{p}j}$ on $\mathbf{p}$, an electron performs a periodic motion in a constant electric field, unless we take into account its scattering on crystal defects and on phonons.

5.12. Electric Conduction in Solids. Basics of Band Theory

The preceding section demonstrated that electrons in crystals are described by the Bloch functions; these are traveling waves, and the mean velocity of these states is nonzero. This may indicate that all crystals conduct electric current. Actually we observe metallic, semiconducting, and dielectric crystals. The reason for this is that a filled band does not contribute to electric current.

Materials in which all bands are either filled or empty are dielectric. Materials with partly filled bands are metals.

The reason for this is that $\mathbf{v}_{\mathbf{p}j} = -\mathbf{v}_{-\mathbf{p}j}$ and $E_{\mathbf{p}j} = E_{-\mathbf{p}j}$. Electrons are fermions; at zero temperature, they fill states up to a certain limiting Fermi energy E_F. In the absence of a field, each occupied state with quasimomentum $\mathbf{p}$ corresponds to an occupied state with quasimomentum $-\mathbf{p}$. The currents of electrons with momenta $\mathbf{p}$ and $-\mathbf{p}$ cancel each other out and the resulting current is zero.

When a constant electric field is imposed, the momentum of an electron changes according to the equation

$$\mathbf{p}(t) = \mathbf{p}_0 + e\mathcal{E}t$$

that is, it grows linearly with time. If the band is filled, the momenta of all electrons in the band change during a time t by the same quantity. Those electrons which thereby reach the boundary of the Brillouin zone are "ejected", by virtue of the periodicity in quasimomentum, to the other boundary of the Brillouin zone, to the point which differs from the original one by a reciprocal lattice vector, so that these electrons continue to move with the momentum of the first Brillouin zone. Therefore, even though the momentum (quasimomentum) of each individual electron changes, the momentum distribution of electrons in the band remains unaltered, the band stays filled, and the total current is zero.

Filled bands thus do not contribute to electric current even when placed in an external field, that is, they do not contribute to electric conduction. Materials with filled bands are dielectrics.

The Fermi level E_F of metals lies inside the filled band, so that the band is only partially filled. When a field is applied, each electron in the band changes its momentum, its velocity and contribution to current. In an ideal crystal without scattering, however, this process in a constant field is periodical: having reached the Brillouin zone boundary, the electron "jumps" to the other boundary and continues its motion again in the first Brillouin zone. The electron's motion is periodical not only in the momentum space but in the coordinate space as well: the sign of its velocity is also periodically reversed and the mean current is zero. This, however, ignores scattering; if scattering on crystal defects is taken into account, the field applied to the crystal produces a stationary distribution of electron momenta in the band, with the number of electrons whose momentum is $\mathbf{p}$ being different from that of electrons with momentum $-\mathbf{p}$: this means that a current is flowing.

The picture remains essentially unchanged at finite temperatures, unless thermal excitation produces partial occupation of empty bands and unoccupied states in formally filled bands. This may happen at kT of the order of the forbidden gap width, that is, of the order of I_0. When band gaps are so wide, melting occurs earlier and so the material remains dielectric even at high temperatures. If, however, the band gap is or the order of or below 1 eV, the situation changes drastically. The upper band which is empty at $T = 0$ (the conduction band) then becomes partially occupied already at room temperatures, while unoccupied states appear in the so-called valence band which is totally occupied at $T = 0$. This is the case of semiconductors which differ from dielectric solids only quantitatively—by the width of the band gap.

Semiconductors at finite temperatures are electrically conductive, and their conduction arises in two bands at the same time: in the conduction band (the electron conduction) and in the valence band (hole conduction).

It has already been shown in connection with phonons that the number of states in a band equals the number of unit cells, N. With the spin taken into account, this number is $2N$. The number of electrons for a monatomic crystal constructed of atoms with Z valence electrons each is NZ. It is clear, therefore, that if Z is an even number, $Z/2$ bands are completely filled, and if Z is odd, one of the bands has to be half-filled and the material is a metal. This is the situation for alkali metals of the first group of the periodic table: Li, Na, and K.

In alkali earth elements of group II, Be, Mg, Cd, and Zn, we have $Z = 2$ so that they could be expected to have only completely filled bands. In fact they are metals since the bands in these crystals overlap and the Fermi level lies, as a result, below the upper boundary of one of the bands while the next band is partially filled (up to the Fermi level). These metals conduct in two bands simultaneously. The elements of group III, Al, Ga, and In, form, as could be predicted, metallic crystals. Elements in group IV have filled bands and represent typical semiconductors.

If a semiconductor has an unoccupied level in the valence band, it will occur close to the upper boundary of the band and one refers to a hole; a hole behaves as a positively charged particle. Indeed, if an electron is absent in a valence state with momentum $\mathbf{p}$, this produces a noncompensated-for contribution $\mathbf{j} = -e\mathbf{v}\mathbf{p}_\text{v}$ to the current ("v" is the index for the valence band). The noncompensated-for current can be interpreted as the current $\mathbf{j} = (-e)\mathbf{v}\mathbf{p}_\text{v} = |e|\mathbf{v}\mathbf{p}_\text{v}$ of a positively charged particle moving at the same velocity $\mathbf{v}$.

Close to the upper boundary of the valence band, the dispersion law can be written in the form

$$E_{\mathbf{p}v} = E_\text{v} - \frac{1}{2}\left(\frac{1}{m_0^*}\right)_{\alpha\beta}(\mathbf{p} - \mathbf{p}_\text{v})_\alpha(\mathbf{p} - \mathbf{p}_\text{v})_\beta + \ldots$$

where p_v is the value of quasimomentum at which the upper boundary E_v of the valence band is reached. In a particular case we may have $p_\text{v} = 0$; this indeed occurs in Ge and Si. The tensor

$$\left(\frac{1}{m_\text{v}^*}\right)_{\alpha\beta}$$

is determined by the expression

$$\left(\frac{1}{m_\text{v}^*}\right)_{\alpha\beta} = -\left(\frac{\partial^2 E_{\mathbf{p}v}}{\partial p_\alpha \partial p_\beta}\right)_{\mathbf{p}=\mathbf{p}_\text{v}}$$

It is known as the inverse tensor of effective hole mass.

For this dispersion law we have

$$v_\alpha = \frac{\partial E_{\mathbf{p}v}}{\partial p_\alpha} = -\left(\frac{1}{m_\text{v}^*}\right)_{\alpha\beta}(\mathbf{p} - \mathbf{p}_\text{v})_\beta$$

and if $m^*_{\alpha\beta}$ is interpreted as a tensor inverse to $(m^{*-1})_{\alpha\beta}$, then

$$(\mathbf{p} - \mathbf{p}_\text{v})_\alpha = -(m_\text{v}^*)_{\alpha\beta} v_\beta$$

With this law of dispersion, the equation of motion in a field becomes

$$\frac{\mathrm{d}p_\alpha}{\mathrm{d}t} = -(m_\mathrm{v}^*)_{\alpha\beta}\dot{v}_\beta = e\boldsymbol{\mathcal{E}}_\alpha + \frac{e}{c}\,[\mathbf{v}\boldsymbol{\mathcal{H}}]_\alpha$$

This is the equation of motion of a particle of a (generally anisotropic) mass m^* and a charge $(-e)$. This confirms the interpretation of a hole as a positively charged particle whose effective mass is m_v^*.

The same is true for the bottom of the upper (conduction) band ("c" is the index for the conduction band), where

$$E_{\mathbf{pc}} = E_\mathrm{c} + \frac{1}{2}\left(\frac{1}{m_\mathrm{c}^*}\right)_{\alpha\beta}(\mathbf{p}-\mathbf{p}_\mathrm{c})_\alpha(\mathbf{p}-\mathbf{p}_\mathrm{c})_\beta$$

because the minimum $E_{\mathbf{pc}}$ is reached at $\mathbf{p} = \mathbf{p}_\mathrm{c}$ and the tensor $(1/m_\mathrm{c}^*)_{\alpha\beta}$, being a symmetrical tensor, is diagonalized in a system of coordinates connected with crystal axes; the principal values of the tensor $(1/m_\mathrm{c}^*)$ and of the inverse to it, tensor m_c^*, are positive. In the conduction band, an electron behaves as a particle with negative charge e and effective mass m_c^*.

A pure conductor has two types of charge carriers: electrons and holes. The same is true of semimetals, characterized by a very small overlapping of bands (in bismuth, about 10^{-5} of the upper band is occupied), but then neither the conductivity nor the holes disappear even at $T \to 0$. Holes exist also in metals with overlapping bands; this was deduced already in the last century from Hall effect data, when the sign of charge carriers in some metals was found to be opposite to that of electrons.

In metals, electric conductivity, or susceptibility, due to conduction electrons in a nonfilled band can be calculated using a kinetic equation of the same type as that used in Chapter 3 for the plasma. The difference lies only in the law of dispersion for electrons and in that the distribution function f can be treated as a function of $\mathbf{p}$, $\mathbf{r}$, and t:

$$\frac{\partial f}{\partial t} + \mathbf{v}\,\frac{\partial f}{\partial \mathbf{r}} + e\boldsymbol{\mathcal{E}}\frac{\partial f}{\partial \mathbf{p}} = -\frac{f - f_{0j}}{\tau} \tag{5.12.1}$$

The collision integral is taken here in its simplest form; τ is the relaxation time and $\tau^{-1} = \nu$ is the collision frequency.

Equation (5.12.1) is a first-order partial differential equation and it can be integrated by general methods. We will be interested only in stationary solutions of this equation for the harmonic field $\boldsymbol{\mathcal{E}}$. For the time being, we put aside the effects of spatial dispersion that we discussed in detail for the plasma in Chapter 3. This statement means that we set $\mathbf{k} = 0$ and that the term with $\partial f/\partial \mathbf{r}$ can be neglected. Assuming the field $\boldsymbol{\mathcal{E}}$ to be weak, we can set $f = f_{0j} + \varphi$, where φ is a small perturbation of the equilibrium distribution function f_{0j}. In contrast to the case of the plasma, f_{0j} is not a Maxwellian function of velocities but the Fermi distribution (E_{F} is the Fermi energy, or rather, the chemical potential):

$$f_{0j}(\mathbf{p}) = \frac{1}{\exp\left(\frac{E_{\mathbf{p}j} - E_{\mathrm{F}}}{kT}\right) + 1}$$

and

$$2\sum_{\mathbf{p}} f_{0j}(\mathbf{p}) = 2\int \frac{V\,\mathrm{d}^3p}{(2\pi\hbar)^3} f_{0j}(\mathbf{p}) = N_j \tag{5.12.2}$$

The factor 2 in the normalization of f_{0j} arises because of the two values assumed by the electron spin, and N_j is the total number of electrons in the jth band. Integration over momentum in (5.12.2) is carried out over the first Brillouin zone.

For φ_ω we obtain from (5.12.1) the equation

$$-\mathrm{i}\omega\varphi_\omega + \nu\varphi_\omega = -E_{\omega\beta}\frac{\partial f_{0j}}{\partial p_\beta}$$

Since the current density is expressed through φ by

$$j_{\omega\alpha} = 2e\int (v_{\mathbf{p}j})_\alpha \varphi_\omega \frac{\mathrm{d}^3p}{(2\pi\hbar)^3}$$

we have

$$j_{\omega\alpha} = -\frac{2\mathrm{i}e^2 E_{\omega\beta}}{\omega + \mathrm{i}\nu}\int \frac{\partial f_{0j}}{\partial p_\beta}(v_{\mathbf{p}j})_\alpha \frac{\mathrm{d}^3p}{(2\pi\hbar)^3}$$

and therefore

$$\begin{aligned}\chi_{\alpha\beta}(\omega) &= \frac{2e^2}{\omega(\omega + \mathrm{i}\nu)}\int (v_{\mathbf{p}j})_\alpha \frac{\partial f_{0j}}{\partial p_\beta}\frac{\mathrm{d}^3p}{(2\pi\hbar)^3} \\ &= \frac{2e^2}{\omega(\omega + \mathrm{i}\nu)}\int \frac{\mathrm{d}f_{0j}}{\mathrm{d}E_{\mathbf{p}j}}(v_{\mathbf{p}j})_\alpha (v_{\mathbf{p}j})_\beta \frac{\mathrm{d}^3p}{(2\pi\hbar)^3}\end{aligned} \tag{5.12.3}$$

In metals, $\mathrm{d}f_{0j}/\mathrm{d}E_{\mathbf{p}j}$ is almost a δ-function because at realistic temperatures ($kT \ll E_{\mathrm{F}} \sim I_0$) the Fermi distribution f_{0j} is almost a step function. Therefore, the contribution to electric conduction in the case of a half-filled band (in metals) or nearly half-filled (when bands overlap) is made only by a narrow region about kT wide around the Fermi level E_{F}. This is physically quite transparent; f changes in a weak field only close to the Fermi surface in the $\mathbf{p}$ space, and electrons get redistributed in the narrow strip about kT wide near E_{F}. If the dispersion law is

$$E_{\mathbf{p}j} = E_{0j} + \frac{1}{2m^*}p^2$$

we obtain

$$\chi_{\alpha\beta}(\omega) = -\frac{2e^2}{\omega(\omega + i\nu)} \int f_{0j}(\mathbf{p}) \frac{\partial^2 E_{\mathbf{p}j}}{\partial p_\alpha \partial p_\beta} \frac{d^3p}{(2\pi\hbar)^3}$$
$$= -\frac{2e^2}{\omega(\omega + i\nu)m^*} \int f_{0j} \frac{d^3p}{(2\pi\hbar)^3} = -\frac{n_j e^2}{m^*\omega(\omega + i\nu)} \quad (5.12.4)$$

where $n_j = N_j/V$ is the electron density; this expression coincides with that for the plasma, with m replaced by the effective mass m^*. It is rather interesting that with this dispersion law, the expression for χ includes the total number of electrons per unit volume, n_j, not the gradient of this density at the Fermi surface. The reason for this is that in the model with effective mass, the Fermi energy is expressed in a very simple way in terms of the electron density n_j.

At low frequencies, when $\omega \ll \nu = \tau^{-1}$, the electric conductivity of a metal is

$$\sigma = -i\omega\chi = \frac{n_j e^2 \tau}{m^*}$$

and is largely determined by the relaxation time τ.

In the general case, χ is often written as

$$\chi(\omega) = -\left(\frac{n}{m}\right)_{\text{eff}} \frac{e^2}{\omega(\omega + i\nu)} \qquad \left(\frac{n}{m}\right)_{\text{eff}} = \int f_{0j}(\mathbf{p}) \frac{\partial^2 E_{\mathbf{p}j}}{\partial p_\alpha \partial p_\beta} \frac{d^3p}{(2\pi\hbar)^3}$$

The dimension of the factor $(n/m)_{\text{eff}}$ is that of the ratio of the electron density to the mass and its value depends on a specific law of dispersion of electrons.

In semiconductors, the Fermi level lies in the forbidden gap and kT is much smaller than the gap width. In this case the Fermi distribution for electrons in the conduction band and for holes in the valence band transforms into the classical distribution

$$f_{0j} \sim \exp\left(-\frac{E_{\mathbf{p}j}}{kT}\right)$$

Together with the quadratic law of dispersion at the boundaries of bands with the effective masses m_v^* and m_c^*, this simply leads to the ordinary formula for the plasma, with m replaced by the effective masses.

The dielectric permittivity of metals has the form

$$\varepsilon = \varepsilon_0 - \frac{4\pi n e^2}{m^*\omega(\omega + i\nu)}$$

where ε_0 is connected with the contribution of other bands. If $\omega^2 < \omega_0^2 = 4\pi n e^2/m^*\varepsilon_0$, the dielectric permittivity is negative, $\varepsilon < 0$, and light is reflected by the metal. If $m^* \sim m$, the plasma frequency $\omega_0 \sim I_0/\hbar$ and metals reflect light up to the ultraviolet range.

If $\varepsilon = 0$, longitudinal waves are produced; their frequencies ω_0 are in the UV part of the spectrum. If the expression for the frequency of longitudinal waves is compared with that for longitudinal phonons, this latter contains M instead of m, so that the corresponding frequencies fall in the infrared range. Surface waves appear when $\varepsilon(\omega) = -1$, at a frequency

$$\omega = \omega_0 \sqrt{\frac{\varepsilon_0}{1+\varepsilon_0}} \sim \frac{\omega_0}{\sqrt{2}}$$

(surface plasmons).

In the next section, we derive the general expression for the linear susceptibility in the band theory. The main shortcoming of the band theory is that correlations between electrons are insufficiently taken into account, since its initial approach is to obtain the equation of motion for one electron in the average self-consistent field of all other electrons.

5.13. Linear Susceptibility in Band Theory

We will now write the expression for the linear susceptibility in the band theory, making use of the general formula (2.6.1). In order to make the formula more specific, we need to find the expressions for the matrix elements of the current density operator.

As follows from Chapter 2, the current density operator is the operator

$$j_\alpha(\mathbf{r}) = -\frac{\mathrm{i}e\hbar}{2m}\left(\frac{\partial}{\partial x'_\alpha}\delta(\mathbf{r}-\mathbf{r}') + \delta(\mathbf{r}-\mathbf{r}')\frac{\partial}{\partial x'_\alpha}\right)$$

Therefore,

$$\begin{aligned}(j_\alpha(\mathbf{r}))_{\mathbf{p}j,\mathbf{p}'j'} &= \frac{1}{N}\int_V \exp\left(-\frac{\mathrm{i}}{\hbar}\mathbf{p}\mathbf{r}'\right) u^*_{\mathbf{p}j}(\mathbf{r}')j_\alpha(\mathbf{r}) \\ &\quad \times \exp\left(\frac{\mathrm{i}}{\hbar}\mathbf{p}'\mathbf{r}'\right) u_{\mathbf{p}'j'}(\mathbf{r}')\,\mathrm{d}^3r' \\ &= \frac{e}{2mN}\exp\left(-\frac{\mathrm{i}}{\hbar}(\mathbf{p}-\mathbf{p}')\mathbf{r}\right)\left[(p_{\alpha'}+p_\alpha)u^*_{\mathbf{p}j}u_{\mathbf{p}'j'}\right. \\ &\quad \left. -\mathrm{i}\hbar\left(u^*_{\mathbf{p}j}\frac{\partial u_{\mathbf{p}'j'}}{\partial x_\alpha} - \frac{\partial u^*_{\mathbf{p}j}}{\partial x_\alpha}u_{\mathbf{p}'j'}\right)\right]\end{aligned} \tag{5.13.1}$$

The expression in brackets is a periodic function that can be expanded in a Fourier series. Correspondingly,

$$(j_\alpha(\mathbf{r}))_{\mathbf{p}j,\mathbf{p}'j'} = \frac{1}{V}\exp\left(-\frac{\mathrm{i}}{\hbar}(\mathbf{p}-\mathbf{p}')\mathbf{r}\right)\sum_{\mathbf{l}}(j^{\mathbf{l}}_\alpha)_{\mathbf{p}j,\mathbf{p}'j'}\exp(\mathrm{i}\mathbf{g}_{\mathbf{l}}\mathbf{r})$$

In view of this expression,

$$\chi_{\alpha\beta}^{\mathbf{l}\mathbf{l}'}(\mathbf{k},\omega) = \frac{1}{V}\iint\limits_{V} \chi_{\alpha\beta}(\mathbf{r},\mathbf{r}';\omega)$$

$$\times \exp[-\mathrm{i}(\mathbf{k}+\mathbf{g}_{\mathbf{l}})\mathbf{r}]\exp[\mathrm{i}(\mathbf{k}+\mathbf{g}_{\mathbf{l}})\mathbf{r}']\,\mathrm{d}^3r\,\mathrm{d}^3r'$$

$$= -\frac{n^{\mathbf{l}\mathbf{l}'}e^2}{m\omega^2}\delta_{\alpha\beta} + \frac{2}{\omega^2 V}\sum_{\mathbf{p}j,\mathbf{p}'j'}(f_j(\mathbf{p}) - f_{j'}(\mathbf{p}'))(j_\alpha^{\mathbf{l}})_{\mathbf{p}j,\mathbf{p}'j'}(j_\beta^{\mathbf{l}'})_{\mathbf{p}j,\mathbf{p}'j'}$$

$$\times (E_{\mathbf{p}'j'} - E_{\mathbf{p}j} - \hbar\omega)^{-1}\delta_{\mathbf{p}-\mathbf{p}',\hbar\mathbf{k}} \tag{5.13.2}$$

where $n^{\mathbf{l}\mathbf{l}'} = (1/\Omega)\int_\Omega n(\mathbf{r})\exp(-\mathrm{i}\mathbf{g}_{\mathbf{l}-\mathbf{l}'}\mathbf{r})\,\mathrm{d}^3r$ are the components of the Fourier distribution of the electron density $n(\mathbf{r})$.

Forgetting for a while the effects of spatial dispersion, we can set $\mathbf{k}=0$, so that $\mathbf{p}'=\mathbf{p}$. We are only interested in the average susceptibility, so that we only need to calculate $\overline{\chi}_{\alpha\beta} = \chi_{\alpha\beta}^{00}$. To calculate χ^{00}, we only need the Fourier components $(j_\alpha^0)_{\mathbf{p}j,\mathbf{p}j'}$; denoting them by $(J_\alpha)_{jj'}$, we obtain

$$\bar{\chi}_{\alpha\beta}(\omega) = -\frac{\bar{n}e^2}{m\omega^2}\delta_{\alpha\beta} + \frac{2}{\omega^2 V}\sum_{\mathbf{p}jj'} f_j(\mathbf{p})$$

$$\times \left\{\frac{(J_\alpha)_{jj'}(J_\beta)_{j'j}}{E_{\mathbf{p}j'} - E_{\mathbf{p}j} - \hbar\omega} + \frac{(J_\beta)_{jj'}(J_\alpha)_{j'j}}{E_{\mathbf{p}j'} - E_{\mathbf{p}j} + \hbar\omega}\right\} \tag{5.13.3}$$

On the other hand, (5.13.1) yields

$$(J_\alpha)_{jj'} = \int\limits_V (j_\alpha(\mathbf{r}))_{\mathbf{p}j\mathbf{p}j'}\,\mathrm{d}^3r$$

$$= \frac{e}{m}\left(p_\alpha\delta_{jj'} - \frac{\mathrm{i}\hbar}{2}\int\limits_\Omega\left(u_{\mathbf{p}j}^*\frac{\partial u_{\mathbf{p}j'}}{\partial x_\alpha} - \frac{\partial u_{\mathbf{p}j}^*}{\partial x_\alpha}u_{\mathbf{p}j'}\right)\mathrm{d}^3r\right)$$

Using (5.9.7) and the notation introduced by (5.11.3), we can rewrite the above equality as

$$(J_\alpha)_{jj'} = e\left(\frac{\partial E_{\mathbf{p}j}}{\partial \mathbf{p}_\alpha}\delta_{jj'} + \frac{\mathrm{i}}{\hbar}(E_{\mathbf{p}j} - E_{\mathbf{p}j'})R_{jj'}^\alpha\right)$$

Substituting the expression for $(J_\alpha)_{jj'}$ into (5.13.3), we arrive at an explicit equation for $\chi_{\alpha\beta}$:

$$\bar{\chi}_{\alpha\beta} = -\frac{\bar{n}e^2}{m\omega^2}\delta_{\alpha\beta} + \frac{2e^2}{\hbar^2\omega^2 V}\sum_{\mathbf{p}jj'} f_j(\mathbf{p})$$

$$\times \left\{\frac{R_{jj'}^\alpha R_{j'j}^\beta (E_{\mathbf{p}j'} - E_{\mathbf{p}j})^2}{E_{\mathbf{p}j'} - E_{\mathbf{p}j} - \hbar\omega} + \frac{R_{jj'}^\beta R_{j'j}^\alpha (E_{\mathbf{p}j'} - E_{\mathbf{p}j})^2}{E_{\mathbf{p}j'} - E_{\mathbf{p}j} + \hbar\omega}\right\} \tag{5.13.4}$$

The summation is in fact carried out over $j \neq j'$, owing to the factor $(E_{\mathbf{p}j'} - E_{\mathbf{p}j})^2$. It may seem that χ is determined only by the interband transitions. However, the matrix elements R_{jj} diagonal in j are singular, so that we can make use of the sum rule and write (5.13.4) in a form that singles out the contribution of intraband transitions.

The sum rules are obtained when commutation relations are used for the electron coordinate and momentum operators. For any wave function $\psi(r)$, the following must hold:

$$\int \psi^*(\mathbf{r})(p_\beta x_\alpha - x_\alpha p_\beta)\psi(\mathbf{r})\, \mathrm{d}^3 r = -\mathrm{i}\hbar\delta_{\alpha\beta} \int \psi^*(r)\psi(r)\, \mathrm{d}^3 r$$

$$\int \psi^*(\mathbf{r})(x_\alpha x_\beta - x_\beta x_\alpha)\psi(\mathbf{r})\, \mathrm{d}^3 r = 0$$

By virtue of the completeness of the set of Bloch functions $\psi_{\mathbf{p}j}$, the last formula can now be rewritten in the form

$$\sum_{\mathbf{p}j} \int \psi^*(\mathbf{r}) x_\alpha \psi_{\mathbf{p}j}\, \mathrm{d}^3 r \int \psi^*_{\mathbf{p}j} x_\beta \psi\, \mathrm{d}^3 r - \sum_{\mathbf{p}j} \int \psi^* x_\beta \psi_{\mathbf{p}j}\, \mathrm{d}^3 r \int \psi^*_{\mathbf{p}j} x_\alpha \psi\, \mathrm{d}^3 r = 0$$

The first commutation relation is rewritten in a similar manner.

Now we substitute the explicit relations for the matrix elements of the operators x_α and p_β. The matrix element of $p_\beta = -\mathrm{i}\hbar\partial/\partial x_\beta$ has in fact been calculated in (5.9.7):

$$\int \psi^*_{\mathbf{p}j} p_\beta \psi_{\mathbf{p}'j'}\, \mathrm{d}^3 r = m \frac{\partial E_{\mathbf{p}j}}{\partial p_\beta} \delta_{jj'}\delta_{\mathbf{pp}'} - \frac{im}{\hbar}(E_{\mathbf{p}j'} - E_{\mathbf{p}j}) R^\beta_{jj'} \delta_{\mathbf{pp}'}$$

Making calculations similar to those that led from (5.11.2) to (5.11.3), we can show that

$$\int \psi_{\mathbf{p}j'} x_\beta \psi_j\, \mathrm{d}^3 r = \mathrm{i}\hbar \frac{\partial c_j(\mathbf{p})}{\partial p_\beta} \delta_{j'j} + R^\beta_{j'j} c_j(\mathbf{p})$$

$$\int \psi^*_j x_\alpha \psi_{\mathbf{p}j'}\, \mathrm{d}^3 r = \mathrm{i}\hbar \frac{\partial c^*_j}{\partial p_\alpha} + c^*_j(\mathbf{p}) R^\alpha_{jj'}$$

We have assumed here that

$$\psi_j = \sum_{\mathbf{p}} c_j(\mathbf{p})\psi_{\mathbf{p}j}$$

where $c_j(\mathbf{p})$ is an arbitrary but sufficiently smooth function of $\mathbf{p}$, which is periodical, as is $f_j(\mathbf{p})$, on the reciprocal lattice. The function $c_j(\mathbf{p})$ can be considered for $\mathbf{p}$ within the first Brillouin zone only.

The commutation relations now give for $\psi = \psi_j$

$$- \mathrm{i}\hbar \sum_{\mathbf{p}} |c_j(\mathbf{p})|^2 \delta_{\alpha\beta} = \mathrm{i}\hbar m \sum_{\mathbf{p}} \frac{\partial |c_j(\mathbf{p})|^2}{\partial p_\alpha} \frac{\partial E_{\mathbf{p}j}}{\partial p_\beta} + \frac{\mathrm{i}}{\hbar} m \sum_{\mathbf{p}j'} |c_j(\mathbf{p})|^2 (E_{\mathbf{p}j} - E_{\mathbf{p}j'})(R^\alpha_{jj'} R^\beta_{j'j} + R^\beta_{jj'} R^\alpha_{j'j}) \quad (5.13.5)$$

and likewise,

$$0 = \sum_{\mathbf{p}j'} |c_j(\mathbf{p})|^2 (R^\alpha_{jj'} R^\beta_{j'j} - R^\beta_{jj'} R^\alpha_{j'j}) - \mathrm{i}\hbar \sum_{\mathbf{p}} \left(\frac{\partial |c_j(\mathbf{p})|^2}{\partial p_\alpha} R^\beta_{jj} - \frac{\partial |c_j(\mathbf{p})|^2}{\partial p_\beta} R^\alpha_{jj} \right) \quad (5.13.6)$$

These relations hold for arbitrary functions $c_j(\mathbf{p})$. Assuming $|c_j(\mathbf{p})|^2 = f_j(\mathbf{p})$, we arrive at the required relations between the matrix elements $R^\alpha_{jj'}$, which hold for arbitrary functions $f_j(\mathbf{p})$.

Assuming

$$(E_{\mathbf{p}j'} - E_{\mathbf{p}j})^2 = (E_{\mathbf{p}j'} - E_{\mathbf{p}j} - \hbar\omega)(E_{\mathbf{p}j'} - E_{\mathbf{p}j} + \hbar\omega) + (\hbar\omega)^2$$

in expression (5.13.4) for $\overline{\chi}_{\alpha\beta}$, we arrive at precisely those combinations $R^\alpha_{jj'}$ which enter (5.13.5) and (5.13.6). Taking this into account, we find

$$\bar{\chi}_{\alpha\beta}(\omega) = -\frac{\bar{n}e^2}{m\omega^2}\delta_{\alpha\beta} + \frac{2e^2}{m\omega^2 V} \sum_{\mathbf{p}j} f_j(\mathbf{p}) \delta_{\alpha\beta} + \frac{2e^2}{\omega^2 V} \sum_{\mathbf{p}j} \frac{\partial f_j(\mathbf{p})}{\partial p_\alpha} \frac{\partial E_{\mathbf{p}j}}{\partial p_\beta} + \frac{2e^2}{V} \sum_{\mathbf{p}jj'} f_j(\mathbf{p}) \left\{ \frac{R^\alpha_{jj'} R^\beta_{j'j}}{E_{\mathbf{p}j'} - E_{\mathbf{p}j} - \hbar\omega} + \frac{R^\beta_{jj'} R^\alpha_{j'j}}{E_{\mathbf{p}j'} - E_{\mathbf{p}j} + \hbar\omega} \right\} + \frac{2e^2 \mathrm{i}}{mV} \sum_{\mathbf{p}j} \left(\frac{\partial f_j(\mathbf{p})}{\partial p_\alpha} R^\beta_{jj} - \frac{\partial f_j(\mathbf{p})}{\partial p_\beta} R^\alpha_{jj} \right) \quad (5.13.7)$$

Taking normalization

$$\frac{2}{V} \sum_{\mathbf{p}j} f_j(\mathbf{p}) = \frac{1}{V} \sum_j N_j = \bar{n}$$

into account, we notice that the first term in (5.13.7) for $\overline{\chi}$ compensates for the second one. The third term corresponds to transitions within one band; it coincides with (5.12.3) for $\nu \to 0$. The fourth term corresponds to interband transitions and the summation here is actually carried out over $j \neq j'$. The fifth term is an addition to the intraband term; it has the same origin as the last term in the righthand side of equation (5.11.3) at $j = j'$. It is small in comparison with the main intraband term, when $\hbar\omega \ll I_0$. Estimates were obtained in connection with equation (5.11.3). Furthermore, it vanishes identically if $f_j(\mathbf{p}) = f_j(-\mathbf{p})$, which does take place, for instance, for the equilibrium Fermi distribution f_{0j}.

Ignoring the last term and going, as usual, from summation to integration, we obtain

$$\bar{\chi}_{\alpha\beta}(\omega) = \frac{2e^2}{\omega^2}\sum_j \int \frac{\partial f_j(\mathbf{p})}{\partial p_\alpha}\frac{\partial E_{\mathbf{p}j}}{\partial p_\beta}\frac{d^3p}{(2\pi\hbar)^3} + 2e^2\sum_{jj'}\int f_j(\mathbf{p})$$
$$\times \left\{\frac{R^\alpha_{jj'}R^\beta_{j'j}}{E_{\mathbf{p}j'} - E_{\mathbf{p}j} - \hbar(\omega + i\delta)} + \frac{R^\beta_{jj'}R^\alpha_{j'j}}{E_{\mathbf{p}j'} - E_{\mathbf{p}j} + \hbar(\omega + i\delta)}\right\}\frac{d^3p}{(2\pi\hbar)^3} \tag{5.13.8}$$

On the contrary, only the first term must be retained in (5.13.2) for the x-ray range, since the second term decreases faster as frequency increases. At high frequencies electrons behave as free, so that

$$\chi_{\alpha\beta}(\mathbf{r}, \mathbf{r}'; \omega) = -\frac{n(\mathbf{r})}{m\omega^2}\delta_{\alpha\beta}\delta(\mathbf{r} - \mathbf{r}')$$

and

$$\chi^{\mathbf{ll}'}_{\alpha\beta}(\mathbf{k}, \omega) = -\frac{e^2}{m\omega^2}n^{\mathbf{ll}'}\delta_{\alpha\beta}$$

The diffraction of x-ray waves in a crystal and all the peculiarities of x-ray optics (Bragg reflection, Borrmann effect, etc.) are determined by the Fourier components of electron density $n^{\mathbf{ll}'}$ (in fact, by the dependence on the difference $\mathbf{l} - \mathbf{l}'$).

5.14. Electron–Phonon Interaction

We have mentioned already that the finite electric conduction of metals is caused by the deceleration of electrons traveling through the crystal. This deceleration is not caused by scattering by atoms of ideal crystal lattice; this motion in the ideal lattice would not be damped and the metal would be the ideal conductor.

The cause of finiteness of electric conduction is the scattering taken into account in Section 5.12 through the relaxation time τ or the collision frequency ν. One of the scattering channels is scattering by impurities. The main scattering factor in pure and structurally perfect crystals, however, is the scattering by thermal vibrations of the lattice (the phonons).

In quantum terms, an electron in a crystal can absorb and emit phonons, thereby changing its momentum (quasimomentum) and energy.

An electron is described by the Schrödinger equation (5.9.1) or (5.11.1), in the spirit of the adiabatic approximation and self-consistent field in the crystal, but when lattice vibrations get taken into account, the potential $W(\mathbf{r})$ is not strictly periodic; it has an addition $\delta W(\mathbf{r})$.

Taking into account the smallness of the lattice vibration amplitude, we have

$$\delta W(\mathbf{r}) = \sum_{\mathbf{n}s\alpha} \frac{\partial W(\mathbf{r})}{\partial R_{\mathbf{n}s\alpha}} u_{\mathbf{n}s\alpha}$$

for atomic displacements $u_{\mathbf{n}s\alpha}$ from the equilibrium positions $R_{\mathbf{n}s\alpha}$.

If the displacements $u_{\mathbf{n}s\alpha}$ are expressed in terms of the normal mode amplitudes $u_{\mathbf{k}j}$ or, in the quantum case, in terms of the phonon creation and annihilation operators $a^{+}_{\mathbf{k}j}$ and $a_{\mathbf{k}j}$ using (5.3.15), we arrive at

$$\delta W(\mathbf{r}) = \mathscr{H}'_{\text{e-ph}}(\mathbf{r}) = \frac{1}{\sqrt{N}} \sum_{\mathbf{k}j} (\gamma_{\mathbf{k}j}(\mathbf{r}) \times \exp(i\mathbf{k}\mathbf{r}) a^{+}_{\mathbf{k}j} + \gamma^{*}_{\mathbf{k}j}(\mathbf{r}) \exp(-i\mathbf{k}\mathbf{r}) a^{+}_{\mathbf{k}j})$$

$$\gamma_{\mathbf{k}j}(\mathbf{r}) = \sum_{\mathbf{n}s\alpha} \frac{\partial W(\mathbf{r})}{\partial R_{\mathbf{n}s\alpha}} \sqrt{\frac{\hbar}{M\omega_{\mathbf{k}j}}}\, e^{j}_{s\alpha}(\mathbf{k}) \exp[i\mathbf{k}(\mathbf{R}_{\mathbf{n}} - \mathbf{r})] \quad (5.14.1)$$

It can be shown, using the periodicity of the crystal at $u_{\mathbf{n}s\alpha} = 0$ that the functions $\gamma_{\mathbf{k}j}(\mathbf{r})$ are periodical, with the lattice period. Indeed, it is obvious that for an arbitrary vector of the Bravais lattice $\mathbf{R}_{\mathbf{m}}$, we have

$$\frac{\partial W(\mathbf{r} + \mathbf{R}_{\mathbf{m}})}{\partial R_{(\mathbf{n}+\mathbf{m})s\alpha}} = \frac{\partial W(\mathbf{r})}{\partial R_{\mathbf{n}s\alpha}}$$

Therefore,

$$\gamma_{\mathbf{k}j}(\mathbf{r} + \mathbf{R}_{\mathbf{m}}) = \sum_{\mathbf{n}s\alpha} \frac{\partial W(\mathbf{r} + \mathbf{R}_{\mathbf{m}})}{\partial R_{\mathbf{n}s\alpha}} \sqrt{\frac{\hbar}{M\omega_{\mathbf{k}j}}}\, e^{j}_{s\alpha}(\mathbf{k}) \exp[i\mathbf{k}(\mathbf{R}_{(\mathbf{n}-\mathbf{m})} - \mathbf{r})]$$

or, replacing summation over $\mathbf{n}$ by summation over $\mathbf{n}' = \mathbf{n} - \mathbf{m}$,

$$\gamma_{\mathbf{k}j}(\mathbf{r} + \mathbf{R}_{\mathbf{m}}) = \sum_{\mathbf{n}'s\alpha} \frac{\partial W(\mathbf{r} + \mathbf{R}_{\mathbf{m}})}{\partial R_{(\mathbf{n}'+\mathbf{m})s\alpha}} \sqrt{\frac{\hbar}{M\omega_{\mathbf{k}j}}}\, e^{j}_{s\alpha}(\mathbf{k}) \exp[i\mathbf{k}(\mathbf{R}_{\mathbf{n}'} - \mathbf{r})] = \gamma_{\mathbf{k}j}(\mathbf{r})$$

The operator $\mathscr{H}'_{\text{e-ph}}$ is the operator of the electron–phonon interaction. It contains terms that correspond to emission and absorption of phonons; indeed, $a^{+}_{\mathbf{k}j}$ and $a_{\mathbf{k}j}$ are the phonon creation and annihilation operators. Using the interaction operator (5.14.1) and perturbation theory, we can find the probability of transition of an electron from a state with a Bloch function $\psi_{\mathbf{p}l}$ to a state $\psi_{\mathbf{p}'l'}$, simultaneously absorbing or emitting a phonon in the $\mathbf{k}j$ mode. The probability of the process with phonon absorption is

$$\begin{aligned} w_{\mathbf{p}'l',\mathbf{p}l} &= \frac{2\pi}{\hbar} |\mathscr{H}'_{\mathbf{p}'l',\mathbf{p}l}|^2 \delta(E_{\mathbf{p}l} - E_{\mathbf{p}'l'} - \hbar\omega_{\mathbf{k}j}) \\ &= \frac{2\pi}{\hbar N} \left| \int \psi^*_{\mathbf{p}'l'}(\mathbf{r}) \gamma^*_{\mathbf{k}j}(\mathbf{r}) \exp(-\mathrm{i}\mathbf{k}\mathbf{r}) \psi_{\mathbf{p}l}(\mathbf{r})\, \mathrm{d}^3 r \right|^2 \\ &\quad \times (n_{\mathbf{k}j} + 1) \delta(E_{\mathbf{p}l} - E_{\mathbf{p}'l'} - \hbar\omega_{\mathbf{k}j}) \end{aligned} \tag{5.14.2}$$

We have used in (5.14.2) the fact that the matrix element of the operator $a^+_{\mathbf{k}j}$ equals $\sqrt{n_{\mathbf{k}j} + 1}$, where $n_{\mathbf{k}j}$ is the number of phonons in the $\mathbf{k}j$ mode.

We can likewise write the expression for the probability of a process with phonon absorption but need to replace $\gamma_{\mathbf{k}j} \mathrm{e}^{\mathrm{i}\mathbf{k}\mathbf{r}}$ for $\gamma^*_{\mathbf{k}j} \mathrm{e}^{-\mathrm{i}\mathbf{k}\mathbf{r}}$, $n_{\mathbf{k}j}$ for $n_{\mathbf{k}j} + 1$ and $\delta(E_{\mathbf{p}l} - E_{\mathbf{p}'l'} + \hbar\omega_{\mathbf{k}j})$ for $(E_{\mathbf{p}l} - E_{\mathbf{p}'l'} - \hbar\omega_{\mathbf{k}j})$.

The calculation of the matrix element yields the law of conservation of quasimomentum. Indeed,

$$\begin{aligned} &\int_V \exp\left(-\frac{\mathrm{i}}{\hbar}\mathbf{p}'\mathbf{r}\right) u^*_{\mathbf{p}'l'} \gamma_{\mathbf{k}j} \exp(-\mathrm{i}\mathbf{k}\mathbf{r}) \exp\left(\frac{\mathrm{i}}{\hbar}\mathbf{p}\mathbf{r}\right) u_{\mathbf{p}l}\, \mathrm{d}^3 r \\ &= \int_\Omega \exp\left(-\frac{\mathrm{i}}{\hbar}(\mathbf{p}' + \hbar\mathbf{k} - \mathbf{p})\mathbf{r}\right) u^*_{\mathbf{p}'l'} \gamma^*_{\mathbf{k}j} u_{\mathbf{p}j}\, \mathrm{d}^3 r \\ &\quad \times \sum_{\mathbf{n}} \exp\left(-\frac{\mathrm{i}}{\hbar}(\mathbf{p}' + \hbar\mathbf{k} - \mathbf{p})\mathbf{R}_{\mathbf{n}}\right) \end{aligned}$$

Since the sum

$$\sum_{\mathbf{n}} \exp\left(-\frac{\mathrm{i}}{\hbar}(\mathbf{p}' + \hbar\mathbf{k} - \mathbf{p})\mathbf{R}_{\mathbf{n}}\right)$$

equals N if $\mathbf{p}' + \hbar\mathbf{k} - \mathbf{p} = \hbar\mathbf{g}_{\mathrm{l}}$ and vanishes if $\mathbf{p}' + \hbar\mathbf{k} - \mathbf{p} \neq \hbar\mathbf{g}_{\mathrm{l}}$, we arrive at the conservation law for quasimomentum:

$$\mathbf{p}' + \hbar\mathbf{k} = \mathbf{p} + \hbar\mathbf{g}_{\mathrm{l}}$$

Even though phonons do not carry momentum, quasimomentum is conserved when phonons interact with electrons. As in most cases, quasimomentum is conserved up to $\hbar\mathbf{g}_{\mathrm{l}}$, where $\mathbf{g}_{\mathrm{l}}$ is an arbitrary reciprocal lattice vector. If $\mathbf{g}_{\mathrm{l}} \neq 0$, the process we are considering involves umklapp.

Let us evaluate the probability for an electron to be scattered with emission or absorption of a phonon. To do this, we need to sum (5.14.2) up over all possible final states:

$$\nu = \frac{1}{\tau} = \sum_{\mathbf{p}'l'} w_{\mathbf{p}'l',\mathbf{p}l} = \frac{2\pi}{\hbar}\overline{|\gamma|^2(n_{\mathbf{k}j}+1)}\,\frac{V}{N}\int \delta(E_{\mathbf{p}l} - E_{\mathbf{p}'l'} - \hbar\omega_{\mathbf{k}j})$$
$$\times\, \mathrm{d}^3p'(2\pi\hbar)^{-3} \sim \frac{I_0}{\hbar}\sqrt{\frac{m}{M}} \sim 10^{14}\mathrm{s}^{-1} \qquad (5.14.3)$$

For the evaluation, we used

$$|\gamma| \sim \frac{\partial W}{\partial R}\sqrt{\frac{\hbar}{M\omega_{\mathbf{k}j}}} \qquad \hbar\omega_{\mathbf{k}j} \sim \sqrt{\frac{m}{M}}\,I_0 \ll E_{\mathbf{p}l} \qquad \frac{\partial W}{\partial R} \sim \frac{I_0}{a_0}$$

so that

$$|\gamma| \sim \frac{I_0}{a_0}\sqrt{\frac{\hbar^2}{\sqrt{mM}\,I_0}} \sim \sqrt[4]{\frac{m}{M}}\sqrt{\frac{\hbar^2 I_0}{ma_0^2}} \sim \sqrt[4]{\frac{m}{M}}\,I_0$$

Therefore, the mean collision frequency is the atomic frequency $I_0/\hbar$ times $(m/M)^{1/2}$, that is, the squared smallness parameter of the adiabatic approximation.

If there were no quantum effects, the collision frequency would be of the order of $I_0/\hbar$. In an ideal lattice, electrons would not be scattered at all, in view of their quantum nature, but the rigorous periodicity of the lattice is violated by small oscillations of atoms at relative amplitudes $(m/M)^{1/4}$. This is a measure of disorder in the lattice, and the square of this ratio determines the collision frequency.

Experiments confirm this estimate: the free path length time between collisions in metals is of the order of 10^{-14} s and the free path length is of the order of $10^2 a_0 \sim \sqrt{m/M}\,a_0$ (electrons move in crystals at velocities of the order of $\partial E/\partial p \sim I_0 a_0 \hbar^{-1} \sim 10^8$ cm/s). This is true for $\mathrm{k}T \sim \hbar\omega_{\mathbf{k}j} = \mathrm{k}T_\mathrm{D}$ (T_D is the Debye temperature). Since at high temperatures $\bar{n}_{\mathbf{k}j} \sim T/T_\mathrm{D}$, the factor $n_{\mathbf{k}j}$ causes the collision frequency ν to be proportional to T; hence, the electric conductivity of metals is proportional to T^{-1}.

At low temperatures ($\hbar\omega_{\mathbf{k}j} \gtrsim \mathrm{k}T$)

$$\bar{n}_{\mathbf{k}j} \sim \left(\exp\left(\frac{\hbar\omega_{\mathbf{k}j}}{\mathrm{k}T}\right) - 1\right)^{-1}$$

so that only low-frequency acoustic phonons are present. As a result, processes with phonon absorption are considerably restricted and the emission of phonons is also suppressed owing to limitations imposed by the Pauli exclusion principle. Because of these restrictions, only transitions that change energy by about $\mathrm{k}T$ near the Fermi level are possible for electrons. If $\mathrm{k}T < \hbar\omega_{\mathbf{k}j}$, processes that emit a phonon with energy $\hbar\omega_{\mathbf{k}j}$ are greatly suppressed. In view of all this, electric conductivity of pure metals at low temperatures varies as T^{-5}.

The general formulas and estimates for the collision frequency, considered together with formulas of Section 5.12, answer the question about electric conduction in typical metals. At typical electron concentrations in partially occupied bands, $n \sim 10^{23}$ cm^{-3} and at temperatures of the order of the Debye (room) temperature, it is about $\sigma = ne^2\tau/m^* \sim I_0\hbar^{-1}(M/m)^{1/2}na_0^3 \sim 10^{17}$ CGS; as temperature is lowered, conductivity may grow by several orders of magnitude.

The range of variation of conductivity in semiconductors is much wider since the concentration of conduction electrons can itself vary in a very wide range, depending on temperature and concentration of the so-called dopants (impurities capable of capturing or releasing electrons or holes) and structural defects. Hence, semiconductors are usually characterized by the concentration of electrons (holes) and their so-called mobility μ, which is closely related to the free path time: $\mu = e\tau/m^*$.

The main factor at room and still lower temperatures is the scattering by long-wavelength acoustic phonons. Indeed, in this case kT is less than the optical phonon energy. Hence, the number of such phonons is low; neither can electrons emit them since their energy is insufficient. At the same time, the thermal velocity of electrons, $(\mathrm{k}T/m^*)^{1/2}$, is much higher than the velocity of sound s down to the lowest achievable temperatures. It is not difficult to verify that owing to the conservation of energy and quasimomentum, the quasimomenta of the emitted or absorbed acoustic phonons cannot considerably exceed that of the electron; hence, their energies are much lower than that of the electron: $\hbar\omega_{\mathbf{k}} = \hbar s|\mathbf{k}| \ll p(p/m^*) = 2E_{\mathbf{p}}$.

Substituting $E_{\mathbf{p}} = E_0 + p^2/2m^*$ into formula (5.14.2) and taking into account that the energy of the emitted (absorbed) photons is small compared with the thermal energy of electrons, we obtain

$$\nu_{\mathbf{p}} \cong \frac{1}{\pi\hbar}\overline{|\gamma_{\mathbf{k}}|^2(n_{\mathbf{k}}+1)}\, m^* p\Omega\hbar^{-3}$$

where Ω is the unit cell volume.

Before we can use this formula for quantitative evaluations, it is necessary to evaluate the quantity $\overline{|\gamma|^2}$, which differs substantially, for long-wavelength phonons, from the general estimate obtained above. The point is that in the limit of infinite wavelength, the acoustic deformation reduces simply to a parallel translation of the crystal as a whole. Clearly, electron energies cannot be changed by translation, that is, the matrix elements of γ in (5.14.2) must tend to zero. The following arguments help to find their values for finite but large wavelengths.

Let us consider a macroscopically large element of crystal volume, comprising a large number of unit cells but whose dimensions are nevertheless small compared with wavelength. An acoustic wave not only displaces this volume as a whole but is also equivalent to a uniform deformation, for example, tension or compression. This deformation signifies changes in the crystal lattice periods, which inevitably changes the energies of electron states. If the relative deformations ε are small (the wave amplitude is low), the increment of electron energy can be expanded in a series in powers of ε; retaining only the first term, we obtain

$$\mathcal{H}'_{\text{e-ph}} = D\varepsilon$$

where D is the so-called deformation potential. It can be easily evaluated: indeed, shifts in electron energy levels are obviously of the order of the characteristic electron energies I_0 for the relative change in interatomic spacings (relative deformation) $\varepsilon \sim 1$. Hence, $D \sim I_0$; recalling that the relative deformation is $\partial u_\alpha / \partial x_\beta$ and comparing the expression for $\mathcal{H}_{\text{e-ph}}$ thus obtained with (5.14.1), we easily arrive at the following expression for γ:

$$|\gamma_{\mathbf{k}j}| = |D| \sqrt{\frac{\hbar}{M\omega_{\mathbf{k}j}}}\, |\mathbf{k}| = |D| \sqrt{\frac{\hbar|\mathbf{k}|}{Ms}}$$

Finally, since $\hbar\omega_{\mathbf{k}j} \ll \mathrm{k}T$ for all phonons of interest for us, the band occupancy numbers are $n_{\mathbf{k}j} \simeq \mathrm{k}T/\omega_{\mathbf{k}j}$. Collecting all these results, we finally obtain the following formula for the scattering frequency of electrons (or holes) in semiconductors:

$$\frac{\nu_{\mathbf{p}} = \frac{1}{\pi\hbar^4}\, m^* D^2 \mathrm{k}T}{\rho s^2\, |\mathbf{p}|}$$

where $\rho = M/\Omega$ is the crystal density. Since the mean thermal momentum of an electron is $\overline{|\mathbf{p}|} = \sqrt{m^*\mathrm{k}T}$, we immediately have the familiar law for mobility: $\mu \sim T^{-3/2}$. The free path length, that is, $l \sim |\mathbf{p}| m^* \nu_{\mathbf{p}}$, is found to be independent of momentum (velocity) and inversely proportional to temperature. Substituting typical values for evaluation (in CGS units) $m^* \sim 10^{-28}$–10^{-27}, $D \sim 3 \times 10^{-11}$, $s \sim 3 \times 10^5$, $\rho \sim 10$, we obtain $l \sim (10^{-4}/T)$ cm. At room temperature, $l \sim 10^{-6}$ cm, that is, of the same order of magnitude as in metals, and can reach about 10^{-4} cm at $T \sim 1$ K.

5.15. Mechanisms of Light Absorption in Solids

The light absorption (including infrared absorption) in pure crystals is caused by

1. Intraband transitions, or by the contribution of free charge carriers (in metals and semiconductors),

2. Interband transitions (the so-called intrinsic absorption) and
3. Absorption by excitons.

The first of these mechanisms was already discussed in Section 5.12. The absorption coefficient for the electromagnetic radiation is determined by the imaginary (anti-Hermitian) component of the tensor χ; the contribution to this tensor due to intraband transitions is given by formulas (5.12.3) and (5.12.4) and by the estimates of the previous sections.

This mechanism does not work in dielectric materials since there are virtually no free electrons nor unoccupied states in filled bands even at finite temperatures, and $\partial f_0/\partial \mathbf{p} = 0$.

As for interband transitions, an analogue here is the atomic linear spectrum. The light absorption coefficient is determined by the anti-Hermitian part of χ, through which losses $\dot{Q}$ are expressed (see Chapter 1). Expression (5.13.8) yields

$$\dot{Q} = \frac{\omega}{4\mathrm{i}} E^*_{\omega\alpha}(\chi_{\alpha\beta} - \chi^*_{\beta\alpha})E_{\omega\beta} = \omega\pi e^2 E^*_{\omega\alpha} \sum_{jj'} \int (f_j(\mathbf{p}) - f_{j'}(\mathbf{p}))R^{\alpha}_{jj'}R^{\beta}_{j'j}\delta(E_{\mathbf{p}j'} - E_{\mathbf{p}j} - \hbar\omega)\,\frac{\mathrm{d}^3 p}{(2\pi\hbar)^3}\,E_{\omega\beta} \quad (5.15.1)$$

In deriving this formula from (5.13.8), we have used that

$$\begin{aligned}
&(R^{\alpha}_{jj'})^* = R^{\alpha}_{j'j}\\
&(E_{\mathbf{p}j'} - E_{\mathbf{p}j} - \hbar(\omega + \mathrm{i}\delta))^{-1} - (E_{\mathbf{p}j'} - E_{\mathbf{p}j} - \hbar(\omega - \mathrm{i}\delta))^{-1}\\
&\quad = 2\pi\mathrm{i}\delta(E_{\mathbf{p}j'} - E_{\mathbf{p}j} - \hbar\omega)
\end{aligned}$$

As we have mentioned in Chapter 2, the same result could be obtained from the general formulas when calculating the probability of the interband transition by perturbation theory, using the perturbation operator $\mathscr{H}' = -e\mathbf{r}\boldsymbol{\mathcal{E}}$. According to the arguments of Sections 5.11 and 5.12, the matrix elements of $\mathbf{r}$ for $j \neq j'$ are then

$$(x_\alpha)_{\mathbf{p}'j',\mathbf{p}j} = R^{\alpha}_{j'j}(\mathbf{p})\delta_{\mathbf{p}'\mathbf{p}}$$

We will use (5.15.1) to evaluate the light absorption coefficient produced by the interband transition. The absorption coefficient $\alpha(\omega)$ for the energy flux I of the electromagnetic field propagating along the z axis is, by definition,

$$\alpha(\omega) = \frac{1}{I}\frac{\mathrm{d}I}{\mathrm{d}z} = \frac{\dot{Q}}{I} = \frac{8\pi\dot{Q}}{c|\mathbf{E}_\omega|^2} = \frac{8\pi}{c|\mathbf{E}_\omega|^2}\,\pi\omega e^2 E^*_{\omega\alpha}\sum_{jj'}\int (f_j(\mathbf{p}) - f_{j'}(\mathbf{p}))R^{\alpha}_{jj'}R^{\beta}_{j'j}\delta(E_{\mathbf{p}j'} - E_{\mathbf{p}j} - \hbar\omega)\,\frac{\mathrm{d}^3 p}{(2\pi\hbar)^3}\,E_{\omega\beta}$$

For a transition in a semiconductor between the valence band (j = v) and the conduction band (j' = c), with the radiation being linearly polarized along the x axis, we obtain

$$\alpha_x(\omega) = \frac{8\pi^2\omega e^2}{c} \int (f_{\rm v}(\mathbf{p}) - f_{\rm c}(\mathbf{p}))|R^x_{\rm cv}|^2 \times \delta(E_{\mathbf{p}\rm c} - E_{\mathbf{p}\rm v} - \hbar\omega)\, \frac{{\rm d}^3 p}{(2\pi\hbar)^3} \tag{5.15.2}$$

We will evaluate this absorption coefficient for the dispersion law

$$E_{\mathbf{p}\rm c} = E_{\rm c} + \frac{p^2}{2m^*_{\rm c}} \qquad E_{\mathbf{p}\rm v} = E_{\rm v} - \frac{p^2}{2m^*_{\rm v}}$$

so that

$$E_{\mathbf{p}\rm c} - E_{\mathbf{p}\rm v} = E_{\rm g} + \frac{p^2}{2m^*}$$

where $E_{\rm g} = E_{\rm c} - E_{\rm v}$ is the gap width and

$$m^* = (m^{*-1}_{\rm c} + m^{*-1}_{\rm v})^{-1}$$

is the normalized effective electron and hole mass.

We see from (5.15.2) that $\alpha \neq 0$ only if $\hbar\omega > E_{\rm g}$. Note now that $f_{\rm v} - f_{\rm c} = -[(1 - f_{\rm v}) + f_{\rm c}] + 1 \simeq 1$ if the populations of the electron and hole levels, $f_{\rm c}$ and $(1 - f_{\rm v})$, are low; this is always the case in a nondegenerate semiconductor if the band gap $E_{\rm g} \gg {\rm k}T$ and the concentration of dopant impurities is moderate. With these qualifications, we can ignore the dependence of $|R_{\rm vc}|^2$ on $\mathbf{p}$ (indeed, we are considering absorption near the absorption band edge, when $\hbar\omega \simeq E_{\rm g}$ and $\mathbf{p} \simeq 0$) and write

$$\begin{aligned}\alpha_x(\omega) &= \frac{8\omega e^2 m^{*3/2}}{\hbar^3 c\sqrt{2}} |R^x_{v\rm c}(0)|^2 \int \sqrt{\xi - E_{\rm g} + \hbar\omega}\; \delta(\xi)\, {\rm d}\xi \\ &= \frac{8\omega e^2 m^{*3/2}}{\hbar^3 c\sqrt{2}} |R^x_{v\rm c}|^2 \sqrt{\hbar\omega - E_{\rm g}}\end{aligned} \tag{5.15.3}$$

To calculate the integral in (5.15.2), we introduced the integration variable

$$\xi = E_{\rm g} + \frac{p^2}{2m^*} - \hbar\omega$$

For the evaluation, we assume

$$m^* \sim m \qquad \hbar\omega \sim E_{\rm g} \sim I_0 \qquad |R_{v\rm c}| \sim a_0$$

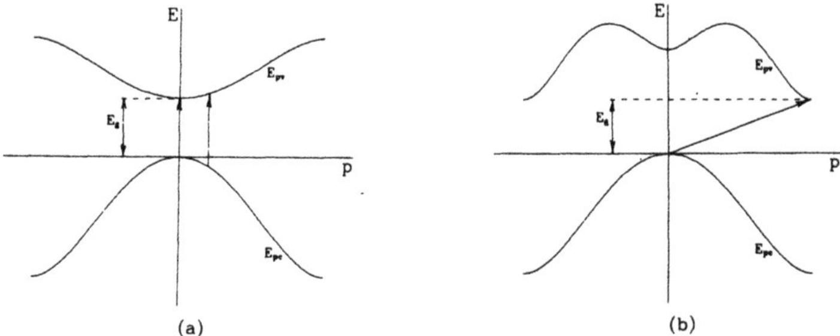

Figure 5.16. Direct (a) and indirect (b) interband transitions.

so that

$$\alpha \sim \frac{\hbar\omega e^2 m^{3/2}}{\hbar^4 c} a_0^2 \sqrt{I_0} \sqrt{\frac{\hbar\omega - E_g}{I_0}} \sim \frac{e^2}{\hbar c} \frac{1}{a_0} \sqrt{\frac{\hbar\omega - E_g}{I_0}}$$
$$\sim \frac{1}{137} 10^8 \sqrt{\frac{\hbar\omega - E_g}{I_0}} \text{ cm}^{-1} \sim 10^6 \sqrt{\frac{\hbar\omega - E_g}{I_0}} \text{ cm}^{-1}$$

Far from the absorption edge, where $\hbar\omega - E_g \sim I_0$, the absorption coefficient must be of the order of 10^6 to 10^7 cm^{-1}, which agrees well with experimental data.

So far we have been discussing the so-called direct, or vertical, transitions, when as a result of smallness of the wave vector $\mathbf{q}$ of the light wave we have $\mathbf{p} \simeq \mathbf{p}'$ (Figure 5.16(a)). It often happens, however (specifically, in Ge and Si) that

$$E_{\mathbf{p}c} = E_c + \frac{1}{2m_c^*} (\mathbf{p} - \mathbf{p}_c)^2$$

where $\mathbf{p}_c \neq 0$. Then the energy gap is $E_g = E_c - E_v$ but the ceiling of the valence band lies at $\mathbf{p} = 0$ and the bottom of the conduction band E_c lies at $\mathbf{p} = \mathbf{p}_c \neq 0$; direct transitions are now impossible at $\hbar\omega \simeq E_g$ owing to the conservation of quasimomentum. Indirect transitions involving phonons become important in this situation.

The probability of interband transition is found in the next (second) order of perturbation theory, by using the composite matrix element of the perturbation operator. This operator is written as a sum of operators of interaction with electromagnetic field and of electron–phonon interaction (5.14.1):

$$\mathscr{H}' = -e\mathcal{E}\mathbf{r} + \mathscr{H}'_{\text{e-ph}}$$

In this case the transition involves absorption or emission of a phonon with energy $\hbar\omega_{\mathbf{k}j}$ and momentum $\hbar\mathbf{k}$. The energy for the electron transition is supplied by the photon while the required quasimomentum is carried in (or out) by a phonon; all this results in an indirect transition shown in Figure 5.16(*b*).

The composite matrix element of second order in perturbation theory describes a virtual transition from a state $E_{\mathbf{p}\mathrm{v}}$ to an intermediate state $E_{\mathbf{p}j}$, which includes absorption of a photon and subsequent emission or absorption of a phonon, and the transition of the electron from the intermediate to the final state $E_{\mathbf{p}\pm\hbar\mathbf{k}c}$:

$$\mathscr{H}^{(2)} = \sum_j \frac{(\mathscr{H}'_{\text{e-ph}})_{cj}(-e\boldsymbol{\mathcal{E}}\mathbf{R}_{j\mathrm{v}})}{(E_{\mathbf{p}v} - E_{\mathbf{p}j} + \hbar\omega)}$$

Squared composite matrix element determining the transition probability includes an extra factor of the order of

$$|\mathscr{H}'_{\text{e-ph}}|^2 I_0^{-2} \sim |\gamma|^2 I_0^{-2} \sim \sqrt{\frac{m}{M}}$$

as compared with the squared matrix element of the direct transition. As a result, the absorption coefficient α for indirect transitions is less by a factor of $\sqrt{M/m} \sim 10^{-2}$ than that for direct transitions, being of the order of 10^{-4}–10^{-3} cm^{-1}.

An indirect transition occurs between electron states for which

$$E_{\mathbf{p}'c} - E_{\mathbf{p}v} = \hbar\omega \pm \hbar\omega_{\mathbf{k}j} \simeq E_{\mathrm{g}}$$

and the momentum changes by

$$\mathbf{p}' - \mathbf{p} \simeq \pm\hbar\mathbf{k} \simeq \mathbf{p}_{\mathrm{c}}$$

Therefore, if a semiconductor is such that $\mathbf{p}_{\mathrm{c}} \neq 0$ when $\hbar\omega \simeq E_{\mathrm{g}}$, then first indirect interband transitions begin; direct transitions occur at higher photon energies $\hbar\omega$ and the absorption coefficient begins to grow much faster (see Figure 5.16), reaching values of the order of 10^5–10^6 cm^{-1}.

If $\mathbf{p}_{\mathrm{c}} = 0$, interband absorption begins with direct transitions (Figure 5.17) but exciton absorption arises even before the threshold $\hbar\omega = E_{\mathrm{g}}$ is reached. An exciton is a bound state of an electron and a hole. Two types of excitons are recognized: strongly bound Frenkel excitons, or small-radius excitons, and weakly bound Wannier–Mott excitons, or large-radius excitons. Wannier–Mott excitons are found in semiconductor crystals with narrow band gaps and high values of dielectric permittivity.

In accordance with the above, an electron and a hole in a crystal behave as oppositely charged particles and therefore undergo Coulomb attraction,

$$V(\mathbf{r}) = -\frac{e^2}{\varepsilon r}$$

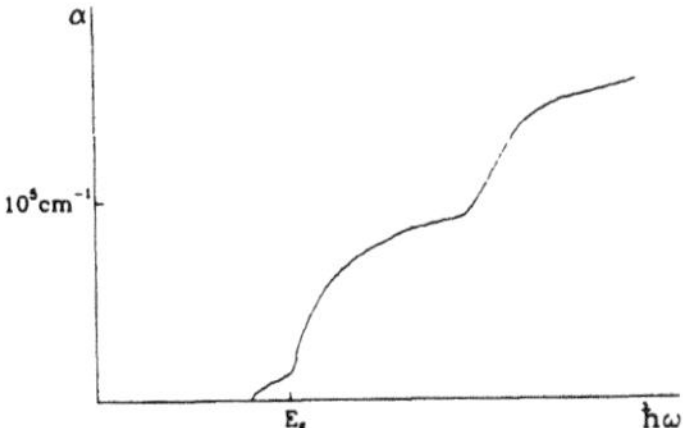

Figure 5.17. Light absorption coefficient α in a semiconductor as a function of light frequency.

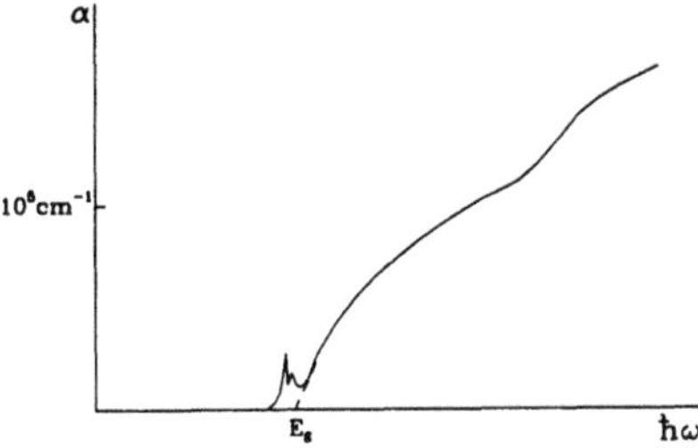

Figure 5.18. Exciton absorption of light close to the intrinsic absorption band of a semiconductor.

where ε is the dielectric permittivity. The Wannier–Mott exciton can be described as a bound hydrogenlike state of an "electron–hole" pair with energy

$$E_{\text{ex}} = -\frac{e^4 m^*}{2\hbar^2 \varepsilon n^2}$$

and effective "Bohr radius"

$$a_{\text{ex}} = \frac{\varepsilon \hbar^2}{m^* e^2}$$

If $\varepsilon = 5$ and $m^* = 0.5m$, the energy E_{ex} for $n = 1$ is about 0.25 eV. Hence, the minimum energy required to form an exciton is

$$E = E_{\text{g}} - \frac{m^* e^4}{2\hbar^2 \varepsilon}$$

which is less than the gap width E_{g}. The spectrum of the exciton absorption is a series of lines below the edge of the main absorption in the crystal (see Figure 5.18). These lines are broadened and the topmost of them merge with the intrinsic absorption band.

The value of a_{ex} at the same values of parameters is greater by an order of magnitude than the interatomic distance in crystals; this explains the term "large-radius excitons."

The other type of excitons (the Frenkel excitons) are typical of molecular crystals. Molecular crystals are characterized by low intermolecular interaction. Hence, the ground state wave function of a crystal is a product of wave functions of molecules in their ground states:

$$\Phi_0 = \prod_{\mathbf{n}} \psi_{\mathbf{n}0}$$

The energy of this state is $E_0 = N\varepsilon_0$ where ε_0 is the ground-state energy of a molecule.

In the same approximation, the wave function of the excited state of a crystal comprising N molecules is a product of the wave functions of individual molecules of which one is in an excited state,

$$\Phi_{\mathbf{n}j} = \psi_{\mathbf{n}j} \prod_{\mathbf{n}' \neq \mathbf{n}} \psi_{\mathbf{n}'0} \tag{5.15.4}$$

where $\psi_{\mathbf{n}j}$ is the wave function of the jth excited state of the molecule. The energy of this state is $E_j = E_0 + (\varepsilon_j - \varepsilon_0)$; this term is N-fold degenerate. However small the energy of intermolecular interaction, the state (5.15.4) will not be stationary and the excitation will migrate from one molecule to another. As a result of the dipole–dipole interaction between molecules, an excited molecule returns to the ground state while another molecule is raised to the excited state; in other words, excitation energy is transferred from one molecule to the other. As for the stationary state $\Phi_{\mathbf{k}j}$, it differs from (5.15.4) but tends to a certain linear combination of (5.15.4)-type states as the intermolecular interaction gets weaker. A correct linear combination can be derived from the requirement discussed earlier that a wave function must correspond to an irreducible representation of the group of translations defined by a vector $\mathbf{k}$. The exact wave function $\Phi_{\mathbf{k}}$ must possess the symmetry property

$$T_{\mathbf{n}}\Phi_{\mathbf{k}} = \exp(i\mathbf{k}\mathbf{R}_{\mathbf{n}})\Phi_{\mathbf{k}}$$

The correct linear combination into which $\Phi_{\mathbf{k}}$ is transformed is

$$\Phi_{\mathbf{k}j} = \frac{1}{\sqrt{N}} \sum_{\mathbf{n}} \exp(i\mathbf{k}\mathbf{R}_{\mathbf{n}})\Phi_{\mathbf{n}j} \tag{5.15.5}$$

If the intermolecular interaction is taken into account, then in general the energy in the state $\Phi_{\mathbf{k}}$ differs from the energy E_j; in perturbation theory, the correction can be found as the mean value of the intermolecular interaction energy in the state (5.15.5). This energy $E_{\mathbf{k}}$ is a function of $\mathbf{k}$ and determines the exciton energy band into which spreads the energy E_j which differs from E_0 by the energy of excitation of one molecule. Hence, the Frenkel exciton is a molecular excitation state propagating through a crystal.

When a crystal in the ground state Φ_0 interacts with the electromagnetic wave whose wave vector is $\mathbf{k}$, excitons with the wave vector $\mathbf{k} = \mathbf{q}$ can be excited. If we use the general theory of linear susceptibility of a medium, outlined in Chapter 2, then $\chi^{(1)}$ is determined by the Fourier components of the matrix elements of the current density operator

$$(j_\alpha^{\mathrm{l}}(\mathbf{q}))_{\mathbf{k}'j',\mathbf{k}j} = \frac{1}{\sqrt{V}} \int (j_\alpha(\mathbf{r}))_{\mathbf{k}'j',\mathbf{k}j} \exp[-\mathrm{i}(\mathbf{q} + \mathbf{g}_\mathrm{l})\mathbf{r}]\, \mathrm{d}^3 r$$

which are nonzero only for $\mathbf{k}' = \mathbf{k} - \mathbf{q}$. Therefore, the part of the tensor $\chi^{(1e)}$ connected with excitons can be written for a crystal in the ground state ($T = 0$) as

$$\chi_{\alpha\beta}^{\mathrm{ll}'}(\mathbf{k}, \omega) = \frac{1}{\omega^2} \sum_{j \neq 0} \left\{ \frac{(j_\alpha^{\mathrm{l}}(\mathbf{k}))_{0j} (j_\beta^{-\mathrm{l}'}(-k))_{j0}}{E_{\mathbf{k}j} - E_0 - \hbar(\omega + \mathrm{i}\delta)} + \frac{(j_\beta^{-\mathrm{l}'}(-k))_{0j} (j_\alpha^{\mathrm{l}}(\mathbf{k}))_{j0}}{E_{\mathbf{k}j} - E_0 + \hbar(\omega + \mathrm{i}\delta)} \right\} \tag{5.15.6}$$

This is an expression of the same type (in the neighborhood of a resonance) as the one obtained above for the susceptibility of ionic lattices, the frequency $\omega_{\mathbf{k}j}$ being equal here to $\hbar^{-1}(E_{\mathbf{k}j} - E_0)$. Therefore, all the earlier remarks in Sections 5.5–5.7 about the vibrations of ionic lattices and their interaction with the field are mostly transferred to the case of electron excitations, that is, excitons. Thus, long-wavelength excitons mix strongly with photons and form exciton polaritons (often referred to simply as polaritons).

In the same manner as for lattices, (5.15.6) is an expression for $\chi^{(\mathrm{e})}$, provided the wave functions $\Phi_{\mathbf{k}j}$ and the energies $E_{\mathbf{k}j}$ are found taking into account completely the transverse and the longitudinal long-wavelength fields. If $\Phi_{\mathbf{k}j}$ and $E_{\mathbf{k}j}$ and the corresponding matrix elements of the current density operator are found, ignoring the long-wavelength fields (the so-called mechanical excitons), we arrive at the true susceptibility $\chi_{\alpha\beta}$. The long-wavelength field can be taken into account via Maxwell's equations. If only longitudinal fields are taken into account in Maxwell's equations or in the propagation function D_0, we arrive at the so-called Coulomb excitons. The law of dispersion for them can be obtained in the quantum-mechanical theory by introducing the long-wavelength longitudinal field into the Hamiltonian, that is, by taking into account the entire Coulomb interaction. In fact, the model description of the Wannier–Mott and Frenkel excitons given above did correspond to the Coulomb excitons.

When the transverse field is also taken into account using Maxwell's equations (as we have shown above, only the long-wavelength transverse field is important in optics), one obtains the dispersion law for exciton polaritons (an analogue of polaritons in the case of electron excitations) which would be obtained in the comprehensive quantum-mechanical theory from the poles of (5.15.6) for $\chi^{(e)}$ when the induced transverse electromagnetic field has been taken into account completely in the Hamiltonian. It is these polaritons that are the actually observed excited states (normal waves) of a crystal in the exciton range of energies.

References

[1] Keldysh L V, Kirzhnits D A and Maradudin A A *The Dielectric Function of Condensed Systems* (New York: North-Holland)

[2] Landau L D and Lifshitz E M 1977 *Quantum Mechanics* (Oxford: Pergamon Press)

[3] Landau L D and Lifshitz E M 1979 *Statistical Physics* (Oxford: Pergamon Press)

[4] Landau L D and Lifshitz E M 1960 *Electrodynamics in Continuous Media* (Oxford: Pergamon Press)

[5] Lifshitz E M and Pitaevsky L P 1979 *Physical Kinetics* (Moscow: Nauka)

[6] Agranovich V M and Ginzburg V L 1983 *Spatial Dispersion in Crystal Optics and the Theory of Excitons* 2nd edition (Berlin: Springer)

[7] Ginzburg V L 1970 *Propagation of Electromagnetic Waves in Plasmas* (New York: Pergamon Press)

[8] Allen L and Eberly J H 1975 *Optical Resonance and Two-Level Atoms* (New York: Wiley)

[9] Kittel C 1971 *Introduction to Solid State Physics* (New York: Wiley)

[10] Ziman J 1972 *Principles of the Theory of Solids* 2nd edition (Cambridge: Cambridge University Press)

[11] Bunker P R 1979 *Molecular Symmetry and Spectroscopy* (New York: Academic Press)

[12] Wigner E P 1959 *Group Theory* (New York: Academic Press)

[13] Bloembergen N 1965 *Nonlinear Optics* (New York: Benjamin)

[14] Shen Y R 1984 *The Principles of Nonlinear Optics* (New York: Wiley)

[15] Flygare W H 1978 *Molecular Structure and Dynamics* (New Jersey: Prentice Hall)

[16] Aleksandrov A F, Bogdankevich L S and Rukhadze A A 1988 *Fundamentals of the Plasma Electrodynamics* (Moscow: Vysshaya Shkola) (in Russian)

[17] Bunkin F V, Kazakov A E and Fedorov M V 1972 Interaction of high-intensity optical radiation with free electrons (non-relativistic case) *Usp. Fiz. Nauk* **107** 559–94

[18] Keldysh L V 1964 Ionization in the field of high-intensity electromagnetic wave *Zh. Eksp. Teor. Fiz.* **47** 1945–57

[19] Delone N B and Fedorov M V 1989 Multiphoton ionization of atoms: novel effects *Usp. Fiz. Nauk* **158** 215–53

[20] Delone N B, Krainov V P and Shepelyansky D L 1983 Highly excited atoms in electromagnetic field *Usp. Fiz. Nauk* **140** 355–92

[21] Casati G, Chirkov B V, Shepelyansky D L and Guarnevi I 1987 Relevance of a classical chaos in quantum mechanics: the hydrogen atom in a monochromatic field *Phys. Reports* **154** 77–123

[22] Akhmanov S A and Khokhlov R V 1972 *Problems of Nonlinear Optics* (New York: Gordon and Breach)

[23] Silin V P and Rukhadze A A 1961 *Electromagnetic Properties of Plasma and Plasma-Like Media* (Moscow: Gosatomizdat) (in Russian)

[24] Klyshko D N 1986 *Fundamental Physics of Quantum Electronics* (Moscow: Nauka) (in Russian)
[25] Apanasevich P A 1977 *Fundamentals of the Theory of Interaction of Light and Matter* (Minsk: Nauka i Tekhnika) (in Russian)
[26] Pines D and Nozières P 1966 *Theory of Quantum Liquids* (New York: Benjamin)
[27] Born M and Huang K 1956 *Dynamical Theory of Crystal Lattices* (London: Oxford Press)

Index

Acoustic branch, 227, 248
Anharmonism, 213, 242, 244, 249, 250, 256

Balance equation, 112, 116

Collision integral, 138, 139, 292
Coordinate representation, 10, 57, 58, 61, 80, 92, 127
Correlation function, 13, 88, 89, 91
Current of free charges, 14, 16

Deformation potential, 304
Diatomic string, 227, 228, 233
Dispersion, 23, 33, 146, 162, 225-228, 230, 231, 234, 235, 243, 247-249, 261, 263, 267, 268, 281, 283, 284, 291, 293, 294, 306, 312
Displacement current, 14, 23
Distribution function, 137, 138, 140, 148, 150, 292

Eigenfunction, 66, 82, 88, 100, 101, 167, 170, 174, 178, 179, 187, 196-198, 207, 278, 287
Electron polarizability tensor of the molecule, 216
Electron radius, 127
Exciton, 305, 308-312
External field, 5, 6, 10, 14, 15, 17, 18, 29, 31-33, 48, 67-69, 72, 73, 75-78, 82, 85-88, 90, 94, 96, 97, 102, 103, 153, 157, 161, 172, 182, 212, 252, 253, 263, 264, 268, 270, 273, 274, 283, 287, 290
Extraneous current, 90-92

Frequency of local vibrations, 232

Gauge invariance, 3, 4, 11, 74
Gauge transformation, 7, 11, 96, 154
Group theory, 163, 165, 258

Interaction representation, 66-71, 75, 78, 82, 86, 88, 103, 115, 117, 122
Intrinsic absorption, 305, 309

Lattice site, 49, 50, 229, 268, 274
Law of dispersion, 40, 134, 146, 150, 242, 259-262, 267, 268, 279, 281, 283, 289, 292, 294, 311

Magnetization, 14-17, 23-27, 92, 256
Mean field, 13, 14, 16, 21, 32, 40, 43, 44, 46, 49, 52, 97, 138, 275
Mean linear and nonlinear susceptibilities, 56
Molecular bonding, 191
Molecular crystal, 191, 219, 220, 228, 277, 286, 310

Nearly-free-electrons approximation, 286
Nonlinear (two-photon) absorption, 111
Nonlinearity parameter, 18
Nonlocality radius, 19, 20, 25, 27, 28, 92, 93, 144, 273, 274

Optical phonon linewidth, 248
Optical wave, 6, 227, 274

Parity, 172, 174, 204
Phonon lifetime, 242, 246, 248, 249, 256, 276
Phonons, 240-243, 245-249, 256, 261, 277, 278, 280, 289, 290, 295, 299-304, 307
Plasma frequency, 133, 134, 295
Polarization current, 14
Primitive unit cell, 42

Quadratic susceptibility of the ionic lattice, 255

Random collisions model, 62, 64
Reciprocal lattice of the crystal, 37, 40, 242
Responses to the external field, 29
Retardation effect, 19, 23, 33, 49

Spatial dispersion, 23, 25, 26, 29, 33, 34, 92, 93, 95-97, 101, 102, 135, 139, 142-147, 149, 150, 153, 262, 264, 292, 296
Susceptibilities for the external field, 32
Symmetry elements, 164, 169
Symmetry group, 163-169, 171, 172, 174, 175, 200, 201, 221, 235
Symmetry of kinetic coefficients, 29, 100
Symmetry relations, 36, 109, 237

Tight-binding approximation, 286
Total current, 16, 24, 25, 129, 290
Total induction, 16
Total polarization, 16, 17, 24
Transport cross section, 131, 132

Wave function, 11, 13, 57-59, 65-67, 76, 79, 99, 100, 114, 154, 155, 157, 158, 163, 166, 168, 169, 171-173, 175, 176, 178, 179, 181-183, 185-188, 194, 195, 197-199, 202-207, 210-212, 214, 219, 221, 276, 277, 280-284, 286, 297, 310, 311

GPSR Compliance
The European Union's (EU) General Product Safety Regulation (GPSR) is a set of rules that requires consumer products to be safe and our obligations to ensure this.

If you have any concerns about our products, you can contact us on

ProductSafety@springernature.com

In case Publisher is established outside the EU, the EU authorized representative is:

Springer Nature Customer Service Center GmbH
Europaplatz 3
69115 Heidelberg, Germany

www.ingramcontent.com/pod-product-compliance
Ingram Content Group UK Ltd.
Pitfield, Milton Keynes, MK11 3LW, UK
UKHW020103200726
13856UKWH00002B/351

* 9 7 8 1 4 8 9 9 1 5 7 1 9 *